Model Free Adaptive Control

Theory and Applications

Model Free Adaptive Control

Theory and Applications

Zhongsheng Hou • Shangtai Jin

CRC Press
Taylor & Francis Group
Boca Raton London New York

CRC Press is an imprint of the
Taylor & Francis Group, an **informa** business

CRC Press
Taylor & Francis Group
6000 Broken Sound Parkway NW, Suite 300
Boca Raton, FL 33487-2742

Printed on acid-free paper
Version Date: 20130808

International Standard Book Number-13: 978-1-4665-9418-0 (Hardback)

Visit the Taylor & Francis Web site at
http://www.taylorandfrancis.com

and the CRC Press Web site at
http://www.crcpress.com

Contents

Preface

During the past half century, modern control theory has developed greatly and many branches and subfields have emerged, for example, linear system theory, optimal control, system identification, adaptive control, robust control, sliding mode control, and stochastic system theory, etc. Meanwhile, lots of remarkable applications of modern control theory have been carried out in many practical fields, such as industrial processes, aerospace, urban transportation, and so on. However, we now face great challenges in studying and applying these control methods, because many new problems have been generated by the theoretical development of the control discipline itself and by practical control requirements of plants in the real world. Modern control theory was established and developed based on a fundamental assumption that the mathematical model or nominal model of a controlled plant is accurately known. As we know, building a model for a practical plant is a tough task, and sometimes it is impossible. Even if the model is properly derived, its accuracy is also in serious doubt. Therefore, difficulty in modeling, poor robustness, lack of safety, as well as a huge gap between theoretical analysis and practical performance, are common phenomena when applying modern control theory in practice. Some important theoretical problems, which are difficult to address within the framework of traditional model-based control theory, including accurate modeling versus model/controller simplification, unmodeled dynamics versus robustness, and unknown uncertainties versus assumption on their known upper bound required by robust design, and so on, hinder the healthy development of the modern control theory. Further, even when the mathematical model of the controlled plant is accurately built, it must be a very complicated one with strong nonlinearity and high order, due to the complexity of a practical system itself. Complex system model definitely leads to a complex controller, which renders unavoidable difficulties in controller design and system analysis. A complicated controller costs much, is hard to apply and maintain, and is unacceptable to engineers.

With the development of information science and technology in recent years, practical processes in many fields, such as chemical industry, metallurgy, machinery, electricity, transportation and logistics, and so on, have been undergoing significant changes. The scale of the enterprises is becoming increasingly large, the production process is becoming more and more complex, and the requirements on product quality are also becoming higher and higher. This challenges the existing control theory and methods since both of these are based on accurate mathematical model of the controlled plant. On the other hand, a huge amount of process data generated by practical plants and industrial processes could be obtained and stored, and it contains all the valuable state information about the process operations implicitly. In this case, how to utilize the collected data and the mined knowledge to develop efficient control and optimization methods for industrial processes when accurate process models are unavailable has become a vital issue faced by the control theory community. Therefore, developing data-driven control theory and methods is an inevitable choice for development of the control theory in the new era and is of great significance in both theory and practice.

Model-free adaptive control (MFAC), as a typical data-driven control method, was proposed by the first author of this book in his doctoral dissertation in 1994. For MFAC, only the measurement I/O data of the controlled plant are directly used for controller design and closed-loop system analysis, without any model dynamics involved. By applying MFAC, adaptive control for an unknown nonlinear system with time-varying parameters and time-varying structure is uniformly realized, and the existing difficulties in modern control methods, such as the dependence of controller design on system model, unmodeled dynamics, traditional robustness issues, and other related theoretical problems, are avoided within the data-driven control framework. Both the theoretical results incessantly developed and improved in the past two decades, and their successful practical applications in motor control, the chemical industry, machinery, and so on, have made MFAC become a novel control theory with a systematic and rigorous framework.

The main contents of the book cover the dynamic linearization approach, model-free adaptive control, model-free adaptive predictive control, and model-free adaptive iterative learning control for discrete-time SISO and MIMO nonlinear systems, with corresponding stability analysis and typical practical applications. Moreover, some more important issues are also studied in this monograph, including model-free adaptive control for complex connected systems, modularized controller designs between model-free adaptive control and other control methods, robustness of model-free adaptive control, and concept of the symmetric similarity for adaptive control system design.

The first author of this book would like to gratefully acknowledge the National Science Foundation of China for long-term support under grant No. 60834001, 61120106009, and 60474038, which helped to complete this monograph.

The first author expresses his sincere appreciation to his students whose contributions enriched and shaped the theory and applications of MFAC, namely Dr. Ronghu Chi, Dr. Weihong Wang, Dr. Chenkun Yin, Dr. Jingwen Yan, Dr. Xuhui Bu, and Dr. Xiangbin Liu, and PhD candidates Yongqiang Li, Heqing Sun, Yuanming Zhu, and Honghai Ji. Special thanks are due to Dr. Ronghu Chi and Dr. Chenkun Yin for their careful proofreading of this monograph. Many thanks go to Professor Jian-Xin Xu of the National University of Singapore, the enlightening conversations with him helped in rigorously expressing the dynamic linearization approach. Thanks go to Associate Professor Cheng Xiang of the National University of Singapore and Professor Mingxuan Sun of Zhejiang University of Technology for their careful proofreading and helpful comments. The first author thanks all the scholars who use MFAC methods in practical fields. Without their successful applications, it would be a hard task for him to keep on researching in this direction with enough confidence and courage.

Zhongsheng Hou
Beijing

Authors

Zhongsheng Hou received his bachelor's and master's degrees from Jilin University of Technology, Changchun, China, in 1983 and 1988, respectively, and his PhD degree from Northeastern University, Shenyang, China, in 1994. He was a postdoctoral fellow with Harbin Institute of Technology, Harbin, China, from 1995 to 1997 and a visiting scholar with Yale University, New Haven, Connecticut, from 2002 to 2003. In 1997, he joined Beijing Jiaotong University, Beijing, China and is currently a full professor and the founding director of the Advanced Control Systems Lab and the dean of Department of Automatic Control in the School of Electronic and Information Engineering.

His research interests lie in the fields of data-driven control, model-free adaptive control, iterative learning control, and intelligent transportation systems. He has been the principal investigator for the state key program and the major international cooperation and exchange program of the National Natural Science Foundation of China. Up to now, he has over 110 peer-reviewed journal papers published and over 120 papers in prestigious conference proceedings. He is the author of the monograph *Nonparametric Model and Its Adaptive Control Theory*, Science Press of China, 1999. Dr. Hou has served as technical committee member for over 50 international and Chinese conferences and as associate editor and guest editor for a number of prestigious international and Chinese journals.

Shangtai Jin received his BS, MS, and PhD degrees from Beijing Jiaotong University, Beijing, China, in 1999, 2004, and 2009, respectively. He is currently a lecturer with Beijing Jiaotong University. His research interests include model-free adaptive control, data-driven control, learning control, and intelligent transportation systems.

Symbols

L	Control input linearization length constant in the PFDL data model
L_y	Controlled output linearization length constant in the FFDL data model
L_u	Control input linearization length constant in the FFDL data model
$\phi_c(k)$	PPD in the CFDL data model for SISO discrete-time nonlinear systems at time instant k
$\phi_c(k,i)$	PPD in the iteration related CFDL data model for SISO discrete-time nonlinear systems at time instant k of the ith iteration
$\boldsymbol{\phi}_c(k)$	PG in the CFDL data model for MISO discrete-time nonlinear systems at time instant k
$\boldsymbol{\phi}_{c,i}(k)$	PG in the CFDL data model of the ith subsystem for interconnected systems at time instant k
$\boldsymbol{\phi}_{p,L}(k)$	L-dimensional PG in the PFDL data model for SISO discrete-time nonlinear systems at time instant k
$\bar{\boldsymbol{\phi}}_{p,L}(k)$	mL-dimensional PPG in the PFDL data model for MISO discrete-time nonlinear systems at time instant k
$\boldsymbol{\phi}_{i,pL}(k)$	L-dimensional PG in the PFDL data model of the ith subsystem for interconnected systems at time instant k
$\boldsymbol{\phi}_{f,L_y,L_u}(k)$	$L_y + L_u$-dimensional PG in the FFDL data model for SISO discrete-time nonlinear systems at time instant k
$\bar{\boldsymbol{\phi}}_{f,L_y,L_u}(k)$	$L_y + mL_u$-dimensional PPG in the FFDL data model for MISO discrete-time nonlinear systems at time instant k
$\boldsymbol{\phi}_S(k)$	PG in the PFDL data model for systems in series connection at time instant k
$\boldsymbol{\phi}_P(k)$	PG in the PFDL data model for systems in parallel connection at time instant k
$\boldsymbol{\phi}_F(k)$	PG in the PFDL data model for systems in feedback connection at time instant k
$\boldsymbol{\Phi}_c(k)$	$m \times m$-dimensional PJM in the CFDL data model for MIMO discrete-time nonlinear systems at time instant k

$\boldsymbol{\Phi}_{p,L}(k)$	$m \times mL$-dimensional PPJM in the PFDL data model for MIMO discrete-time nonlinear systems at time instant k
$\boldsymbol{\Phi}_{f,L_y,L_u}(k)$	$mL_y \times mL_u$-dimensional PPJM in the FFDL data model for MIMO discrete-time nonlinear systems at time instant k
$\boldsymbol{U}_L(k)$	Vector in the PFDL data model for SISO discrete-time nonlinear systems, consisting of all control input signals within a moving time window $[k-L+1,k]$, that is, $\boldsymbol{U}_L(k) = [u(k),\ldots,u(k-L+1)]^T$
$\boldsymbol{U}_{i,L_i}(k)$	Vector in the PFDL data model of the ith subsystem for interconnected systems, consisting of all control input signals within a moving time window $[k-L+1,k]$, that is, $\boldsymbol{U}_{i,L_i}(k) = [u_i(k),\ldots,u_i(k-L+1)]^T$
$\bar{\boldsymbol{U}}_L(k)$	Vector in the PFDL data model for MIMO (MISO) discrete-time nonlinear systems, consisting of all control input signals within a moving time window $[k-L_y+1,k]$, that is, $\bar{\boldsymbol{U}}_L(k) = [\boldsymbol{u}^T(k),\ldots,\boldsymbol{u}^T(k-L+1)]^T$
$\boldsymbol{H}_{L_y,L_u}(k)$	Vector in the FFDL data model for SISO discrete-time nonlinear systems, consisting of all control input signals within a moving time window $[k-L_u+1,k]$, and all system output signals within a moving time window $[k-L_y+1,k]$, that is, $\boldsymbol{H}_{L_y,L_u}(k) = [y(k),\ldots,y(k-L_y+1),u(k),\ldots,u(k-L_u+1)]^T$
$\bar{\boldsymbol{H}}_{L_y,L_u}(k)$	Vector in the FFDL data model for MIMO discrete-time nonlinear systems, consisting of all control input signals within a moving time window $[k-L_u+1,k]$, and all system output signals within a moving time window $[k-L_y+1,k]$, that is, $\bar{\boldsymbol{H}}_{L_y,L_u}(k)=[\boldsymbol{y}^T(k),\ldots,\boldsymbol{y}^T(k-L_y+1),\boldsymbol{u}^T(k),\ldots,\boldsymbol{u}^T(k-L_u+1)]^T$
$\breve{\boldsymbol{H}}_{L_y,L_u}(k)$	Vector in the FFDL data model for MISO discrete-time nonlinear systems, consisting of all control input signals within a moving time window $[k-L_u+1,k]$, and all system output signals within a moving time window $[k-L_y+1,k]$, that is, $\breve{\boldsymbol{H}}_{L_y,L_u}(k) = [y^T(k),\ldots,y^T(k-L_y+1),\boldsymbol{u}^T(k),\ldots,\boldsymbol{u}^T(k-L_u+1)]^T$
R	Real number set
R^n	n-dimensional real vector space
$R^{n\times m}$	$n \times m$-dimensional real matrix space
Z^+	Positive integer set
I	Identity matrix
q^{-1}	Unit delay operator
Δ	Difference operator
sign(x)	Sign function
round($\cdot$)	Round function
$\lvert\cdot\rvert$	Absolute value
$\lVert\cdot\rVert_v$	Consistent norm
$\hat{a}(k)$	Estimation value of variable a at time k

$\tilde{a}(k)$	Estimation error of variable a, that is, $\tilde{a}(k) = \hat{a}(k) - a(k)$
$s(A)$	Spectral radius of A
A^{-1}	Inversion matrix of A
A^T	Transpose of A
A^*	Adjoint of A
$\det(A)$	Determinant of A
$\sigma_1(A)$	Condition number of A
$\lambda_{max}[A]$	Largest eigenvalue of A
$\lambda_{min}[A]$	Smallest eigenvalue of A
$\nabla J(\boldsymbol{\theta})$	Gradient of function $J(\boldsymbol{\theta})$ with respect to $\boldsymbol{\theta}$
$\nabla^2 J(\boldsymbol{\theta})$	Hessian matrix of function $J(\boldsymbol{\theta})$ with respect to $\boldsymbol{\theta}$

Acronyms

CFDL	Compact form dynamic linearization
CFDL–MFAC	Compact form dynamic linearization based model-free adaptive control
CFDL–MFAILC	Compact form dynamic linearization based model-free adaptive iterative learning control
CFDL–MFAPC	Compact form dynamic linearization based model-free adaptive predictive control
DDC	Data-driven control
FFDL	Full form dynamic linearization
FFDL–MFAC	Full form dynamic linearization based model-free adaptive control
FFDL–MFAPC	Full form dynamic linearization based model-free adaptive predictive control
IFT	Iterative feedback tuning
ILC	Iterative learning control
MBC	Model-based control
MFAC	Model-free adaptive control
MFAILC	Model-free adaptive iterative learning control
MFAPC	Model-free adaptive predictive control
MIMO	Multiple-input and multiple-output
MISO	Multiple-input and single-output
PFDL	Partial form dynamic linearization
PFDL–MFAC	Partial form dynamic linearization based model-free adaptive control
PFDL–MFAPC	Partial form dynamic linearization based model-free adaptive predictive control
PG	Pseudo gradient
PID	Proportional-integral-differential control
PJM	Pseudo-Jacobian matrix

PPD	Pseudo partial derivative
PPG	Pseudo partitioned gradient
PPJM	Pseudo partitioned Jacobian matrix
SISO	Single-input and single-output
VRFT	Virtual reference feedback tuning

Chapter 1

Introduction

In this chapter, the existing problems and challenges of the model-based control (MBC) theory are briefly reviewed, followed by a brief discussions of the available data-driven control (DDC) methods with definition, classifications, traits, insights, and applications. Finally, the outline of this book is given.

1.1 Model-Based Control

The introduction of state-space models by Kalman in 1960 [1,2] gave birth to modern control theory and methods. Modern control theory was founded and developed on a basis that the mathematical model or the nominal model of the controlled system is exactly known, thus, it is also called MBC theory. With the prosperous development of MBC theory including linear system theory, system identification theory, optimal control theory, adaptive control theory, robust control theory, filter and estimation theories, and so on, many successful applications of MBC are found in practical fields, especially in the fields of aerospace, national defense, and industry. On the other hand, the scale of industrial plants and enterprises becomes increasingly large, the production technology and processes also become more and more complex, and the requirements on product quality become higher and higher. All these, which were not exposed fully before, bring great challenges to the theoretical studies and practical applications of MBC theory.

1.1.1 Modeling and Identification

At present, almost all the control methods for linear and nonlinear systems are model-based. For the MBC methods, the first step is to build the mathematical model of the plant, then to design the controller based on the plant model obtained with the faith that the plant model represents the true system, and finally to analyze the closed-loop control system still in virtue of the mathematical model. The "certainty equivalence principle" is the cornerstone of modern control theory, and the plant model is the starting point and landmark for the controller design and analysis, as well as the evaluation criterion and control destination for the MBC methods.

There are two kinds of methods for modeling a plant: the first principles method and the system identification method. Modeling a plant using first principles means to establish the dynamic equations of the controlled plant according to some physical or chemical principles, and to determine the model parameters by a series of experiments. Modeling a plant by system identification is to develop an input–output plant model, which lies in a specified model set covering the true system and can approximate the true system in terms of bias or error, using the online or offline measurement data. It is recognized that modeling by first principles or system identification is just an approximation of the true system with some error, due to the complexities in system structure and operation environment. In other words, the unmodeled dynamics and other uncertainties always exist in the modeling process. Consequently, the application of the controllers designed on the inaccurate mathematical model may bring various practical problems. The corresponding closed-loop control system may have weak robustness and inherently possible lack of safety because of the unmodeled dynamics and external disturbances [3–6].

To preserve the obvious advantages of MBC design while increasing robustness against model errors, much effort has been put into the development of robust control theory. Various ways of describing model errors in the configuration of closed-loop systems have been considered. These include additive, multiplicative descriptions and the assumption on *a priori* bounds for noise, modeling errors, or uncertainties. However, the descriptions of these uncertainties upon which robust control design methods are based, cannot be obtained quantitatively or qualitatively by physical modeling or identification modeling. Even for the upper bound of the uncertainty, so far no identification method is able to supply its quantification. In other words, the descriptions of these uncertainties are not consistent with the results delivered by physical modeling or identification modeling [7]. Therefore, theoretically speaking, it is difficult to apply MBC methods to synthesize the controller for practical systems, and the control performance and safety may not be guaranteed when the model-based controller is connected to the practical plants [8].

For control system design, a very intuitive way is to first put a significant amount of effort to obtain a very accurate mathematical model (including uncertainties) for the unknown system by first-principles modeling or identification techniques, and then design a model-based controller based on this model for practical applications. However, there are both practical and theoretical difficulties in the establishment of a perfect plant model and the controller. First, unmodeled dynamics and the robustness are inevitable twinborn problems and they cannot be solved simultaneously within the conventional MBC theoretical framework. Up to now, there is no efficient tool or method, in mathematics theory or system identification, to produce an accurate plant model for complex nonlinear systems that exist everywhere in the real world. Accurate modeling sometimes is more difficult than control system design itself. If the plant dynamics is with time-varying structure or fast-varying parameters, it would be hard to determine the plant model, to design and analyze the control system using analytical mathematics tools. Second, the more accurate the model is, the more effort or cost we should spend, and the more complicated the controller is as well. In a sequel, the complex controller must lead to weak robustness and low reliability of the controlled system. Moreover, great difficulties would be brought in the implementation and application of the control system. If the system dynamics is of very high order, it is not suitable for practical controller design, since such a model would definitely lead to a controller with very high order, which possibly makes control system design, analysis, and application complicated and brings many difficulties in system monitoring and maintenance. In practice, reduction of the model or controller must be performed additionally to obtain a simple and practical control system. It seems paradoxical to build an accurate high-order model to target high performance for control system design, and then to perform model simplification for a low-order controller. The last but not least difficulty is the "persistent excitation" condition for modeling. It is a great challenge for input signals to satisfy the "persistent excitation" condition during modeling and closed-loop control. Without the persistently exciting inputs, an accurate model cannot be produced. In such a circumstance, most MBC theoretical results of a closed-loop control system, such as stability and convergence, cannot be guaranteed as what they claimed when they are used in practice [4–6,8]. "Persistent excitation" condition and control performance are also paradoxical, and cannot be handled within the framework of traditional MBC theory.

1.1.2 Model-Based Controller Design

For modern control theory, controller design is based on the mathematical model of a plant. Typical linear control system design methodologies are zero-pole assignment, linear quadratic regulator (LQR) design, optimal control design, and so on. For nonlinear systems, the inevitable controller design methods are the Lyapunov-based methods with backstepping controller design, feedback linearization, and so on. These controller design methodologies are recognized to be typical MBC system design methods. The key skill of MBC controller design and analysis is performed through mathematical analysis of the error dynamics of the closed-loop system, and the plant model is included in every phase of the control system design and applications, such as operation monitoring, evaluation, and diagnosis. The architecture of MBC theory is shown in Figure 1.1. This diagram shows that the system model and assumptions are the starting point for controller design, and also the destination for MBC control system analysis.

Since the unmodeled dynamics and other uncertainties always exist in modeling and closed-loop control process, a model-based controller may not work well

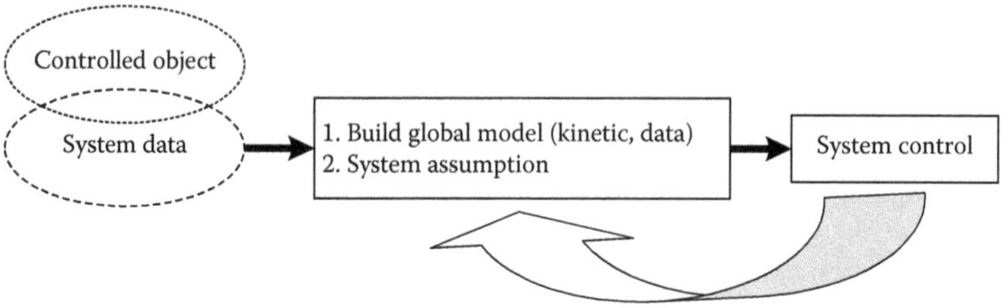

Figure 1.1 Architecture of MBC theory.

in practice, and even leads to bad performance or causes the closed-loop system unstable. An arbitrarily small modeling error could lead to a very bad closed-loop performance [9]. Rohr's counterexample in adaptive control is a warning bell for scholars to contemplate MBC theory and methods. The Rohr's counterexample has demonstrated that reported stable adaptive control systems, based on some assumptions made about the system model and the certainty equivalence principle, may show a certain unexpected behavior in the presence of unmodeled dynamics [10,11]. Thus, the correctness and applicability of MBC control system design methods are challenged.

Even when the model is accurate enough, the results of a theoretical analysis, such as the stability, convergence, and robustness of a closed-loop control system proven by rigorous mathematical derivation, are not always valuable if the additional assumptions made about the system are not reasonable. Taking adaptive control as an example, adaptive control methods often say that under assumptions A, B, C, D, and E, and with the use of algorithm F, all signals remain bounded as time goes to infinity, and then some other results occur. However, the uncertainties, including disturbances and the parameter shifting due to aging when the plant goes into operation, are inevitable for the adaptive control in practice. Those external factors and unmodeled dynamics may lead to an unstable closed-loop system. In other words, the stated conclusion in adaptive control does not rule out the possibility that at some time before time goes to infinity, the controller connected to the plant could make the closed-loop system unstable [6].

There are two kinds of uncertainties. One is the parametric uncertainties and the other is the nonparametric uncertainties. To enhance the robustness, many modification techniques and robust adaptive control design methods were proposed, such as normalization, dead-zone method, projection method, σ-modification and sliding mode adaptive method, and so on. The robustness issue of adaptive control systems is an intractable topic that attracts much research attention.

For MBC control system design, the modeling accuracy and correctness of the assumptions imposed on the plant mathematical model together determine the control system performance, reliability, and safety, since the available system model kinetics is embedded in the control system. If the system model is unavailable or the assumptions do not hold, then no conclusion on controller design and analysis could be obtained, and further we have no way to discuss applications. The MBC control method comes from the system model and ends up in the system model. To some extent, it may be called model theory rather than control theory.

1.2 Data-Driven Control

With the development of information science and technology, many industrial processes, for instance, chemical industry, metallurgy, machinery, electronics, electricity, transportation, and logistics, have been undergoing significant changes. As the scale of the enterprises becomes increasingly large, the production technology, equipment, and production process also become more and more complex, and the requirements on product quality become higher and higher. Hence, modeling these processes using the first-principles or identification method becomes more and more difficult, which leads us to conclude that using the traditional MBC theory to deal with the control issues in these kinds of enterprises would be impractical. On the other hand, however, many industrial processes generate and store a huge amount of process data, which contain all the valuable state information of the process operations and the equipments. In this case, it is of great significance to utilize these process data, gained online or offline, directly for controller design, monitoring, prediction, diagnosis, and assessment of the industrial processes when the accurate process models are unavailable. Therefore, the establishment and development of the DDC theory and methods are of significant importance in both control theory and applications.

Up to now, there exist a few DDC methods, such as proportion integral differential (PID) control, model-free adaptive control (MFAC), iterative feedback tuning (IFT), virtual reference feedback tuning (VRFT), iterative learning control (ILC), and so on. Although the studies on DDC are still in their embryonic stage, they attract much attention in the control theory community. The Institute for Mathematics and Its Applications (IMA) in the University of Minnesota held a workshop titled "IMA Hot Topics Workshop: Data-Driven Control and Optimization" in 2002. The Natural Science Foundation of China (NSFC) held a workshop titled "Data-Based Control, Decision, Scheduling, and Fault Diagnostics" in November 2008. A special issue with the same title as above in *Acta Automatica Sinica* was published in June 2009 that included 20 papers in these four directions [12]. NSFC and Beijing Jiaotong University jointly held another workshop on this topic titled "International Workshop on Data Based Control, Modeling and Optimization" in November 2010. The Chinese Automation Congresses, held by the Chinese Automation Association in 2009 and 2011, respectively, also listed this hot topic as one of main forums. Further, *IEEE Transactions on Neural Networks*, *Information Sciences*, and *IEEE Transactions on Industrial Informatics* also launched their call for papers for their special issues on this topic in 2010 and 2011, respectively, and the *IEEE Transactions on Neural Networks* has published its special issue in December 2011 [13].

1.2.1 Definition and Motivation of Data-Driven Control

There are three literal definitions of DDC in the literature so far, found by searching the Internet:

Definition 1.1 [14]

Data-driven control is the control theory and method, in which the controller is designed merely using online or offline I/O data of the controlled system or using knowledge from the data processing without explicitly or implicitly using information from the mathematical model of the controlled process, and whose stability, convergence, and robustness can be guaranteed by rigorous analysis under certain reasonable assumptions.

Definition 1.2 [15]

Data-driven control design is the synthesis of a controller using measurement data acquired on the actual system to be controlled, without explicitly using (non)parametric models of the system to be controlled during adaptation.

Definition 1.3 [16]

Measured data are used directly to minimize a control criterion. Only one optimization in which the controller parameters are the optimization variables is used to calculate the controller.

In Definition 1.1, the DDC controller design merely uses the plant measurement I/O data, not including any dynamics information and structure information of the controlled system. In Definition 1.2, the DDC controller design may include implicit use of the structure information of the controlled plant. In Definition 1.3, the DDC controller structure is assumed predetermined, and the controller's parameters are obtained via offline optimization. From the above three definitions, we could find that the motivation of DDC methodologies is to design the controller directly using the measurement input–output data of the controlled plant.

Summarizing all the three definitions above, a more generic definition of DDC is proposed as follows:

Definition 1.4 [17]

Data-driven control is the control theory and method, in which the controller is directly designed by using online or offline I/O data of the controlled system or

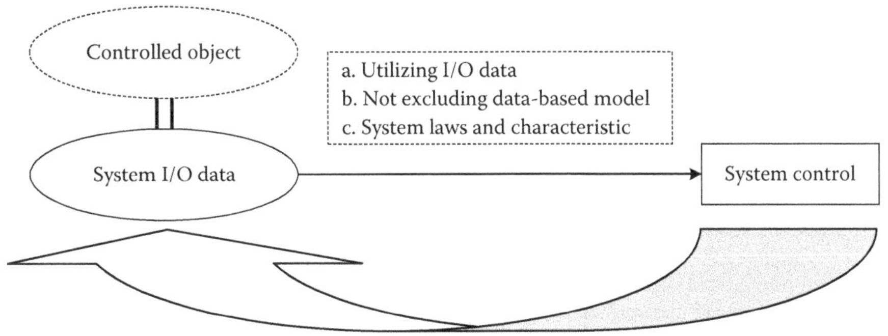

Figure 1.2 Architecture of DDC methodologies.

knowledge from the data processing without explicitly using the information from the mathematical model of the controlled process, and whose stability, convergence, and robustness can be guaranteed by rigorous analysis under certain reasonable assumptions.

The only difference between Definitions 1.4 and 1.1 lies in the fact that the former includes the methods that may implicitly use the information from the mathematical model of the controlled process. With Definition 1.4, direct adaptive control, subspace predictive control, and so on are in the scope of DDC.

The architecture diagram of the DDC methodologies is shown in Figure 1.2.

1.2.2 Object of Data-Driven Control Methods

The control system consists of two main parts: the controlled plant and the controller. The controlled plants can be categorized into the following four classes:

C1. The accurate first-principles or the identified model is available.
C2. The first-principles or the identified model is inaccurate with uncertainties.
C3. The first-principles or the identified model is known but complicated with very high order and very strong nonlinearities.
C4. The first-principles or the identified model is difficult to be established or is unavailable.

Generally speaking, C1 and C2 have been well addressed by the modern control theory, also called MBC theory. For C1, although controller design for general nonlinear systems is complicated, many well-developed approaches have been presented to deal with the linear or nonlinear system, such as zero-pole assignment, Lyapunov controller design methods, backstepping design methods, feedback linearization, and so on. For C2, adaptive control and robust control were developed to tackle uncertainties when the uncertainties could be parameterized or their bound is

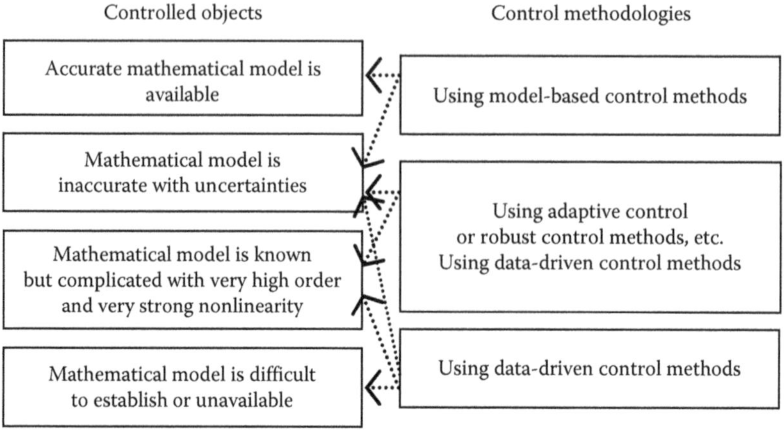

Figure 1.3 Controlled objects of DDC.

assumed known or available. Of course, many other well-developed modern control branches are also devoted to handling these two classes of controlled objects.

For C3, although the first-principles or the identified model is available with quite a high accuracy, its controller design and analysis is still quite a difficult task since the model is possibly with thousands of states, equations, high orders, and strong nonlinearity. As we know, the high-order nonlinear system model definitely leads to a controller with high order or strong nonlinearity. A complex high-order nonlinear controller must bring about a huge difficulty in implementation, performance analysis, practical application, and maintenance. In this case, the procedure of model reduction or controller reduction is inevitable. Usually, a complex high-order nonlinear model is not suitable for the controller design, analysis, and applications. Based on such an observation, we conclude that there is no available method to deal with these kinds of control problems both for C3 and C4.

For these four classes of controlled plants, only half of them or less are well addressed by the existing MBC methods, but there is not any good approach to deal with the other half and more up to now. System I/O data, however, can be always obtained regardless of the type of controlled objects. Therefore, it is recommended to apply DDC control methods. If the system model is unavailable or is with large uncertainty, DDC is an inevitable choice. The relationship between control methods and controlled objects is illustrated in Figure 1.3.

1.2.3 Necessity of Data-Driven Control Theory and Methods

A complete control theory should cover all the methodologies and theories for all the aforementioned controlled objects. From this point of view, MBC methods and DDC methods should be two inevitable parts of the complete control theory; namely, a perfect control theory must consist of MBC theory and DDC theory.

In the following, we will show the necessity of DDC theory and methods from three aspects: theory studies, applications, and historical development of control theory.

From the angle of basic theory: (1) Some difficult issues concerning unmodeled dynamics and robustness are inevitable for MBC theory and methods. On one hand, MBC methods are incapable of dealing with the control issue without the available system model. On the other hand, the unmodeled dynamics and robustness cannot be avoided for modeling and control. This is a dilemma. (2) The complexity in the system mathematical model determines the complexity in controller structure. Generally, a complicated high-order nonlinear system model definitely leads to a complicated high-order nonlinear controller. Therefore, it is necessary to consider model simplification, order reduction, and robustness of the MBC controller. (3) To implement a robust control, the upper bound of the uncertainties must be quantitatively given and described *a priori*; however, it can hardly be achieved in theory by existing modeling approaches.

From the angle of application, low-cost automation systems and equipments satisfying certain control criterions of decision makers are required in most of the practical fields, such as chemical process, industrial producing process, and so on. It costs, however, too much to build the first-principles model or global model mathematically. Especially for batch processes, it is impossible to precisely build the mathematical model for each subsystem in each batch in the name of improving production and quality. For a complicated system, building its global model is impossible since the system is with intrinsic complexity and external disturbances. Even in the local model, there is also discrepancy. Therefore, the MBC method may fail when it is used in practice. Rich theoretical achievements in control theory research versus poor control methods to tackle the practical control problems is the primary challenge to the management and control of many complicated systems in the field of process industry. In addition, the profound mathematical tools and the professional skills required by the MBC method make engineers feel unconfident and powerless when they use it to design and maintain control systems, particularly for complicated systems. The gap between control theory and its application becomes larger and larger. This strongly restricts the healthy development of control theory.

The historical development of control theory is in a spiral fashion, undergoing the following stages: simple regulation devices and PID control without the need of a mathematical model, the classical control theory based on the transfer function model, the MBC theory based on the state-space model, the control theory and methods depending on professional knowledge of the controlled plants including ruler-based model, neural-networks-based model and expert systems, and the DDC theory aiming at removing the dependence on the plant model. By DDC theory and methods, the controller is designed directly using I/O data of the controlled plant, which definitely accords with the developing trend of control theory.

According to the integrity of the theory, the existing control theory can be categorized into three classes: (1) control theory and methods fully depending on a mathematical model, such as control approaches in aerospace, optimal control, linear and nonlinear control, the coordinated control and decomposition technique for large systems and the zero-pole assignment technique, and so on; (2) control theory and methods partially depending on a mathematical model, including robust control, sliding mode control, adaptive control, fuzzy control, expert control, neural networks (NN) control, intelligent control, and so on; and (3) DDC theory and methods merely using the I/O data of the control plant, for example, PID control, ILC, and others. Establishing DDC theory and methods meets the integrity requirement of the control theory framework.

It is worth noting that DDC methods and MBC methods are not mutually exclusive, and they cannot be replaced with each other. DDC and MBC methods do have their own advantages and disadvantages. Based on this observation, DDC and MBC methods could separately develop and also work together in a complementary fashion. The primary difference between MBC and DDC is that the former is the model-based control design approach in case the accurate model is available and the latter is the data-based control design approach in case the accurate model is unavailable. The advantage of DDC theory and methods lies in the fact that the dependence on the mathematical model of a controlled plant is removed for control system design. Some important issues that are inevitable in traditional MBC methods, including unmodeled dynamics versus robustness, accurate modeling versus model simplification or controller reduction, robust control design versus the unavailable description on the upper bound of the uncertainties, good theoretical results versus bad practical performance, and so on, do not exist any longer within the framework of DDC.

1.2.4 Brief Survey on Data-Driven Control Methods

So far, there have been over 10 kinds of different DDC methods in the literature. According to the type of data usage, these methods could be summarized as three classes: the online data based DDC method; the offline data based DDC method, and the hybrid online/offline data based DDC method. According to controller structure, these methods could be divided into two classes: DDC methods with a known controller structure and DDC methods with an unknown controller structure. In the following, we briefly survey the existing DDC methods by virtue of these two observations.

1.2.4.1 DDC Classification According to Data Usage

Online Data Based DDC methods include Simultaneous Perturbation Stochastic Approximation (SPSA) based DDC, Model-Free Adaptive Control, and Unfalsified Control (UC), etc.

Simultaneous Perturbation Stochastic Approximation Based DDC Method — The simultaneous perturbation stochastic approximation (SPSA) based direct controller approximation method was first proposed by J. C. Spall in 1992 [18]. The method is solely using the closed-loop measured data, rather than the mathematical model of a class of discrete-time nonlinear plants, to tune the controller parameters. The SPSA-based control methods assume that the nonlinear dynamics of the controlled plant is unknown, and the controller is a function approximator, whose structure is fixed and the parameters are tunable. NN, polynomial, and so on, may be adopted as the approximators. The controller design is to find optimal controller parameters by minimizing a control performance index by using I/O data at each time instant. To solve the above optimal problem in case the system model is unknown, the SPSA algorithm is applied to estimate the gradient information of cost function with respect to the control input [19,20]. SPSA-based control algorithm has also been utilized in traffic control [21] and industrial control [22].

Model-Free Adaptive Control — The MFAC approach was first proposed by Zhongsheng Hou in 1994 for a class of general discrete-time nonlinear systems [23–26]. Instead of identifying a more or less nonlinear model of a plant, a virtual equivalent dynamical linearization data model is built at each dynamic operation point of the closed-loop system using a new dynamic linearization technique with a novel concept called pseudo partial derivative (PPD). With the help of the equivalent virtual data model, adaptive control of the nonlinear system is carried out by a proposed controller. The time-varying PPD is estimated merely using the I/O measurement data of a controlled plant. The dynamic linearization techniques include compact-form dynamic linearization (CFDL) data model, partial-form dynamic linearization (PFDL) data model, and full-form dynamic linearization (FFDL) data model. Compared with traditional adaptive control schemes, the MFAC approach has several merits, which make it more suitable for many practical control applications. First, MFAC is a pure DDC method, since it only depends on the real-time measurement I/O data of the controlled plant, but not on any model information of the plant. This implies that a general controller for a class of practical industrial processes could be designed independently. Second, MFAC does not require any external testing signals or any training process, which are necessary for the neural-networks-based nonlinear adaptive control; therefore, it is a low-cost controller. Third, MFAC is simple and easily implemented with small computational burden and has strong robustness. Fourth, under certain practical assumptions, the monotonic convergence and bounded-input bounded-output (BIBO) stability of the CFDL data model based MFAC and PFDL data model based MFAC can be guaranteed. It is a highlighted feature compared with other DDC approaches. Finally, the simplest CFDL data model based MFAC scheme has been successfully implemented in many practical applications, for example, the chemical industry [27,28], linear motor control [29], injection modeling process [30], pH value control [31], and so on.

Unfalsified Control— The unfalsified control (UC) method was first proposed by M. G. Safonov in 1995 [32]. It recursively falsifies control parameter sets that fails to satisfy the performance specification in order to derive the proper parameter and corresponding controller. The whole process is operated only by using the I/O data rather than the mathematical model of the controlled plant. UC belongs to a type of switching control method in essence, but is different from the traditional switching control. UC can falsify the controller that cannot stabilize the control system before being inserted into the feedback loop, so the transient performance is pretty good. UC includes three elements: an invertible controller candidate set, a cost detectable performance specification, and a controller switching mechanism [33,34]. The input–output stability of the unfalsified adaptive switching control system in a noisy environment is obtained in Ref. [35], and other modifications could be found in Ref. [36]. The UC method has been successfully used in the fields of missile guidance, robot arm control, and industrial process control [37].

Offline Data Based DDC methods include PID Control, Iterative Feedback Tuning, Correlation-Based Tuning, and Virtual Reference Feedback Tuning, etc.

PID Control — The body of literature on the PID control methods is large. The techniques are well developed and widely used in practical applications. Up to now, 95% of the control methods utilized in the industrial process are PID [38]. It is worth pointing out that PID control may be considered as the earliest DDC method, although parameter tuning methods and techniques for PID control are still under development.

Iterative Feedback Tuning — The IFT method was first proposed by H. Hjalmarsson in 1994 [39]. It is a data-driven controller tuning scheme, which iteratively optimizes the parameter of a feedback controller by using an offline gradient estimation of a control performance criterion with respect to the control input. At each iteration, the gradient is estimated by using a finite set of data obtained partly from the normal operating condition of the closed-loop system, and partly from a special experiment whose reference signal is a specified signal. Under suitable assumptions, the algorithm converges to a local minimum of the control performance criterion [40]. Some extension results of IFT to nonlinear systems could be found in Refs. [41–44]. The industrial and experimental applications of IFT are summarized in Refs. [40,45].

Correlation-Based Tuning — The correlation-based tuning (CbT) method was proposed by K. Karimi, L. Miskovic, and D. Bonvin in 2002 [46]. It is a kind of data-driven iterative controller tuning method. The underlying idea is inspired by the well-known correlation approach in system identification. The controller parameters are tuned iteratively either to decorrelate the closed-loop output error between the designed and achieved closed-loop systems with the external reference signal (decorrelation procedure) or to reduce this correlation (correlation reduction). It is necessary to point out that CbT and IFT are closely related methods, but they differ in two aspects:

the underlying control objective and the way of obtaining the gradient estimate. CbT is extended to the multiple-input and multiple-output (MIMO) systems in Ref. [47] and applied to the suspension system in Refs. [48,49].

Virtual Reference Feedback Tuning — The VRFT method was proposed by G. O. Guardabassi and S. M. Savaresi in 2000 [50]. It is a direct data-driven method to optimize the parameters of the controller with a prespecified controller structure for a linear time-invariant (LTI) system. VRFT transforms the controller design problem into a controller parameter identification problem by introducing a virtual reference signal. VRFT and IFT belong to the same class of controller design methods but their features are quite different. IFT is a gradient descent based iterative algorithm. VRFT is a "one-shot" batch method that searches for the global minimum of the performance index, without the need for iteration or initialization. It makes use of a set of input–output data collecting in an open or closed loop, and does not require specific elaborate experiments [51]. In Ref. [52], the VRFT approach is extended to a nonlinear setup. In Ref. [53], VRFT is extended to the MIMO system. VRFT has been successfully applied to some practical systems, such as a vertical-type one-link arm [54], an active suspension system [55], and velocity controllers in self-balancing industrial manual manipulators [56], and so on.

Online/Offline Data Based DDC methods include Iterative Learning Control and Lazy Learning (LL), etc.

Iterative Learning Control — The ILC was first proposed by M. Uchiyama in 1978 in Japanese [57], which did not draw enough attention. Since Ref. [58] was published in English in 1984, ILC has been extensively studied with significant progress in theory and has been widely applied in many fields. For a system that repeats the same task over a finite interval, ILC is an ideal control method, since it targets control performance improvement by learning from the repetitive operations via output tracking errors and control input signals at previous iterations. ILC is of a very simple controller structure, which is considered as an integrator in the iteration domain. ILC as a kind of DDC control method requires little prior knowledge of the system and can guarantee learning error convergence when the iteration number goes to infinity. Recent research results [59–63] of ILC are summarized comprehensively and systematically, and the contraction mapping method forms the theoretical basis of most ILC studies [64,65]. In addition, ILC has been widely applied in many practical fields [66,67]. Compared with other DDC approaches, ILC uses the collected online/offline data in a more abundant and systematic way. It is worth noting that ILC does not tune controller parameters using the I/O data, but finds the optimal control input signal directly.

Lazy Learning — Lazy learning (LL) algorithm is one of the supervised machine learning algorithms. S. Schaal and C. G. Atkeson first applied LL algorithms to the control field in 1994 [68]. Similar to other supervised machine learning algorithms, the goal of LL algorithms is to find the relationship between input and output from a collection of input/output data, called the training set. LL-based control, using the historical data, builds online a local linear dynamic model for the nonlinear plant, and then designs a local controller at every time instant. Owing to the real-time updating of the historical dataset, LL-based control can be considered as an intrinsically adaptive control method. Its computational cost, however, is high. Besides this main shortcoming of LL-based control, there is also a lack of theoretical analysis of the stability [69,70]. In the literature, there are some other methods similar to LL, such as just-in-time learning (JITL) [71], instance-based learning [72], local weighted model [73], model-on-demand [74,75], and so on.

1.2.4.2 DDC Classification According to Controller Structure Design

In this section, we will use another criterion, that is, whether the structure of the controller is known or not, to classify DDC approaches in order to make the readers understand them well.

DDC Methods with Known Controller Structure include PID, IFT, VRFT, UC, SPSA-based control, CbT, etc.

For this kind of DDC method, controller design is carried out by using a plant's I/O measurement data on the basis of a known controller structure with unknown parameters. The controller parameters are obtained by some procedure of optimization, such as batch and recursive algorithms. In other words, the controller design problem for this kind of DDC method is transformed into controller parameter identification with the help of the assumption that the controller structure is known *a priori* and linear in controller parameters. Typical ones are PID, IFT, VRFT, UC, SPSA-based control, CbT, and so on. No explicit information of the plant model and the dynamics model is involved in these methods. The focus, however, is how to determine the controller structure. Generally speaking, it is quite difficult to construct a proper controller with unknown parameters for a particular plant, especially the one with a general nonlinear structure. Sometimes, the difficulty is equal to that of accurately building a plant model. Another obstacle to apply this kind of DDC method is the lack of a stability analysis methodology for a closed-loop system.

DDC Methods with Unknown Controller Structure include model-related DDC methods and model-unrelated DDC methods.

Model-Related DDC Methods — It seems that this kind of DDC method is merely dependent on the plant I/O measurement data, but in essence the plant model structure and the dynamics information are implicitly involved in the controller design. Control system design and theoretical analysis approaches for these kinds of DDC methods are similar to those of MBC design. However, model-related DDC methods are also of significance for the control system design because strong robustness can be achieved when they are used in practice. Typical ones are direct adaptive control, subspace predictive control, and so on.

Model-Unrelated DDC Methods — For these kinds of DDC methods, the controller is designed only using the plant measurement I/O data without involving any explicit or implicit model information. Model-unrelated DDC methods can uniformly fulfill the control task for linear and nonlinear systems. Another outstanding feature of these kinds of DDC methods is that it has a systematic framework for controller design and provides a systematic stability analysis means. Typical ones are ILC and MFAC. Compared with other DDC methods, the effectiveness or rationality of the controller structure or controller design in model-unrelated DDC methods is theoretically guaranteed.

1.2.5 Summary of Data-Driven Control Methods

Toward a comprehensive understanding of DDC methods for the readers, some brief remarks are listed here:

1. Theoretically speaking, ILC, SPSA, UC, and MFAC are originally developed for nonlinear systems by directly using the I/O data of the controlled plants. Other methods, such as IFT and VRFT, are proposed for LTI systems and then extended to nonlinear systems.
2. SPSA, MFAC, UC, and LL have adaptive features, while other DDC methods are nonadaptive control methods. The adaptability of SPSA may be affected by variation of plant structure or parameters.
3. Essentially, SPSA, IFT, and VRFT are controller parameter identification approaches. Specifically, VRFT is a one-shot direct identification method, that is, only one experiment is needed for I/O data collection, and controller parameters are directly identified through offline optimization. The other two are iterative identification methods.
4. Both MFAC and LL are based on dynamic linearization. Specifically, MFAC gives a systematic dynamic linearization framework, including several virtual data models and a series of controller design strategies, with contraction-mapping-like stability analysis and error convergence of closed-loop systems. No framework is formed yet for the LL-based control method.
5. Most of the DDC methods except for PID, ILC, and VRFT need to estimate the gradient using measured I/O data. SPSA, IFT, and gradient-based UC estimate a gradient of a certain cost function with respect to the controller parameters. The dynamic linearization based MFAC and LL, however, calculate the gradient of the system output with respect to the control input online at each instant.
6. SPSA, UC, and MFAC use online measured I/O data, and PID, IFT, and VRFT use offline measured I/O data, and ILC and LL use both online and offline data. It is worth noting that ILC provides a systematic way to use the online/offline I/O measurement data. Moreover, ILC approximates directly

to the desired control signal in the iteration domain rather than tuning controller parameter.

7. For ILC, a perfect systematic framework is put forward in both the controller design and performance analysis. MFAC has the similar feature, but the other DDC methods need further study.
8. Almost all the DDC methods mentioned above are designed based on controller parameter tuning approaches except ILC. Some of them are online tuning, such as MFAC, UC, and SPSA, while the others are offline tuning. The key of the DDC methods is that the controller structure does not depend on the plant model, although controller structure is assumed known *a priori* in some DDC methods. Comparatively, MFAC and LL go one step ahead since their controller structure is based on the dynamic linearization data model and certain optimization criterions that are theoretically supported, while the assumption in the other DDC methods that the controller structure is known inevitably leads to a problem of how to determine the controller structure. Sometimes, the difficulty in determining a proper controller structure for a given plant is equivalent to that of building an accurate plant model.
9. Moreover, as we know, controller parameter tuning is an optimization problem in mathematics, but the optimization in DDC controller design is quite different from traditional optimization because the system model during DDC controller design is unknown, while the model in MBC is known *a priori*. From this point of view, an outstanding feature of MFAC, SPSA, and IFT is that the technique to calculate or estimate the gradient information using the I/O data of the controlled plant is developed when the objective function is unknown. Specifically, MFAC and IFT use the deterministic approach, and SPSA uses the stochastic approximation approach.
10. The key to distinguish DDC methods and MBC methods lies in whether or not its control design is only dependent on I/O data of the controlled system, that is, whether or not the information about dynamic model structure or dynamics (including other forms of expression such as NN, fuzzy rules, expert knowledge, etc.) is embedded in the controller. If only I/O data of the controlled system is utilized, but neither any information about the model structure nor dynamics equality is involved in the control design, it is a DDC method; otherwise, it is an MBC method.

1.3 Preview of the Book

There are 12 chapters in this book. Chapters 2 through 10 focus on the design and analysis of the MFAC theory and methodology, Chapter 11 is dedicated to MFAC applications, and Chapter 12 presents the conclusions.

Chapter 1 first surveys the issues on modeling, identification, and control within the framework of MBC theory and then gives the definition, classification, and highlights on the existing DDC methods.

Chapter 2 introduces the online parameter estimation algorithms, which intends to make the monograph self-contained.

Chapter 3 presents a novel dynamic linearization method for a class of discrete time single-input single-output (SISO), multiple-input and single-output (MISO), and MIMO nonlinear systems, which serves as the theoretical foundation for the design and analysis of MFAC. By this dynamic linearization method, the CFDL data model, the PFDL data model, and the FFDL data model are introduced.

Chapters 4 and 5 study the design, stability analysis, and simulation verification of the dynamic linearization based MFAC schemes for discrete-time SISO, MISO, and MIMO nonlinear systems, respectively.

Chapter 6 describes the design, stability analysis, and simulation verification of model-free adaptive predictive control (MFAPC) schemes for discrete-time SISO nonlinear systems.

Chapter 7 presents the model-free adaptive iterative learning control (MFAILC) scheme for a class of discrete-time SISO nonlinear systems, and shows that the scheme guarantees monotonic convergence of the maximum learning error in the iteration domain.

Chapter 8 extends the MFAC control system design to complicated interconnected nonlinear systems, and the modularized controller design between MFAC and other existing control methods, such as adaptive control and ILC, are also presented in this chapter.

Chapter 9 discusses robustness issues for the MFAC scheme by considering external disturbance and data dropout.

Chapter 10 introduces a conceptual description of the symmetric similarity structure and design principle. The similarity among different control methods, especially for MFAC and MFAILC, adaptive control and ILC, is analyzed.

Chapter 11 lists several applications of the MFAC scheme to practical plants, including a three water tank system, a linear motor system, a traffic system, the welding process, and a wind turbine.

Chapter 12 concludes this book and points out several future research directions closely related to MFAC theory and implementation.

Chapter 2

Recursive Parameter Estimation for Discrete-Time Systems

2.1 Introduction

Mathematical model of a system is the exact description of a plant to be studied. System modeling for a plant consists of two parts: specifying the system model structure and determining the model parameters. There are two ways to get the model parameters if the model structure has been determined. One way is to calibrate the parameter precisely for the first-principles model of the plant by using physical or chemical experiments. The other way is to identify the unknown parameters by using system I/O data according to some optimization process to a certain criterion function of parameter estimation. This procedure is usually called parameter estimation.

The parameter estimation algorithm includes two classes of schemes: the online scheme and the offline scheme. The offline estimation algorithm, also called the batch estimation algorithm, means that the parameter is calculated in a one-shot way by some optimization algorithm on some criterion function using a certain amount of the system I/O data collected *a priori*. Its main advantages are that the real time is not strictly required and a high precision can be achieved by applying complex and advanced algorithms. The online parameter estimation algorithm is also known as the recursive parameter estimation algorithm. Compared to the offline scheme, the online scheme needs to measure the system I/O data persistently, and to recursively update and correct the parameter estimation by making full use of the online data measured. In other words, data measurement and parameter estimation are executed simultaneously. A high operating speed is indispensable for online estimation, which is highly desired for the real-time process and adaptive control system design, since one iterative computation should be accomplished within one sampling period. In practical applications, the online estimation algorithm should have a simple and stable updating structure besides the real-time requirement.

The recursive parameter estimation algorithm usually has the following general structure:

$$\hat{\boldsymbol{\theta}}(k) = \boldsymbol{g}(\hat{\boldsymbol{\theta}}(k-1), D(k), k), \tag{2.1}$$

where $\hat{\boldsymbol{\theta}}(k)$ is the estimation of the parameter $\boldsymbol{\theta}$ at time k; $D(k)$ denotes the available data set of the dynamic system till time k, which generally includes the present and past measurements of system outputs and control inputs,

$$D(k) = \left\{Y(k), U(k)\right\} = \left\{y(k), y(k-1), \ldots; u(k), u(k-1), \ldots\right\},$$

and $\boldsymbol{g}(\ldots)$ is a function. Different functions $\boldsymbol{g}(\ldots)$ mark different parameter estimation algorithms.

A typical form of parameter estimation algorithm widely used in practice is

$$\hat{\boldsymbol{\theta}}(k) = \hat{\boldsymbol{\theta}}(k-1) + M(k-1)\bar{\boldsymbol{\phi}}(k-1)\bar{e}(k), \tag{2.2}$$

where $M(k-1)$ denotes an algorithm gain (possibly a matrix), $\bar{\boldsymbol{\phi}}(k-1)$ denotes a regression vector of some kind composed of elements from $Y(k-1)$ and $U(k-1)$, and $\bar{e}(k)$ denotes some kind of modeling error.

To discuss conveniently with a clear outline, we restrict our attention to the deterministic SISO system in this chapter. This chapter is organized as follows. Section 2.2 introduces two typical parameter estimation algorithms for linearly parameterized systems, that is, the projection algorithm and the least-squares algorithm. Parameter estimation algorithms of nonlinearly parameterized systems in the existing literature, as well as the revised projection algorithm and the revised least-squares algorithm, are presented in Section 2.3. Section 2.4 presents the conclusions.

2.2 Parameter Estimation Algorithm for Linearly Parameterized Systems

A large class of linear and nonlinear deterministic dynamical systems can be simply described by the following linearly parameterized model:

$$y(k) = \boldsymbol{\phi}^T(k-1)\boldsymbol{\theta}_0, \tag{2.3}$$

where $\boldsymbol{\phi}(k-1)$ denotes a regression vector that is a linear or nonlinear function of system I/O data before time k. $\boldsymbol{\theta}_0$ is an unknown time-invariant parameter vector.

The two classical parameter estimation algorithms presented in this section are from Ref. [76]. They are cited here so as to draw forth the revised parameter estimation algorithms for nonlinearly parameterized systems in Section 2.3, and to facilitate citation and discussion in the later chapters of the book.

2.2.1 Projection Algorithm

To estimate the time-invariant parameter $\boldsymbol{\theta}_0$ in system (2.3), the projection algorithm is constructed as

$$\hat{\boldsymbol{\theta}}(k) = \hat{\boldsymbol{\theta}}(k-1) + \frac{\boldsymbol{\phi}(k-1)}{\boldsymbol{\phi}^T(k-1)\boldsymbol{\phi}(k-1)}(y(k) - \boldsymbol{\phi}^T(k-1)\hat{\boldsymbol{\theta}}(k-1)), \tag{2.4}$$

with a given initial value $\hat{\boldsymbol{\theta}}(0)$.

The following theorem implies that the projection algorithm is the optimal solution of an optimization problem.

Theorem 2.1

Given $\hat{\boldsymbol{\theta}}(k-1)$ and $y(k)$, the minimum of

$$J(\hat{\boldsymbol{\theta}}(k)) = \frac{1}{2} \| \hat{\boldsymbol{\theta}}(k) - \hat{\boldsymbol{\theta}}(k-1) \|^2 \tag{2.5}$$

subject to

$$y(k) = \boldsymbol{\phi}^T(k-1)\hat{\boldsymbol{\theta}}(k) \tag{2.6}$$

can be achieved by implementing projection algorithm (2.4).

To avoid dividing by zero in algorithm (2.4), the following modified updating algorithm is usually used for practical applications:

$$\hat{\boldsymbol{\theta}}(k) = \hat{\boldsymbol{\theta}}(k-1) + \frac{\alpha\boldsymbol{\phi}(k-1)}{c + \boldsymbol{\phi}^T(k-1)\boldsymbol{\phi}(k-1)}(y(k) - \boldsymbol{\phi}^T(k-1)\hat{\boldsymbol{\theta}}(k-1)), \tag{2.7}$$

where $c > 0$ and $0 < \alpha < 2$.

Denote

$$\tilde{\boldsymbol{\theta}}(k) = \hat{\boldsymbol{\theta}}(k) - \boldsymbol{\theta}_0, \tag{2.8}$$

$$e(k) = y(k) - \boldsymbol{\phi}^T(k-1)\hat{\boldsymbol{\theta}}(k-1) = -\boldsymbol{\phi}^T(k-1)\tilde{\boldsymbol{\theta}}(k-1). \tag{2.9}$$

Elementary properties of the projection algorithm are summarized as below.

Theorem 2.2

For projection algorithm (2.7) and model (2.3), it follows that

i. $\|\hat{\boldsymbol{\theta}}(k) - \boldsymbol{\theta}_0\| \le \|\hat{\boldsymbol{\theta}}(k-1) - \boldsymbol{\theta}_0\| \le \|\hat{\boldsymbol{\theta}}(0) - \boldsymbol{\theta}_0\|, \quad k \ge 1;$ (2.10)

ii. $\lim_{N\to\infty} \sum_{k=1}^{N} \frac{e^2(k)}{c + \boldsymbol{\phi}^T(k-1)\boldsymbol{\phi}(k-1)} < \infty,$ (2.11)

and this implies that

a. $\lim_{k\to\infty} \frac{e(k)}{(c + \boldsymbol{\phi}^T(k-1)\boldsymbol{\phi}(k-1))^{1/2}} = 0,$ (2.12)

b. $\lim_{N\to\infty} \sum_{k=1}^{N} \frac{\boldsymbol{\phi}^T(k-1)\boldsymbol{\phi}(k-1)e^2(k)}{(c + \boldsymbol{\phi}^T(k-1)\boldsymbol{\phi}(k-1))^2} < \infty,$ (2.13)

c. $\lim_{N\to\infty} \sum_{k=1}^{N} \left\|\hat{\boldsymbol{\theta}}(k) - \hat{\boldsymbol{\theta}}(k-1)\right\|^2 < \infty,$ (2.14)

d. $\lim_{N\to\infty} \sum_{k=t}^{N} \left\|\hat{\boldsymbol{\theta}}(k) - \hat{\boldsymbol{\theta}}(k-t)\right\|^2 < \infty,$ (2.15)

e. $\lim_{k\to\infty} \left\|\hat{\boldsymbol{\theta}}(k) - \hat{\boldsymbol{\theta}}(k-t)\right\| = 0,$ for any finite t. (2.16)

2.2.2 Least-Squares Algorithm

2.2.2.1 Least-Squares Algorithm for Time-Invariant Parameter Estimation

For estimating the time-invariant parameter in model (2.3), the recursive least-squares algorithm described as follows can also be applied:

$$\hat{\boldsymbol{\theta}}(k) = \hat{\boldsymbol{\theta}}(k-1) + \frac{P(k-2)\boldsymbol{\phi}(k-1)}{1+\boldsymbol{\phi}^T(k-1)P(k-2)\boldsymbol{\phi}(k-1)}(y(k) - \boldsymbol{\phi}^T(k-1)\hat{\boldsymbol{\theta}}(k-1)), \quad (2.17)$$

$$P(k-1) = P(k-2) - \frac{P(k-2)\boldsymbol{\phi}(k-1)\boldsymbol{\phi}^T(k-1)P(k-2)}{1+\boldsymbol{\phi}^T(k-1)P(k-2)\boldsymbol{\phi}(k-1)}, \quad (2.18)$$

where $\hat{\boldsymbol{\theta}}(0)$ is the given initial estimation of $\boldsymbol{\theta}$ and $P(-1)$ is any given positive definite matrix P_0.

Theorem 2.3

Algorithm (2.17)–(2.18) minimizes the following quadratic cost function:

$$J_N(\boldsymbol{\theta}) = \frac{1}{2}\sum_{k=1}^{N}(y(k) - \boldsymbol{\phi}^T(k-1)\boldsymbol{\theta})^2 + \frac{1}{2}(\boldsymbol{\theta} - \hat{\boldsymbol{\theta}}(0))^T P_0^{-1}(\boldsymbol{\theta} - \hat{\boldsymbol{\theta}}(0)). \quad (2.19)$$

The following well-known matrix inversion lemma is used to prove Theorem 2.3.

Lemma 2.1 (Matrix Inversion Lemma)

If

$$P^{-1}(k-1) = P^{-1}(k-2) + \boldsymbol{\phi}(k-1)\boldsymbol{\phi}^T(k-1)\alpha(k-1), \quad (2.20)$$

where the scalar $\alpha(k-1) > 0$, then

$$P(k-1) = P(k-2) - \frac{P(k-2)\boldsymbol{\phi}(k-1)\boldsymbol{\phi}^T(k-1)P(k-2)\alpha(k-1)}{1+\boldsymbol{\phi}^T(k-1)P(k-2)\boldsymbol{\phi}(k-1)}, \quad (2.21)$$

and also

$$P(k-1)\boldsymbol{\phi}(k-1) = \frac{P(k-2)\boldsymbol{\phi}(k-1)}{1+\boldsymbol{\phi}^T(k-1)P(k-2)\boldsymbol{\phi}(k-1)\alpha(k-1)}, \quad (2.22)$$

$$P(k-2)\boldsymbol{\phi}(k-1) = \frac{P(k-1)\boldsymbol{\phi}(k-1)}{1-\boldsymbol{\phi}^T(k-1)P(k-1)\boldsymbol{\phi}(k-1)\alpha(k-1)}, \quad (2.23)$$

$$\boldsymbol{\phi}^T(k-1)P(k-1)\boldsymbol{\phi}(k-1) = \frac{\boldsymbol{\phi}^T(k-1)P(k-2)\boldsymbol{\phi}(k-1)}{1+\boldsymbol{\phi}^T(k-1)P(k-2)\boldsymbol{\phi}(k-1)\alpha(k-1)}, \quad (2.24)$$

$$\boldsymbol{\phi}^T(k-1)P(k-2)\boldsymbol{\phi}(k-1) = \frac{\boldsymbol{\phi}^T(k-1)P(k-1)\boldsymbol{\phi}(k-1)}{1-\boldsymbol{\phi}^T(k-1)P(k-1)\boldsymbol{\phi}(k-1)\alpha(k-1)}. \tag{2.25}$$

The basic convergence properties of the least-squares algorithm are summarized in the following theorem.

Theorem 2.4

For least-squares algorithm (2.17)–(2.18) and model (2.3), it follows that

i. $$\|\hat{\boldsymbol{\theta}}(k)-\boldsymbol{\theta}_0\|^2 \le \sigma_1 \|\hat{\boldsymbol{\theta}}(0)-\boldsymbol{\theta}_0\|^2, \quad k \ge 1, \tag{2.26}$$

where σ_1 is the condition number of $P(-1)^{-1}$, that is, $\sigma_1 = \lambda_{\max}(P^{-1}(-1))/(\lambda_{\min}(P^{-1}(-1)))$.

ii. $$\lim_{N\to\infty}\sum_{k=1}^{N}\frac{e^2(k)}{1+\boldsymbol{\phi}^T(k-1)P(k-2)\boldsymbol{\phi}(k-1)} < \infty, \tag{2.27}$$

and this implies that

a. $$\lim_{k\to\infty}\frac{e(k)}{\left(1+\sigma_1\boldsymbol{\phi}^T(k-1)\boldsymbol{\phi}(k-1)\right)^{1/2}} = 0, \tag{2.28}$$

b. $$\lim_{N\to\infty}\sum_{k=1}^{N}\frac{\boldsymbol{\phi}^T(k-1)P(k-2)\boldsymbol{\phi}(k-1)e^2(k)}{\left(1+\boldsymbol{\phi}^T(k-1)P(k-2)\boldsymbol{\phi}(k-1)\right)^2} < \infty, \tag{2.29}$$

c. $$\lim_{N\to\infty}\sum_{k=1}^{N}\|\hat{\boldsymbol{\theta}}(k)-\hat{\boldsymbol{\theta}}(k-1)\|^2 < \infty, \tag{2.30}$$

d. $$\lim_{N\to\infty}\sum_{k=t}^{N}\|\hat{\boldsymbol{\theta}}(k)-\hat{\boldsymbol{\theta}}(k-t)\|^2 < \infty, \tag{2.31}$$

e. $$\lim_{k\to\infty}\|\hat{\boldsymbol{\theta}}(k)-\hat{\boldsymbol{\theta}}(k-t)\| = 0, \quad \text{for any finite } t. \tag{2.32}$$

2.2.2.2 *Modified Least-Squares Algorithm for Time-Varying Parameter Estimation*

Case 1: Least squares with exponential data weighting—According to the fact that the most recent data are assumed to be more informative than past data, the basic idea in this case is that we exponentially discard old data. The algorithm presented here is derived by using the following exponentially weighted cost function:

$$S_N(\boldsymbol{\theta}) = \alpha(N-1)S_{N-1}(\boldsymbol{\theta}) + \left(y(N) - \boldsymbol{\phi}^T(N-1)\boldsymbol{\theta}\right)^2, \tag{2.33}$$

where $0 < \alpha(\cdot) < 1$ and N is the length of the data collected. Note that $\alpha(k) \equiv 1$ gives the standard least-squares cost function.

Analogous to the method used in Section 2.2.2.1, it can be readily shown that minimization of the cost function $S_N(\boldsymbol{\theta})$ leads to the following recursive algorithm:

$$\begin{aligned}\hat{\boldsymbol{\theta}}(k) &= \hat{\boldsymbol{\theta}}(k-1)\\ &\quad + \frac{P(k-2)\boldsymbol{\phi}(k-1)}{\alpha(k-1) + \boldsymbol{\phi}^T(k-1)P(k-2)\boldsymbol{\phi}(k-1)}(y(k) - \boldsymbol{\phi}^T(k-1)\hat{\boldsymbol{\theta}}(k-1)),\end{aligned} \tag{2.34}$$

$$P(k-1) = \frac{1}{\alpha(k-1)}\left(P(k-2) - \frac{P(k-2)\boldsymbol{\phi}(k-1)\boldsymbol{\phi}^T(k-1)P(k-2)}{\alpha(k-1) + \boldsymbol{\phi}^T(k-1)P(k-2)\boldsymbol{\phi}(k-1)}\right), \tag{2.35}$$

where $\hat{\boldsymbol{\theta}}(0)$ is the given initial estimation of $\boldsymbol{\theta}$, $P(-1)$ is any given positive definite matrix, such as $P(-1) = cI$, c is a positive constant, and I is the identity matrix.

In Ref. [77], a good candidate for $\alpha(k)$ is offered by

$$\alpha(k) = \alpha_0\alpha(k-1) + (1 - \alpha_0), \tag{2.36}$$

with typical values $\alpha(0) = 0.95$ and $\alpha_0 = 0.99$.

Case 2: Least squares with covariance resetting—In practice, the ordinary least-squares algorithm has a very fast initial convergence rate, but the algorithm gain reduces dramatically when the matrix P becomes smaller after a few iterations (typically 10 or 20); as a result, the updating ability becomes weaker. To prevent the undesired phenomenon, one feasible way is to reset the covariance matrix P at various time instants. This will revitalize the algorithm and is helpful in maintaining an overall fast rate of convergence. This is particularly useful when estimating time-varying parameters. In this case, it is the time to reset P when one suspects that a significant parameter change occurs.

The least-squares algorithm with covariance resetting is described as follows:

$$\hat{\boldsymbol{\theta}}(k) = \hat{\boldsymbol{\theta}}(k-1) + \frac{P(k-2)\boldsymbol{\phi}(k-1)}{1 + \boldsymbol{\phi}^T(k-1)P(k-2)\boldsymbol{\phi}(k-1)}(y(k) - \boldsymbol{\phi}^T(k-1)\hat{\boldsymbol{\theta}}(k-1)), \tag{2.37}$$

$$P(-1) = k_0 I, \quad k_0 > 0, \tag{2.38}$$

where $\hat{\boldsymbol{\theta}}(0)$ is the given initial value. Let $Z_s = \{k_1, k_2, \ldots\}$ be the time series when resetting occurs. For $k \notin Z_s$, an ordinary recursive least-squares algorithm update is used, that is,

$$P(k-1) = P(k-2) - \frac{P(k-2)\boldsymbol{\phi}(k-1)\boldsymbol{\phi}^T(k-1)P(k-2)}{1+\boldsymbol{\phi}^T(k-1)P(k-2)\boldsymbol{\phi}(k-1)}. \tag{2.39}$$

Otherwise, for $k = k_i \in Z_s$, $P(k_i - 1)$ is reset as follows:

$$P(k_i - 1) = k_i I, \quad 0 < k_{\min} \leq k_i \leq k_{\max} < \infty. \tag{2.40}$$

Theorem 2.5

For algorithm (2.37)–(2.40) and subject to model (2.3), it follows that

i. $\|\hat{\boldsymbol{\theta}}(k) - \boldsymbol{\theta}_0\|^2 \leq \|\hat{\boldsymbol{\theta}}(k-1) - \boldsymbol{\theta}_0\|^2 \leq \|\hat{\boldsymbol{\theta}}(0) - \boldsymbol{\theta}_0\|^2, \quad \forall k \geq 1,$ (2.41)

ii. $\lim_{N\to\infty} k_{\min} \sum_{k=1}^{N} \frac{e^2(k)}{1 + k_{\max}\boldsymbol{\phi}^T(k-1)\boldsymbol{\phi}(k-1)} < \infty,$ (2.42)

and this implies that

a. $\lim_{k\to\infty} \frac{e(k)}{\left(1 + k_{\max}\boldsymbol{\phi}^T(k-1)\boldsymbol{\phi}(k-1)\right)^{1/2}} = 0,$ (2.43)

b. $\lim_{k\to\infty} \|\hat{\boldsymbol{\theta}}(k) - \hat{\boldsymbol{\theta}}(k-t)\| = 0, \quad$ for any finite t. (2.44)

Remark 2.1

From the conclusion in Theorem 2.5, one can readily learn that if P is reset at every time instant; the algorithm simply reduces to the projection algorithm. Thus, scheme (2.37)–(2.40) with covariance resetting allows one to take advantage of both the least-squares and projection algorithms.

Case 3: Least squares with covariance modification—Another way to deal with the time-varying problem is simply using the ordinary least-squares algorithm and adding a modified term into the covariance matrix when parameter changes are detected. This actually gives a similar effect as the previous algorithm since P is kept from converging to zero.

The least-squares algorithm with covariance modification is

$$\hat{\boldsymbol{\theta}}(k) = \hat{\boldsymbol{\theta}}(k-1) + \frac{P(k-2)\boldsymbol{\phi}(k-1)}{1+\boldsymbol{\phi}^T(k-1)P(k-2)\boldsymbol{\phi}(k-1)}(y(k) - \boldsymbol{\phi}^T(k-1)\hat{\boldsymbol{\theta}}(k-1)), \quad (2.45)$$

$$\bar{P}(k-1) = P(k-2) - \frac{P(k-2)\boldsymbol{\phi}(k-1)\boldsymbol{\phi}^T(k-1)P(k-2)}{1+\boldsymbol{\phi}^T(k-1)P(k-2)\boldsymbol{\phi}(k-1)}, \quad (2.46)$$

$$P(k-1) = \bar{P}(k-1) + Q(k-1), \quad (2.47)$$

where $0 \leq Q(k-1) < \infty$.

The convergence of algorithm (2.45)–(2.47) follows as in Theorem 2.5.

2.3 Parameter Estimation Algorithm for Nonlinearly Parameterized Systems

In different parameterized models, parameters can be shown either in the linear form or in the nonlinear form. In fact, most practical plants can be described by nonlinearly parameterized models. In this section, parameter estimation algorithms for nonlinearly parameterized systems are discussed further.

2.3.1 Projection Algorithm and Its Modified Form for Nonlinearly Parameterized Systems

2.3.1.1 Existing Projection Algorithm

Consider the parameter estimation problem for nonlinearly parameterized systems with the following form:

$$y(k+1) = f(Y(k), U(k), \boldsymbol{\theta}), \quad (2.48)$$

where $y(k)$ is the system output at time k; $Y(k)$ and $U(k)$ denote the measurements of the system outputs and inputs until time k, respectively; $\boldsymbol{\theta}$ is an unknown time-invariant or slowly time-varying parameter; and $f(\cdots)$ is a nonlinear function with a known structure.

A projection algorithm for estimating the parameter in system model (2.48) is [78,79]

$$\hat{\boldsymbol{\theta}}(k) = \hat{\boldsymbol{\theta}}(k-1) + \frac{\delta\boldsymbol{\phi}(k-1)}{\|\boldsymbol{\phi}(k-1)\|^2}(y(k) - f(Y(k-1), U(k-1), \hat{\boldsymbol{\theta}}(k-1))), \quad (2.49)$$

where δ is an appropriate positive constant, and

$$\boldsymbol{\phi}(k-1) = \left.\frac{\partial f(Y(k-1),U(k-1),\boldsymbol{\theta})}{\partial \boldsymbol{\theta}(k)}\right|_{\boldsymbol{\theta}=\hat{\boldsymbol{\theta}}(k-1)}.$$

In fact, algorithm (2.49) can be obtained from the following derivation.

Assume that $y(k)$ and $\hat{\boldsymbol{\theta}}(k-1)$ are known. Taking the first-order Taylor expansion for $f(Y(k-1),U(k-1),\boldsymbol{\theta})$ at $\boldsymbol{\theta} = \hat{\boldsymbol{\theta}}(k-1)$, it follows that

$$f(Y(k-1),U(k-1),\boldsymbol{\theta}) \cong f(Y(k-1),U(k-1),\hat{\boldsymbol{\theta}}(k-1)) + \boldsymbol{\phi}^T(k-1)(\boldsymbol{\theta} - \hat{\boldsymbol{\theta}}(k-1)). \tag{2.50}$$

The meaning of the above equation is to approximate the original nonlinear model using a first-order dynamic linear model around operating point $\hat{\boldsymbol{\theta}}(k-1)$. Then, substituting (2.50) into the following criterion function:

$$J(\boldsymbol{\theta}) = \left(y(k) - f(Y(k-1),U(k-1),\boldsymbol{\theta})\right)^2, \tag{2.51}$$

minimizing (2.51), and applying the matrix inversion lemma gives algorithm (2.49), where $\hat{\boldsymbol{\theta}}(k)$ is the optimal solution and the constant δ is added to make the algorithm more general. To avoid dividing by zero in algorithm (2.49), a small positive constant μ should be added in the denominator.

2.3.1.2 Modified Projection Algorithm

The parameter estimation algorithm (2.49), derived by virtue of criterion function (2.51), has the ability to track the time-varying parameter since it minimizes the error between the model output and the actual output at time k. This algorithm, however, may be very sensitive to some abnormal or abrupt measured data (possibly caused by sensor malfunction, man-made influence, noise and disturbance, etc.); therefore, its robustness is not strong enough. In addition, the parameter estimation algorithm for nonlinear systems, aiming at adaptive control system design, must meet real-time requirements for online control and avoid complex calculation. Thus, linear expansion of the nonlinear system at each operating point is an effective way to settle the real-time issue. During the expansion, the distance between two operating points, namely, the linearization scope, should not be too large. Otherwise, estimation accuracy must be worsened, or even the algorithm becomes unstable. It has been proved in practical applications that algorithm (2.49) based on criterion function (2.51) always brings about a large or fast variation of parameter estimation due to some abnormal data, or even becomes unstable. Based on this consideration, we put forward a new criterion function as follows:

$$J(\boldsymbol{\theta}) = \left(y(k+1) - f(Y(k),U(k),\boldsymbol{\theta})\right)^2 + \mu\|\boldsymbol{\theta} - \hat{\boldsymbol{\theta}}(k)\|^2, \qquad (2.52)$$

where $\mu > 0$ is a weighting factor. The term $\mu\|\boldsymbol{\theta} - \hat{\boldsymbol{\theta}}(k)\|^2$ is introduced to restrain the large variation of the estimation value.

Taking first-order Taylor expansion for $f(Y(k),U(k),\boldsymbol{\theta})$ at $\boldsymbol{\theta} = \hat{\boldsymbol{\theta}}(k)$ leads to

$$f(Y(k),U(k),\boldsymbol{\theta}) \cong f(Y(k),U(k),\hat{\boldsymbol{\theta}}(k)) + \boldsymbol{\phi}^T(k)(\boldsymbol{\theta} - \hat{\boldsymbol{\theta}}(k)), \qquad (2.53)$$

where $\boldsymbol{\phi}(k) = \left.\frac{\partial f(Y(k),U(k),\boldsymbol{\theta})}{\partial \boldsymbol{\theta}}\right|_{\boldsymbol{\theta}=\hat{\boldsymbol{\theta}}(k)}$. Substituting (2.53) into criterion function (2.52), solving $J'(\boldsymbol{\theta}) = 0$, and applying the matrix inversion lemma, we can obtain the following modified projection algorithm:

$$\hat{\boldsymbol{\theta}}(k+1) = \hat{\boldsymbol{\theta}}(k) + \frac{\eta_k \boldsymbol{\phi}(k)}{\mu + \|\boldsymbol{\phi}(k)\|^2}(y(k+1) - f(Y(k),U(k),\hat{\boldsymbol{\theta}}(k))). \qquad (2.54)$$

Remark 2.2

The factor η_k is added into algorithm (2.54) to make the algorithm more general, and it will facilitate the convergence proof of the proposed algorithm.

Remark 2.3

Since the penalty factor μ is used to smooth the variation of the parameter estimation, it will enhance the algorithm's robustness to the abnormal data. By appropriate selection of the factor, it will also restrict the substitution range of the nonlinear system (2.48) by the dynamic linear system (2.53). Moreover, it can be seen from algorithm (2.54) that only if $\mu > 0$, the singularity of the proposed algorithm can be avoided.

Remark 2.4

When the system is linearly parameterized, that is,

$$y(k+1) = \boldsymbol{\phi}^T(k)\boldsymbol{\theta}_0,$$

the modified projection algorithm (2.54) becomes projection algorithm (2.7), while the meaning of μ is different. In algorithm (2.54), μ is a penalty factor to smooth the variation of the parameter estimation.

The basic property of the above algorithm is as given below.

Theorem 2.6

For nonlinearly parameterized system (2.48), assume the following conditions: (1) the system has continuous partial derivatives with respect to $\boldsymbol{\theta}$; (2) observed values of system output $y(k)$ are bounded by b_1 for any k; (3) $f(\cdots)$ is uniformly bounded by b_2 for all independent variables at any time k; (4) the estimation sequence $\{\hat{\boldsymbol{\theta}}(k)\}$ obtained from algorithm (2.54) satisfies $\inf_k\{(\boldsymbol{\phi}^*(k))^T\boldsymbol{\phi}(k)\} = \gamma > 0$, where $\boldsymbol{\phi}^*(k)$ denotes the value of the partial derivative $\partial f(Y(k),U(k),\boldsymbol{\theta})/(\partial\boldsymbol{\theta})$ at a certain point between $\hat{\boldsymbol{\theta}}(k)$ and $\hat{\boldsymbol{\theta}}(k+1)$. By appropriately selecting μ and η_k, algorithm (2.54) gives

$$\lim_{k\to\infty}\left|y(k+1) - f(Y(k),U(k),\hat{\boldsymbol{\theta}}(k+1))\right| = 0. \tag{2.55}$$

Proof

Applying Cauchy's mean value theorem to $f(Y(k),U(k),\hat{\boldsymbol{\theta}}(k+1))$ at $\hat{\boldsymbol{\theta}}(k+1) = \hat{\boldsymbol{\theta}}(k)$ gives

$$f(Y(k),U(k),\hat{\boldsymbol{\theta}}(k+1)) = f(Y(k),U(k),\hat{\boldsymbol{\theta}}(k)) + \left(\boldsymbol{\phi}^*(k)\right)^T\Delta\hat{\boldsymbol{\theta}}(k+1), \tag{2.56}$$

where $\Delta\hat{\boldsymbol{\theta}}(k+1) \triangleq \hat{\boldsymbol{\theta}}(k+1) - \hat{\boldsymbol{\theta}}(k)$.

Substituting algorithm (2.54) into (2.56) and subtracting it from the observed system outputs yields

$$\begin{aligned}&\left|y(k+1) - f(Y(k),U(k),\hat{\boldsymbol{\theta}}(k+1))\right|\\&\leq \left|y(k+1) - f(Y(k),U(k),\hat{\boldsymbol{\theta}}(k))\right|\left|1 - \frac{\eta_k\left(\boldsymbol{\phi}^*(k)\right)^T\boldsymbol{\phi}(k)}{\mu + \|\boldsymbol{\phi}(k)\|^2}\right|.\end{aligned} \tag{2.57}$$

Using assumptions (2), (3), and (2.57) gives

$$|y(k+1) - f(Y(k),U(k),\hat{\boldsymbol{\theta}}(k+1))| \le (b_1 + b_2)\left|1 - \frac{\eta_k (\boldsymbol{\phi}^*(k))^T \boldsymbol{\phi}(k)}{\mu + \|\boldsymbol{\phi}(k)\|^2}\right|. \quad (2.58)$$

Assumption (4) means that, for any $0 < \varepsilon < \gamma$, there must exist an integer N such that the following inequality holds for any $k > N$:

$$\frac{(\boldsymbol{\phi}^*(k))^T \boldsymbol{\phi}(k)}{\mu + \|\boldsymbol{\phi}(k)\|^2} \ge \frac{\gamma - \varepsilon}{\mu + \bar{b}^2} > 0, \quad (2.59)$$

where $\bar{b}$ is a constant satisfying $\|(\partial f(Y(k),U(k),\boldsymbol{\theta}))/(\partial \boldsymbol{\theta})\| \le \bar{b}$ (according to assumption (1)).

On the other hand, we may always choose a proper μ by virtue of assumption (1), such that

$$\frac{(\boldsymbol{\phi}^*(k))^T \boldsymbol{\phi}(k)}{\mu + \|\boldsymbol{\phi}(k)\|^2} \le 1. \quad (2.60)$$

Using (2.59) and (2.60), and selecting η_k to satisfy $0 < \eta_k \le (\mu + \bar{b}^2)/(\gamma - \varepsilon)$, (2.58) yield

$$|y(k+1) - f(Y(k),U(k),\hat{\boldsymbol{\theta}}(k+1))| \le \left(1 - \eta_k \frac{\gamma - \varepsilon}{\mu + \bar{b}^2}\right)(b_1 + b_2). \quad (2.61)$$

Moreover, if η_k also satisfies

$$\eta_k \ge \left(1 - \frac{\varepsilon}{b_1 + b_2}\right)\frac{\mu + \bar{b}^2}{\gamma - \varepsilon},$$

then (2.61) implies that

$$\left|y(k+1) - f(Y(k),U(k),\hat{\boldsymbol{\theta}}(k+1))\right| \le \varepsilon,$$

which gives the conclusion. ■

Remark 2.5

Assumption (4) in the above theorem is not strict. For instance, a wide range of systems with time-varying parameters is often described as follows:

$$y(k+1) = \boldsymbol{\varphi}^T(k)\boldsymbol{\theta}(k),$$

where $\boldsymbol{\varphi}(k)$ denotes a regressive vector consisting of available I/O data at time k. When the I/O data are not zero, the relationship $\inf_k\{(\boldsymbol{\phi}^*(k))^T\boldsymbol{\phi}(k)\} = \inf_k\{\boldsymbol{\varphi}^T(k)\boldsymbol{\varphi}(k)\}$ is always guaranteed. In such circumstance, assumption (4) is satisfied immediately.

2.3.2 Least-Squares Algorithm and Its Modified Form for Nonlinearly Parameterized Systems

For nonlinearly parameterized system (2.48), the recursive least-squares algorithm and its modified form are discussed in this section.

2.3.2.1 Existing Least-Squares Algorithm

The criterion function of the least-squares algorithm is

$$J_N(\boldsymbol{\theta}) = \frac{1}{2}\sum_{k=1}^{N}(y(k) - \hat{y}(k,\boldsymbol{\theta}))^2, \tag{2.62}$$

where $\hat{y}(k,\boldsymbol{\theta}) = f(Y(k-1),U(k-1),\boldsymbol{\theta})$ and $\boldsymbol{\theta}$ is an unknown time-invariant or slowly time-varying parameter.

Rewrite (2.62) as follows:

$$J_{N+1}(\boldsymbol{\theta}) = J_N(\boldsymbol{\theta}) + \frac{1}{2}(y(N+1) - \hat{y}(N+1,\boldsymbol{\theta}))^2. \tag{2.63}$$

Resorting to a similar idea in Section 2.3.1.1, taking the first-order Taylor expansion for $\hat{y}(N+1,\boldsymbol{\theta})$ at $\boldsymbol{\theta} = \hat{\boldsymbol{\theta}}(N)$ gives

$$\hat{y}(N+1,\boldsymbol{\theta}) \cong \hat{y}(N+1,\hat{\boldsymbol{\theta}}(N)) + (\hat{y}'(N+1,\hat{\boldsymbol{\theta}}(N)))^T(\boldsymbol{\theta} - \hat{\boldsymbol{\theta}}(N)), \tag{2.64}$$

where $\hat{y}'(N+1,\hat{\boldsymbol{\theta}}) = (\partial f(Y(N),U(N),\boldsymbol{\theta}))/(\partial\boldsymbol{\theta})\big|_{\boldsymbol{\theta}=\hat{\boldsymbol{\theta}}(N)}$. Substituting (2.64) into (2.63), and denoting

$$\boldsymbol{\phi}(N) \triangleq \hat{y}'(N+1,\hat{\boldsymbol{\theta}}(N)),$$

$$Z(N+1) \triangleq y(N+1) - \hat{y}(N+1,\boldsymbol{\theta}) + \boldsymbol{\phi}^T(N)\boldsymbol{\theta},$$

we have

$$J_{N+1}(\boldsymbol{\theta}) = J_N(\boldsymbol{\theta}) + \frac{1}{2}(Z(N+1) - \boldsymbol{\phi}^T(N)\boldsymbol{\theta})^2. \tag{2.65}$$

Differentiating (2.65) with respect to θ gives

$$\nabla J_{N+1}(\boldsymbol{\theta}) = \nabla J_N(\boldsymbol{\theta}) - \boldsymbol{\phi}(N)(Z(N+1) - \boldsymbol{\phi}^T(N)\boldsymbol{\theta}). \tag{2.66}$$

Taking the first-order Taylor expansion for $\nabla J_N(\boldsymbol{\theta})$ at $\boldsymbol{\theta} = \hat{\boldsymbol{\theta}}(N)$ yields

$$\nabla J_N(\boldsymbol{\theta}) \cong \nabla J_N(\hat{\boldsymbol{\theta}}(N)) + \nabla^2 J_N(\hat{\boldsymbol{\theta}}(N))(\boldsymbol{\theta} - \hat{\boldsymbol{\theta}}(N)). \tag{2.67}$$

Substituting (2.67) into (2.66) and using the condition that $\hat{\boldsymbol{\theta}}(N)$ is the minimum of $J_N(\boldsymbol{\theta})$ gives

$$\nabla J_{N+1}(\boldsymbol{\theta}) \cong \nabla^2 J_N(\hat{\boldsymbol{\theta}}(N))(\boldsymbol{\theta} - \hat{\boldsymbol{\theta}}(N)) - \boldsymbol{\phi}(N)(Z(N+1) - \boldsymbol{\phi}^T(N)\boldsymbol{\theta}).$$

Letting $\hat{\boldsymbol{\theta}}(N+1)$ be the optimal solution that minimizes $J_{N+1}(\boldsymbol{\theta})$, we obtain from the above equation

$$\hat{\boldsymbol{\theta}}(N+1) = \hat{\boldsymbol{\theta}}(N) + P(N)\boldsymbol{\phi}(N)(Z(N+1) - \boldsymbol{\phi}^T(N)\hat{\boldsymbol{\theta}}(N)), \tag{2.68}$$

where

$$P(N) = (\nabla^2 J_N(\hat{\boldsymbol{\theta}}(N)) + \boldsymbol{\phi}(N)\boldsymbol{\phi}^T(N))^{-1}. \tag{2.69}$$

To derive the recursive formula of $P(N)$, differentiating both sides of Equation (2.66) with respect to $\boldsymbol{\theta}$ leads to

$$\nabla^2 J_{N+1}(\boldsymbol{\theta}) = \nabla^2 J_N(\boldsymbol{\theta}) + \boldsymbol{\phi}(N)\boldsymbol{\phi}^T(N).$$

This means that

$$P^{-1}(N) = \nabla^2 J_{N+1}(\hat{\boldsymbol{\theta}}(N)).$$

Using the matrix inversion lemma, (2.69) gives

$$P(N) = P(N-1) - \frac{P(N-1)\boldsymbol{\phi}(N)\boldsymbol{\phi}^T(N)P(N-1)}{1+\boldsymbol{\phi}^T(N)P(N-1)\boldsymbol{\phi}(N)}.$$

Thus, we can obtain the recursive least-squares algorithm for nonlinearly parameterized systems as follows:

$$\hat{\boldsymbol{\theta}}(k+1) = \hat{\boldsymbol{\theta}}(k) + P(k)\boldsymbol{\phi}(k)(y(k+1) - f(Y(k),U(k),\hat{\boldsymbol{\theta}}(k))), \tag{2.70}$$

$$P(k) = P(k-1) - \frac{P(k-1)\boldsymbol{\phi}(k)\boldsymbol{\phi}^T(k)P(k-1)}{1+\boldsymbol{\phi}^T(k)P(k-1)\boldsymbol{\phi}(k)}, \tag{2.71}$$

$$\boldsymbol{\phi}(k) = \left.\frac{\partial f(Y(k),U(k),\boldsymbol{\theta})}{\partial \boldsymbol{\theta}}\right|_{\boldsymbol{\theta}=\hat{\boldsymbol{\theta}}(k)}. \tag{2.72}$$

Remark 2.6

When the system becomes

$$y(k+1) = \boldsymbol{\phi}^T(k)\boldsymbol{\theta},$$

where $\boldsymbol{\phi}(k)$ is a certain function of available I/O data at time k, algorithm (2.49) and (2.70)–(2.72) have the same forms as the projection algorithm and the least-squares algorithm in Section 2.2, respectively.

2.3.2.2 Modified Least-Squares Algorithm

As we know, the least-squares algorithm and its modifications have been widely used in parameter identification. When the recursive least-squares algorithm is applied, certain conditions, however, are required for the observed data of the system, such as persistent excitation of signals and so on. Moreover, it is of significant importance that the initial matrix must be appropriately selected so as to increase the convergence speed and overcome the problems arising from the ill-conditioned matrix simultaneously.

To improve the performance of the least-squares algorithm, a revised criterion function (2.62) is proposed as

$$J_N(\boldsymbol{\theta}) = \frac{1}{2}\sum_{k=1}^{N}(y(k) - \hat{y}(k,\boldsymbol{\theta}))^2 + \frac{1}{2}\mu(\boldsymbol{\theta} - \hat{\boldsymbol{\theta}}(N-1))^2. \tag{2.73}$$

Following a similar derivation in Section 2.3.2.1, a modified recursive least-squares algorithm for nonlinearly parameterized system (2.48) is as given below:

$$\hat{\boldsymbol{\theta}}(k+1) = \hat{\boldsymbol{\theta}}(k) + P(k)\boldsymbol{\phi}(k)(y(k+1) - f(Y(k),U(k),\hat{\boldsymbol{\theta}}(k))) \\ - \mu P(k)\Delta\hat{\boldsymbol{\theta}}(k), \tag{2.74}$$

$$P(k) = P(k-1) - \frac{P(k-1)\boldsymbol{\phi}(k)\boldsymbol{\phi}^T(k)P(k-1)}{1+\boldsymbol{\phi}^T(k)P(k-1)\boldsymbol{\phi}(k)}, \tag{2.75}$$

$$\boldsymbol{\phi}(k) = \left.\frac{\partial f(Y(k),U(k),\boldsymbol{\theta})}{\partial\boldsymbol{\theta}}\right|_{\boldsymbol{\theta}=\hat{\boldsymbol{\theta}}(k)}, \tag{2.76}$$

where $\Delta\hat{\boldsymbol{\theta}}(k) = \hat{\boldsymbol{\theta}}(k) - \hat{\boldsymbol{\theta}}(k-1)$.

Remark 2.7

Compared with the ordinary least-squares algorithm (2.70)–(2.72) for a nonlinear system, the modified algorithm (2.74)–(2.76) is still with a simple structure and has an extra new term $\mu P(k)\Delta\hat{\boldsymbol{\theta}}(k)$ added into (2.74) without remarkably increasing the computational burden. When $\mu = 0$, it turns into the ordinary least-squares algorithm.

The convergence analysis on algorithm (2.74)–(2.76) for the general case (2.48) is omitted here. Refer to Refs. [80,81] for details. In the following, we will focus on a special class of nonlinear systems with linear parameters for a clear demonstration of the basic properties of the proposed algorithm. Consider

$$y(k+1) = \boldsymbol{\phi}^T(k)\boldsymbol{\theta}, \tag{2.77}$$

where $\boldsymbol{\theta} = (\theta_1,\ldots,\theta_n)^T$ is an unknown time-invariant or slowly time-varying parameter vector, $\boldsymbol{\phi}(k) = \left[f_1(Y(k),U(k)),\ldots,f_n(Y(k),U(k))\right]^T$, $f_i(\cdots), i = 1,2,\ldots,n$ are known nonlinear functions independent of $\boldsymbol{\theta}$. Model (2.77) is widely used in practice, and many nonlinear systems are special cases of it.

The criterion function is given as

$$J_N(\boldsymbol{\theta}) = \frac{1}{2}\sum_{k=1}^{N}(y(k) - \boldsymbol{\phi}^T(k-1)\boldsymbol{\theta})^2 + \frac{1}{2}(\boldsymbol{\theta} - \hat{\boldsymbol{\theta}}(N-1))^T P^{-1}(0)(\boldsymbol{\theta} - \hat{\boldsymbol{\theta}}(N-1)), \tag{2.78}$$

where $P^{-1}(0) = \mu I$, μ is a constant.

Let

$$\mathbf{\Phi}_N = [\boldsymbol{\phi}(1), \boldsymbol{\phi}(2), \ldots, \boldsymbol{\phi}(N)]^T,$$

$$\boldsymbol{Y}_N = [y(1), y(2), \ldots, y(N)]^T.$$

Then criterion function (2.78) is rewritten in the following vector form:

$$J_N(\boldsymbol{\theta}) = \frac{1}{2}\left\|\boldsymbol{Y}_N - \mathbf{\Phi}_{N-1}\boldsymbol{\theta}\right\|^2 + \frac{1}{2}(\boldsymbol{\theta} - \hat{\boldsymbol{\theta}}(N-1))^T P^{-1}(0)(\boldsymbol{\theta} - \hat{\boldsymbol{\theta}}(N-1)). \quad (2.79)$$

Minimizing criterion function (2.79), we can obtain the modified least-squares algorithm as follows:

$$\Delta\hat{\boldsymbol{\theta}}(k) = (P^{-1}(0) + \mathbf{\Phi}_{k-1}^T\mathbf{\Phi}_{k-1})^{-1}\mathbf{\Phi}_{k-1}^T(\boldsymbol{Y}_k - \mathbf{\Phi}_{k-1}\hat{\boldsymbol{\theta}}(k-1)). \quad (2.80)$$

In such circumstances, the ill-conditioned problem of the matrix $\mathbf{\Phi}_{k-1}^T\mathbf{\Phi}_{k-1}$ can be avoided by selecting an appropriate $P(0)$. To avoid directly computing the inverse matrix in algorithm (2.80), the least-squares algorithm in a recursive form for the nonlinearly parameterized system (2.77) can be given as below, following the similar derivation in Section 2.2:

$$\begin{aligned}\hat{\boldsymbol{\theta}}(k) &= \hat{\boldsymbol{\theta}}(k-1) + P(k-1)\boldsymbol{\phi}(k-1)(y(k) - \boldsymbol{\phi}^T(k-1)\hat{\boldsymbol{\theta}}(k-1)) \\ &\quad - P(k-1)P^{-1}(0)\Delta\hat{\boldsymbol{\theta}}(k-1),\end{aligned} \quad (2.81)$$

where $P(k-1) = (P^{-1}(0) + \mathbf{\Phi}_{k-1}^T\mathbf{\Phi}_{k-1})^{-1}$ and

$$P(k-1) = P(k-2) - \frac{P(k-2)\boldsymbol{\phi}(k-1)\boldsymbol{\phi}^T(k-1)P(k-2)}{1 + \boldsymbol{\phi}^T(k-1)P(k-2)\boldsymbol{\phi}(k-1)}. \quad (2.82)$$

To estimate slowly time-varying parameters, an effective way is to introduce proper data weighting so as to diminish the impact of old data. Applying this simple idea to (2.78), a criterion function with a forgetting factor is proposed as follows:

$$J_N(\boldsymbol{\theta}) = \frac{1}{2}\sum_{i=1}^{N} r^{N-i}(y(i) - \boldsymbol{\phi}^T(i-1)\boldsymbol{\theta})^2 + (\boldsymbol{\theta} - \hat{\boldsymbol{\theta}}(k-1))^T P^{-1}(0)(\boldsymbol{\theta} - \hat{\boldsymbol{\theta}}(k-1)), \quad (2.83)$$

where $r \in [0,1]$ is a weighting factor.

Minimizing criterion function (2.83), the modified least-squares algorithm with data weighting for slowly time-varying parameter estimation is obtained:

$$\Delta\hat{\boldsymbol{\theta}}(k) = (P^{-1}(0) + \boldsymbol{\Phi}_{k-1}^T\boldsymbol{\Omega}_{k-1}\boldsymbol{\Phi}_{k-1})^{-1}\boldsymbol{\Phi}_{k-1}^T\boldsymbol{\Omega}_{k-1}(\boldsymbol{Y}_k - \boldsymbol{\Phi}_{k-1}\hat{\boldsymbol{\theta}}(k-1)), \tag{2.84}$$

where $\boldsymbol{\Omega}_{k-1} = diag\{1, r, r^2, \ldots, r^{k-1}\}$. Note that only if $P(0)$ is invertible, the inverted matrix $(P^{-1}(0) + \boldsymbol{\Phi}_{k-1}^T\boldsymbol{\Omega}_{k-1}\boldsymbol{\Phi}_{k-1})^{-1}$ always exists. Therefore, the algorithm is still effective even if $\boldsymbol{\Phi}_{k-1}^T\boldsymbol{\Omega}_{k-1}\boldsymbol{\Phi}_{k-1}$ becomes ill conditioned. It is obvious that, if $r = 1$, algorithm (2.84) is the same as algorithm (2.80).

The recursive form of algorithm (2.84) is

$$\begin{aligned}\hat{\boldsymbol{\theta}}(k) &= \hat{\boldsymbol{\theta}}(k-1) + P(k-1)\boldsymbol{\phi}(k-1)(y(k) - \boldsymbol{\phi}^T(k-1)\hat{\boldsymbol{\theta}}(k-1)) \\ &\quad - rP(k-1)P^{-1}(0)\Delta\hat{\boldsymbol{\theta}}(k-1),\end{aligned} \tag{2.85}$$

where

$$P(k-1) = (P^{-1}(0) + \boldsymbol{\Phi}_{k-1}^T\boldsymbol{\Omega}_{k-1}\boldsymbol{\Phi}_{k-1})^{-1}. \tag{2.86}$$

Now, we discuss the recursive form of (2.86). For convenience, the initial matrix is set to be a diagonal one as $P^{-1}(0) = \mu I$. Using the Cholesky factorization [82], we have

$$P^{-1}(0) = \sum_{i=1}^{n} \mu \boldsymbol{e}_i \boldsymbol{e}_i^T, \tag{2.87}$$

where n represents the dimension of $P^{-1}(0)$; $\boldsymbol{e}_i$, $i = 1, \ldots, n$, denotes the ith unit vector, that is, all the entries are zero except the ith one. Rewrite (2.86) as

$$P(k-1) = \left(rP^{-1}(k-2) + \boldsymbol{\phi}(k-1)\boldsymbol{\phi}^T(k-1) + (1-r)P^{-1}(0)\right)^{-1}. \tag{2.88}$$

Substituting (2.87) into (2.88) gives

$$P(k-1) = \left(rP^{-1}(k-2) + \boldsymbol{\phi}(k-1)\boldsymbol{\phi}^T(k-1) + (1-r)\sum_{i=1}^{n} \mu \boldsymbol{e}_i \boldsymbol{e}_i^T\right)^{-1}. \tag{2.89}$$

Let

$$\bar{P}(0,k-2) = P(k-2)/r,$$

$$\bar{P}(1,k-2) = \left(\bar{P}^{-1}(0,k-2) + (1-r)\mu \boldsymbol{e}_1\boldsymbol{e}_1^T\right)^{-1}$$

$$= \bar{P}(0,k-2) - \frac{(1-r)\mu\bar{\boldsymbol{p}}_1(0,k-2)\bar{\boldsymbol{p}}_1^T(0,k-2)}{1+(1-r)\mu\bar{p}_{11}(0,k-2)},$$

$$\vdots$$

$$\bar{P}(i,k-2) = \bar{P}(i-1,k-2) + \frac{(1-r)\mu\bar{\boldsymbol{p}}_i(i-1,k-2)\bar{\boldsymbol{p}}_i^T(i-1,k-2)}{1+(1-r)\mu\bar{p}_{ii}(i-1,k-2)},$$

where $\bar{\boldsymbol{p}}_i(i-1,k-2)$ denotes the ith column of $\bar{P}(i-1,k-2)$, and $\bar{p}_{ii}(i-1, k-2)$ denotes the ith element of $\bar{\boldsymbol{p}}_i(i-1,k-2)$. Then

$$P(k-1) = \bar{P}(n,k-2) - \frac{\bar{P}(n,k-2)\boldsymbol{\phi}(k-1)\boldsymbol{\phi}^T(k-1)\bar{P}(n,k-2)}{1+\boldsymbol{\phi}^T(k-1)\bar{P}(n,k-2)\boldsymbol{\phi}(k-1)}. \tag{2.90}$$

According to the above discussions, the modified least-squares algorithm with data weighting for slowly time-varying parameter estimation in recursive form is described as follows:

$$\begin{aligned}\hat{\boldsymbol{\theta}}(k) = {} & \hat{\boldsymbol{\theta}}(k-1) + P(k-1)\boldsymbol{\phi}(k-1)(y(k) - \boldsymbol{\phi}^T(k-1)\hat{\boldsymbol{\theta}}(k-1)) \\ & - rP(k-1)P^{-1}(0)\Delta\hat{\boldsymbol{\theta}}(k-1),\end{aligned} \tag{2.91}$$

$$P(k-1) = \bar{P}(n,k-2) - \frac{\bar{P}(n,k-2)\boldsymbol{\phi}(k-1)\boldsymbol{\phi}^T(k-1)\bar{P}(n,k-2)}{1+\boldsymbol{\phi}^T(k-1)\bar{P}(n,k-2)\boldsymbol{\phi}(k-1)}, \tag{2.92}$$

where

$$\bar{P}(i,k-2) = \bar{P}(i-1,k-2) - \frac{(1-r)\mu\bar{\boldsymbol{p}}_i(i-1,k-2)\bar{\boldsymbol{p}}_i^T(i-1,k-2)}{1+(1-r)\mu\bar{p}_{ii}(i-1,k-2)}, \quad i=1,\ldots,n, \tag{2.93}$$

$$\bar{P}(0,k-2) = \frac{P(k-2)}{r}. \tag{2.94}$$

Remark 2.8

If $r = 0$, algorithm (2.91)–(2.94) becomes

$$\hat{\boldsymbol{\theta}}(k) = \hat{\boldsymbol{\theta}}(k-1) + P(k-1)\boldsymbol{\phi}(k-1)(y(k) - \boldsymbol{\phi}^T(k-1)\hat{\boldsymbol{\theta}}(k-1)), \tag{2.95}$$

$$P(k-1) = P(0) - \frac{P(0)\boldsymbol{\phi}(k-1)\boldsymbol{\phi}^T(k-1)P(0)}{1+\boldsymbol{\phi}^T(k-1)P(0)\boldsymbol{\phi}(k-1)}. \tag{2.96}$$

This is the ordinary least-squares algorithm. When $r = 1$, the algorithm becomes (2.81) and (2.82), that is, the time-invariant parameter estimation algorithm.

In the rest of the section, the properties of the modified recursive least-squares algorithm (2.81) and (2.82) and of the modified recursive least squares with data weighting (2.85) and (2.86) will be discussed in detail.

Theorem 2.7

For model (2.77), the modified recursive least-squares algorithm (2.81) and (2.82) guarantees

i. $\left\|\hat{\boldsymbol{\theta}}(k) - \boldsymbol{\theta}\right\|^2 \le k_c \left\|\hat{\boldsymbol{\theta}}(0) - \boldsymbol{\theta}\right\|^2, \quad \forall k > 0,$ (2.97)

where k_c is the condition number of $P^{-1}(0)$;

ii. $\lim_{k\to\infty} \left\|P(k)\boldsymbol{\phi}(k)e(k)\right\| = 0,$ (2.98)

where $e(k) = y(k) - \boldsymbol{\phi}^T(k-1)\hat{\boldsymbol{\theta}}(k-1)$;

iii. $\lim_{k\to\infty} \left\|\hat{\boldsymbol{\theta}}(k) - \hat{\boldsymbol{\theta}}(k-l)\right\| = 0, \quad$ for any finite l. (2.99)

Proof

i. Define $\tilde{\boldsymbol{\theta}}(k) = \hat{\boldsymbol{\theta}}(k) - \boldsymbol{\theta}$, and $\Delta\hat{\boldsymbol{\theta}}(k) = \hat{\boldsymbol{\theta}}(k) - \hat{\boldsymbol{\theta}}(k-1)$. Obviously, it has

$$\tilde{\boldsymbol{\theta}}(k) = \tilde{\boldsymbol{\theta}}(k-1) + \Delta\hat{\boldsymbol{\theta}}(k). \tag{2.100}$$

Substituting (2.80) into (2.100) and using $\boldsymbol{Y}_k = \boldsymbol{\Phi}_{k-1}\boldsymbol{\theta}$ and the relationship $P(k-1) = (P^{-1}(0) + \boldsymbol{\Phi}_{k-1}^T\boldsymbol{\Phi}_{k-1})^{-1}$ gives

$$\begin{aligned}\tilde{\boldsymbol{\theta}}(k) &= \tilde{\boldsymbol{\theta}}(k-1) + \left(P^{-1}(0) + \boldsymbol{\Phi}_{k-1}^T\boldsymbol{\Phi}_{k-1}\right)^{-1}\boldsymbol{\Phi}_{k-1}^T(\boldsymbol{Y}_k - \boldsymbol{\Phi}_{k-1}\hat{\boldsymbol{\theta}}(k-1)) \\ &= \tilde{\boldsymbol{\theta}}(k-1) - P(k-1)\boldsymbol{\Phi}_{k-1}^T\boldsymbol{\Phi}_{k-1}\tilde{\boldsymbol{\theta}}(k-1).\end{aligned} \tag{2.101}$$

Using $P(k-1) = (P^{-1}(0) + \mathbf{\Phi}_{k-1}^T \mathbf{\Phi}_{k-1})^{-1}$ again, (2.101) becomes

$$\begin{aligned}\tilde{\boldsymbol{\theta}}(k) &= \left(I - P(k-1)\mathbf{\Phi}_{k-1}^T \mathbf{\Phi}_{k-1}\right)\tilde{\boldsymbol{\theta}}(k-1) \\ &= P(k-1)P^{-1}(0)\tilde{\boldsymbol{\theta}}(k-1). \end{aligned} \tag{2.102}$$

Let

$$V(k) = \tilde{\boldsymbol{\theta}}^T(k)P^{-1}(k)\tilde{\boldsymbol{\theta}}(k),$$

and compute the difference of $V(k)$ and use (2.102); then we have

$$V(k) - V(k-1) = \tilde{\boldsymbol{\theta}}^T(k-1)(P^{-1}(0)P(k-1)P^{-1}(0) - P^{-1}(k-1))\tilde{\boldsymbol{\theta}}(k-1). \tag{2.103}$$

From (2.82),

$$P(k) = P(0) - \sum_{i=1}^{k} \frac{P(i-1)\boldsymbol{\phi}(i)\boldsymbol{\phi}^T(i)P(i-1)}{1 + \boldsymbol{\phi}^T(i)P(i-1)\boldsymbol{\phi}(i)}. \tag{2.104}$$

Using (2.104) and $P(k-1) = (P^{-1}(0) + \mathbf{\Phi}_{k-1}^T \mathbf{\Phi}_{k-1})^{-1}$, (2.103) becomes

$$\begin{aligned} V(k) - V(k-1) &= \tilde{\boldsymbol{\theta}}^T(k-1) \\ &\times \left(-\sum_{i=1}^{k-1} \frac{P^{-1}(0)P(i-1)\boldsymbol{\phi}(i)\boldsymbol{\phi}^T(i)P(i-1)P^{-1}(0)}{1 + \boldsymbol{\phi}^T(i)P(i-1)\boldsymbol{\phi}(i)} - \mathbf{\Phi}_{k-1}^T \mathbf{\Phi}_{k-1} \right) \tilde{\boldsymbol{\theta}}(k-1), \end{aligned} \tag{2.105}$$

which implies that $V(k)$ is a nonincreasing function; therefore it gives

$$\tilde{\boldsymbol{\theta}}^T(k)P^{-1}(k)\tilde{\boldsymbol{\theta}}(k) \le \tilde{\boldsymbol{\theta}}^T(0)P^{-1}(0)\tilde{\boldsymbol{\theta}}(0). \tag{2.106}$$

According to $P(k) = (P^{-1}(0) + \mathbf{\Phi}_k^T \mathbf{\Phi}_k)^{-1}$ and the definition of $\mathbf{\Phi}_k$, it follows that

$$P^{-1}(k) = P^{-1}(k-1) + \boldsymbol{\phi}(k-1)\boldsymbol{\phi}^T(k-1),$$

and then we have

$$\sigma_{\min}\left[P^{-1}(k)\right] \ge \sigma_{\min}\left[P^{-1}(k-1)\right] \ge \sigma_{\min}\left[P^{-1}(0)\right],$$

where $\sigma_{\min}[\cdot]$ denotes the smallest singular value of a matrix. It immediately gives

$$\begin{aligned}\sigma_{\min}\left[P^{-1}(0)\right]\left\|\tilde{\boldsymbol{\theta}}(k)\right\|^2 &\le \sigma_{\min}\left[P^{-1}(k)\right]\left\|\tilde{\boldsymbol{\theta}}(k)\right\|^2\\ &\le \tilde{\boldsymbol{\theta}}^T(k)P^{-1}(k)\tilde{\boldsymbol{\theta}}(k)\\ &\le \cdots \le \tilde{\boldsymbol{\theta}}^T(0)P^{-1}(0)\tilde{\boldsymbol{\theta}}(0)\\ &\le \sigma_{\max}\left[P^{-1}(0)\right]\left\|\tilde{\boldsymbol{\theta}}(0)\right\|^2, \end{aligned} \tag{2.107}$$

where $\sigma_{\max}[\cdot]$ denotes the largest singular value of a matrix. This brings us to conclusion (i).

ii. Since $V(k)$ is a nonnegative and nonincreasing function, and

$$\lim_{k\to\infty}\left|\sum_{i=1}^{k}(V(i)-V(i-1))\right| = \lim_{k\to\infty}\left|V(k)-V(0)\right| < \infty,$$

then we have

$$\lim_{k\to\infty}(V(k)-V(k-1)) = 0. \tag{2.108}$$

By using (2.108) and noting that every term in the sum of the right-hand side of (2.105) is negative definite, we have

$$\lim_{k\to\infty}\frac{\tilde{\boldsymbol{\theta}}^T(k-1)P^{-1}(0)P(k-1)\boldsymbol{\phi}(k)\boldsymbol{\phi}^T(k)P(k-1)P^{-1}(0)\tilde{\boldsymbol{\theta}}(k-1)}{1+\boldsymbol{\phi}^T(k)P(k-1)\boldsymbol{\phi}(k)} = 0, \tag{2.109}$$

and

$$\lim_{k\to\infty}\tilde{\boldsymbol{\theta}}^T(k-1)\boldsymbol{\Phi}_{k-1}^T\boldsymbol{\Phi}_{k-1}\tilde{\boldsymbol{\theta}}(k-1) = 0. \tag{2.110}$$

By multiplying the left-hand and right-hand sides of (2.109) by $\boldsymbol{\phi}^T(k-1)$ and $\boldsymbol{\phi}(k-1)$, respectively, and using the symmetric properties of $P(0)$ and $P(k-1)$, we have

$$\lim_{k\to\infty}\frac{\left\|P^{-1}(0)P(k-1)\boldsymbol{\phi}(k)e(k)\right\|^2}{1+\boldsymbol{\phi}^T(k)P(k-1)\boldsymbol{\phi}(k)} = 0. \tag{2.111}$$

Equation (2.111) gives that

$$\lim_{k\to\infty}\frac{\left\|P^{-1}(0)P(k-1)\boldsymbol{\phi}(k)e(k)\right\|^2}{1+\boldsymbol{\phi}^T(k)P(k-1)\boldsymbol{\phi}(k)} \geq \lim_{k\to\infty}\frac{\sigma_{\min}^2\left[P^{-1}(0)\right]}{1+\boldsymbol{\phi}^T(k)P(k-1)\boldsymbol{\phi}(k)}\left\|P(k-1)\boldsymbol{\phi}(k)e(k)\right\|^2 \geq 0. \tag{2.112}$$

Since $\sigma_{\min}[P^{-1}(0)] \neq 0$, combining (2.111) and (2.112) yields

$$\lim_{k\to\infty}\left\|P(k-1)\boldsymbol{\phi}(k)e(k)\right\|^2 = 0. \tag{2.113}$$

Applying (2.82) and (2.113), we have

$$\begin{aligned}\lim_{k\to\infty}\left\|P(k)\boldsymbol{\phi}(k)e(k)\right\|^2 &\leq \lim_{k\to\infty}\left\|P(k-1)\boldsymbol{\phi}(k)e(k)\right\|^2 \\ &+ \lim_{k\to\infty}\frac{\left\|P(k-1)\boldsymbol{\phi}(k)\boldsymbol{\phi}^T(k)\right\|^2}{1+\boldsymbol{\phi}^T(k)P(k-1)\boldsymbol{\phi}(k)}\left\|P(k-1)\boldsymbol{\phi}(k)e(k)\right\|^2 \\ &= 0.\end{aligned} \tag{2.114}$$

This is conclusion (ii).

iii. Using (2.100) and (2.101), we have

$$\lim_{k\to\infty}\Delta\hat{\boldsymbol{\theta}}(k) = \lim_{k\to\infty}P(k-1)\boldsymbol{\Phi}_{k-1}^T\boldsymbol{\Phi}_{k-1}\hat{\boldsymbol{\theta}}(k-1). \tag{2.115}$$

Then, using (2.110) and (2.115), it follows that

$$\lim_{k\to\infty}\Delta\hat{\boldsymbol{\theta}}(k) = 0. \tag{2.116}$$

By using (2.116) and the Schwarz inequality, the conclusion (iii) is reached by

$$\begin{aligned}\lim_{k\to\infty}\left\|\hat{\boldsymbol{\theta}}(k)-\hat{\boldsymbol{\theta}}(k-l)\right\| &= \lim_{k\to\infty}\left\|\hat{\boldsymbol{\theta}}(k)-\hat{\boldsymbol{\theta}}(k-1)+\hat{\boldsymbol{\theta}}(k-1)-\cdots\right. \\ &\qquad\left.+\hat{\boldsymbol{\theta}}(k-l+1)-\hat{\boldsymbol{\theta}}(k-l)\right\| \\ &\leq \lim_{k\to\infty}\left\|\hat{\boldsymbol{\theta}}(k)-\hat{\boldsymbol{\theta}}(k-1)\right\|+\cdots+\lim_{k\to\infty}\left\|\hat{\boldsymbol{\theta}}(k-l+1)\right. \\ &\qquad\left.-\hat{\boldsymbol{\theta}}(k-l)\right\| \\ &= 0.\end{aligned}$$

■

Theorem 2.8

If the same initial values and the same measured outputs are used in least-squares algorithm (2.17) and (2.18) and modified least-squares algorithm (2.81) and (2.82), then we have

$$\left\|\tilde{\boldsymbol{\theta}}(k)\right\|^2 \le \left\|\tilde{\boldsymbol{\theta}}'(k)\right\|^2, \tag{2.117}$$

where $\tilde{\boldsymbol{\theta}}(k)$ and $\tilde{\boldsymbol{\theta}}'(k)$ denote the errors of parameter estimation at time k by ordinary least-squares algorithm and the modified one, respectively.

Proof

From (2.102), $\tilde{\boldsymbol{\theta}}(k)$ can be written as

$$\tilde{\boldsymbol{\theta}}(k) = P(k-1)P^{-1}(0)\tilde{\boldsymbol{\theta}}(k-1) = \left[\prod_{i=1}^{k-1} P(i)P^{-1}(0)\right]\tilde{\boldsymbol{\theta}}(0). \tag{2.118}$$

On the other hand, from (2.17), matrix inversion lemma and $P^{-1}(N-1) = P^{-1}(N-2) + \boldsymbol{\phi}(N-1)\boldsymbol{\phi}^T(N-1)$, it follows that

$$\begin{aligned}\tilde{\boldsymbol{\theta}}'(k) &= \tilde{\boldsymbol{\theta}}'(k-1) + P(k-1)\boldsymbol{\phi}(k-1)(y(k) - \boldsymbol{\phi}^T(k-1)\hat{\boldsymbol{\theta}}'(k-1))\\ &= (I - P(k-1)\boldsymbol{\phi}(k-1)\boldsymbol{\phi}^T(k-1))\hat{\boldsymbol{\theta}}'(k-1)\\ &= P(k-1)P^{-1}(k-2)\hat{\boldsymbol{\theta}}'(k-1)\\ &= \left[\prod_{i=1}^{k-1} P(i)P^{-1}(i-1)\right]\hat{\boldsymbol{\theta}}'(0).\end{aligned} \tag{2.119}$$

If $\tilde{\boldsymbol{\theta}}(0) = \tilde{\boldsymbol{\theta}}'(0)$, then

$$\begin{aligned}\left\|\tilde{\boldsymbol{\theta}}(k)\right\|^2 - \left\|\tilde{\boldsymbol{\theta}}'(k)\right\|^2 = \tilde{\boldsymbol{\theta}}^T(0)\Bigg\{&\left[\prod_{i=1}^{k-1} P(i)P^{-1}(0)\right]^T\left[\prod_{i=1}^{k-1} P(i)P^{-1}(0)\right]\\ &-\left[\prod_{i=1}^{k-1} P(i)P^{-1}(i-1)\right]^T\left[\prod_{i=1}^{k-1} P(i)P^{-1}(i-1)\right]\Bigg\}\tilde{\boldsymbol{\theta}}(0).\end{aligned} \tag{2.120}$$

Since $P(i)$ is positive definite for any $i \geq 0$, and satisfies

$$P^{-1}(i) = P^{-1}(i-1) + \boldsymbol{\phi}(i-1)\boldsymbol{\phi}^T(i-1),$$

we obtain the following conclusion:

$$\left\|\tilde{\boldsymbol{\theta}}(k)\right\|^2 \leq \left\|\tilde{\boldsymbol{\theta}}'(k)\right\|^2.$$

■

Corollary 2.1

Applying the modified recursive least-squares algorithm (2.85) and (2.86) with forgetting factor $r \in [0,1]$ to model (2.77) gives

i. $\lim_{k\to\infty}\left\|\tilde{\boldsymbol{\theta}}(k)\right\| \leq k_c\left\|\tilde{\boldsymbol{\theta}}(0)\right\|, \quad \forall k > 0,$

   where k_c denotes the condition number of $P(0)$.

ii. $\lim_{k\to\infty}(y(k) - \boldsymbol{\phi}^T(k-1)\hat{\boldsymbol{\theta}}(k-1)) = 0.$

Proof

Analogous to the proof of Theorem 2.7. Refer to Ref. [83] for details. ■

2.4 Conclusions

There is a large body of literature on parameter estimation. In this chapter, two classes of parameter estimation algorithms, which are widely used in adaptive control system design, were briefly introduced, to help in understanding them and to facilitate citation in other chapters of this book. More details of parameter estimation algorithms can be found in Refs. [7,84–94].

Chapter 3

Dynamic Linearization Approach of Discrete-Time Nonlinear Systems

3.1 Introduction

The controller design and analysis for discrete-time LTI systems have been fully developed, and they have found many successful applications in practice. The linear model, however, is an approximation of a real system, which is almost always nonlinear. Therefore, it is of great significance to study modeling and control for nonlinear systems. Nevertheless, owing to the inherent complexity of nonlinearity, the study of nonlinear systems is much more difficult than that of the linear ones. Existing works in this field focus mainly on system analysis and controller design for certain special classes of discrete-time nonlinear systems, such as the Hammerstein model [95], the bilinear model [96], the Wiener model [97], and so on.

Nonlinear autoregressive moving average with the exogenous input (NARMAX) model [89,98,99] is a general description of discrete-time nonlinear nonaffine systems. However, the controller design for such a model is quite difficult due to the intrinsic nonlinearity with respect to control input. A common approach to deal with this kind of general nonlinear systems is to convert the original NARMAX model into a linearized one, so as to be easily studied within the relatively well-developed linear system framework. There are several typical linearization methods, such as feedback linearization [100–113], Taylor's linearization [114–118], piecewise linearization [119], orthogonal-function-approximation-based linearization

[120–122], and so on. However, all of these linearization methods have their own limitations. Feedback linearization intends to find a direct channel between the system output and the control input, for which the accurate mathematical model of the controlled system is needed. As we know, obtaining the accurate model is very difficult and sometimes impossible for a practical plant. Taylor's linearization uses Taylor expansion around the operating point without high-order terms to obtain a linearized model, which offers an approximate expression of the controlled system. Taylor's linearization has found many successful applications in practice, but also encounters some difficulties in theoretical analysis for control system design due to the omitted high-order terms. Piecewise linearization executes Taylor expansion in a piecewise manner to improve linearization accuracy. More information, however, such as the switching instant and the dwell time of the piecewise linearized dynamics of a controlled plant, is needed. Orthogonal-function-approximation-based linearization utilizes a set of orthogonal basis functions to approximate the nonlinear model of a controlled plant. A large number of parameters, however, will appear in the linearized model, and the number of the parameters will increase exponentially as more orthogonal basis functions are involved, which leads to a heavy computational burden for parameter identification algorithms and complexity in controller design.

The use of the aforementioned linearization methods either needs an accurate dynamics description of the plant model or leads to a more or less negative influence on controller design or system analysis. In other words, these linearization methods are not control-design-oriented. For instance, the orthogonal-function-approximation-based linearization method would definitely lead to a model with many parameters or a very high order. As a result, it will yield a very complicated controller. It is well known that a too complex controller would not be suitable for practical applications, since it is difficult to be designed and maintained and is not robust. To design a controller with lower order, a model simplification or controller simplification procedure is inevitable. A control-design-oriented linearization should have characteristics such as a simple structure, a moderate amount of adjustable parameters, convenient utilization of I/O data directly during the controller design, and so on [14,123].

In this chapter, a novel dynamic linearization approach, as a kind of control-design-oriented linearization method, is developed for general discrete-time nonlinear systems based on the new concepts of the pseudo partial derivative (PPD), the pseudo gradient (PG) or the pseudo Jacobian matrix (PJM). By using this approach, a series of dynamic linearization data models in the form of I/O data increment are presented, respectively. They are the CFDL data model, the PFDL data model, and the FFDL data model [23–26,124–126].

This chapter is organized as follows. Sections 3.2 and 3.3 introduce three kinds of dynamic linearization data models and the theoretical proofs, for a class of SISO and MIMO discrete-time nonlinear systems, respectively. Section 3.4 summarizes the features of these linearization data models and presents the conclusions.

3.2 SISO Discrete-Time Nonlinear Systems

3.2.1 Compact Form Dynamic Linearization

Consider a class of SISO nonlinear discrete-time systems described by

$$y(k+1) = f(y(k),\ldots,y(k-n_y),u(k),\ldots,u(k-n_u)), \tag{3.1}$$

where $u(k) \in R$ and $y(k) \in R$ are the control input and the system output at time instant k, respectively, n_y and n_u are two unknown positive integers, and $f(\cdots) : R^{n_u+n_y+2} \mapsto R$ is an unknown nonlinear function.

Many nonlinear system models, such as the Hammerstein model, the bilinear model, and so on, can be shown to be the special cases of model (3.1).

Some assumptions are made on system (3.1) before the CFDL method is elaborated.

Assumption 3.1

The partial derivative of $f(\cdots)$ with respect to the (n_y+2)th variable is continuous, for all k with finite exceptions.

Assumption 3.2

System (3.1) satisfies the generalized *Lipschitz* condition, for all k with finite exceptions, that is,

$$|y(k_1+1) - y(k_2+1)| \leq b|u(k_1) - u(k_2)|$$

for $u(k_1) \neq u(k_2)$ and any $k_1 \neq k_2$, $k_1,k_2 \geq 0$, where $y(k_i+1) = f(y(k_i),\ldots, y(k_i-n_y),u(k_i),\ldots,u(k_i-n_u))$, $i = 1,2$, and b is a positive constant.

From a practical point of view, these assumptions imposed on the plant are reasonable and acceptable. Assumption 3.1 is a typical constraint for general nonlinear systems in the field of control system design. Assumption 3.2 imposes an upper bound on the change rate of the system output driven by the change of the control input. From an energy viewpoint, the energy change inside a system cannot go to infinity if the energy change of the control input is at a finite level. Many practical systems satisfy this assumption, such as temperature control systems, pressure control systems, liquid level control systems, and so on.

Without loss of generality, the statement "for all k with finite exceptions" is omitted in the subsequent chapters for clarity.

To facilitate the presentation of the following theorem, define $\Delta y(k+1) = y(k+1) - y(k)$ as the output change and $\Delta u(k) = u(k) - u(k-1)$ as the input change between two consecutive time instants.

Theorem 3.1

Consider nonlinear system (3.1) satisfying Assumptions 3.1 and 3.2. If $|\Delta u(k)| \neq 0$, then there exists a time-varying parameter $\phi_c(k) \in R$, called the *pseudo partial derivative* (PPD), such that system (3.1) can be transformed into the following CFDL data model:

$$\Delta y(k+1) = \phi_c(k)\Delta u(k), \tag{3.2}$$

with bounded $\phi_c(k)$ for any time k.

Proof
From the definition of $\Delta y(k+1)$ and system (3.1), we have

$$\begin{aligned}\Delta y(k+1) = {} & f\left(y(k),\ldots,y(k-n_y),u(k),\ldots,u(k-n_u)\right)\\ & - f\left(y(k),\ldots,y(k-n_y),u(k-1),u(k-1),\ldots,u(k-n_u)\right)\\ & + f\left(y(k),\ldots,y(k-n_y),u(k-1),u(k-1),\ldots,u(k-n_u)\right)\\ & - f\left(y(k-1),\ldots,y(k-n_y-1),u(k-1),\ldots,u(k-n_u-1)\right).\end{aligned} \tag{3.3}$$

Denote

$$\begin{aligned}\psi(k) = {} & f\left(y(k),\ldots,y(k-n_y),u(k-1),u(k-1),\ldots,u(k-n_u)\right)\\ & - f\left(y(k-1),\ldots,y(k-n_y-1),u(k-1),\ldots,u(k-n_u-1)\right).\end{aligned}$$

By virtue of Assumption 3.1 and Cauchy's mean value theorem [127], (3.3) can be rewritten as

$$\Delta y(k+1) = \frac{\partial f^*}{\partial u(k)}\Delta u(k) + \psi(k), \tag{3.4}$$

where $\partial f^*/\partial u(k)$ denotes the partial derivative value of $f(\cdots)$ with respect to the (n_y+2)th variable at a certain point between

$$[y(k),\ldots,y(k-n_y),u(k-1),u(k-1),\ldots,u(k-n_u)]^T$$

and

$$[y(k),\ldots,y(k-n_y),u(k),u(k-1),\ldots,u(k-n_u)]^T.$$

For every fixed time k, consider the following equation with a variable $\eta(k)$:

$$\psi(k) = \eta(k)\Delta u(k). \tag{3.5}$$

Since $|\Delta u(k)| \neq 0$, there must exist a unique solution $\eta^*(k)$ to Equation (3.5).

Let $\phi_c(k) = \eta^*(k) + \partial f^*/\partial u(k)$; then (3.4) can be written as $\Delta y(k+1) = \phi_c(k)\Delta u(k)$; this gives the main conclusion of the theorem. The concern of the boundedness of $\phi_c(k)$ is guaranteed directly by using Assumption 3.2. ■

Remark 3.1

Obviously, PPD is a time-varying parameter, even if system (3.1) is an LTI system. From the proof, we can see that PPD is related to the input and output signals till current time instant. For notation simplicity, we denote it as $\phi_c(k)$ without listing all the time indices before current time k. $\phi_c(k)$ can be considered as a differential signal in some sense and it is bounded for any k. If the sampling period and $\Delta u(k)$ are not too large, $\phi_c(k)$ may be regarded as a slowly time-varying parameter.

Further, from the proof of Theorem 3.1, we can see that all the possible complicated behavior characteristics, such as nonlinearities, time-varying parameters or time-varying structure, and so on, of the original dynamic nonlinear system are compressed and fused into a single time-varying scalar parameter $\phi_c(k)$. Therefore, the dynamics of PPD $\phi_c(k)$ may be too complicated to be described mathematically. However, its numerical behavior may be simple and easily estimated. In other words, even though the time-varying parameter, structure, and delay are explicit in the first-principles model, which are hard to handle in the framework of the model-based control system design, the numerical change of the PPD behavior may not be sensitive to these time-varying factors.

Finally, PPD is merely a concept in the mathematical sense. The existence of PPD is theoretically guaranteed by rigorous analysis from the proof of the above theorem, but generally, PPD cannot be analytically formulated. It is determined jointly by the mean value of the partial derivative at some point within an interval and a nonlinear remaining term. Since the mean value in Cauchy's mean value theorem cannot be explicitly figured out in an analytical form even for a known simple nonlinear function, PPD cannot be computed analytically.

For a simple nonlinear system $y(k+1) = f(u(k))$, PPD just represents the derivative value of the nonlinear function $f(\cdot)$ at a certain point between

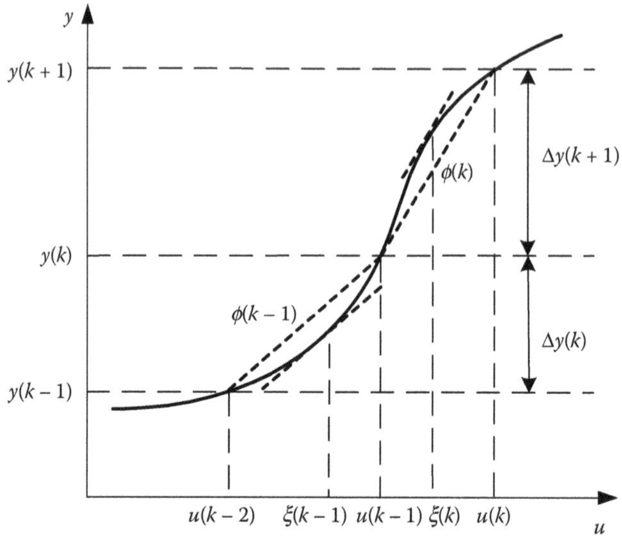

Figure 3.1 Geometric interpretation of PPD.

$u(k-1)$ and $u(k)$. Its geometric interpretation is shown in Figure 3.1. The proposed dynamic linearization data model is denoted by the dashed piecewise line, which is just built along the dynamic operation points of the closed-loop system. The boundedness of PPD means that the nonlinear function does not change abruptly; that is, it has a bounded derivative value. This condition could be satisfied for many practical systems since the input and output variables in practice are always energy related.

Remark 3.2

The dynamic linearization data model only depends on the measurement of I/O data of the closed-loop controlled plant. Neither structure information nor parameter information of the plant dynamic model is explicitly or implicitly involved in applying the proposed dynamic linearization approach. It is a data-driven method. Therefore, the dynamic linearization approach developed in this chapter is applicable to controller design for most of the practical plants, no matter whether system parameters and model structure are time-varying or not.

Dynamic linearization data model (3.2) is a kind of accurate, equivalent, I/O data based model at each sampling point. It is a control-design-oriented linearization model, which means that it is proposed only for the controller design and may not be applicable for other purposes, such as diagnosis, monitoring, and so on. It is also different in essence from other existing linearization methods. Comparatively, Taylor's linearization leads to an approximate linearized model since this method ignores high-order terms when Taylor expansion is applied. Dynamic linearization data model (3.2) is merely related to I/O data of the controlled system, and is independent of system structure and parameters. Comparatively, feedback linearization is based on a known, accurate system structure and parameter. Dynamic linearization data model (3.2) describes a direct mapping from control input increment to system output increment, and one will find that this new linearization approach can bring great convenience to control system design in the next chapter. Hence, it is regarded as a control-design-oriented linearization method. Comparatively, the orthogonal-function-approximation-based linearization focuses on projection and decomposition of the nonlinearity in a certain orthogonal space from a viewpoint of mathematical transformation, but does not intend to care about control system design, and furthermore, the selection of the orthogonal basis and the orthogonal space greatly influences the accuracy and validity of the model transformation.

Remark 3.3

For a linear system as we know, a simple criterion for the controllability of $x(k+1) = Ax(k) + Bu(k)$ is to inspect the rank of the controllability matrix $[B\ AB\ldots A^{n-1}\ B]$. Since all the states in the past and in the future can be completely determined by the coefficient matrices in this linear model, the existence of the feasible control inputs can be guaranteed by the nonsingularity of the controllability matrix; thus, we can say that the controllability test for a linear system, whose state-space model is completely known, is *a priori*.

In the data-driven approach, the only available information about the system is the measured I/O data till current time since the states and the state space model are completely unknown, while the measured data in the future are not available at current time. Therefore, it is not easy to discuss the controllability of the original nonlinear system or the equivalent dynamic linearization data model from the viewpoint of the classical definition of controllability in the state space. To overcome this difficulty, we introduce the concept of "output controllability" as follows. A system is said to be output controllable at a given I/O data point at current time instant k, if the output of the system can be driven to a feasible specified setting point within finite time by a sequence of control inputs.

For the unknown nonlinear system (3.1), the output controllability may be discussed with the help of the novel concept of PPD. It is evident that the system (3.1) is output controllable for a feasible specified setting point if PPD $\phi_c(k)$ is neither zero nor infinite for all k. In Chapter 4, this statement will be verified through rigorous mathematics analysis. For data-driven controller design, the system model (3.1) is not known, the available information about the system is only the measured data until time k, and thus, the output controllability test in a data-driven framework is *a posteriori*. The magnitude of $\phi_c(k)$, which is used to determine whether the system is output controllable or not, will only be estimated using the available operational I/O measured data till current time k. In this sense, the controllability discussed here is quite different from that of the model-based approach. It is worth pointing out that, although the PPD $\phi_c(k)$ cannot be determined analytically like inspecting the controllability matrix for a linear system, its boundedness and nonzero usually can be verified using the collected closed-loop measured data.

From Theorem 3.1, one can see that the boundedness of PPD $\phi_c(k)$ is guaranteed by the assumption of the generalized *Lipschitz* condition made about the system; that is, I/O data of the close-loop system should satisfy a certain condition. Obviously, the generalized *Lipschitz* condition is closely related to the output controllability of the system, and hence is a critical condition for the theoretical analysis.

Theorem 3.1 requires that $|\Delta u(k)| \neq 0$ is satisfied for any k. As a matter of fact, if the case $\Delta u(k) = 0$ happens at a certain sampling time, a slightly different dynamic linearization can be applied after shifting $\sigma_k \in Z^+$ time instants till $u(k) \neq u(k - \sigma_k)$ holds. More details are provided in the following theorem. The other way dealing with this case is the PFDL method or the FFDL method, which will be elaborated in following sections.

Theorem 3.2

For nonlinear system (3.1) satisfying Assumptions 3.1 and 3.2, if there exists an integer $k_0 \geq 1$ such that

$$\Delta u(j) \begin{cases} = 0, & \text{if } j = 1,\ldots,k_0 - 1 \\ \neq 0, & \text{if } j = k_0 \end{cases}$$

then for any integer $k \geq k_0$, a bounded integer σ_k can be always found such that

$$\Delta u(k - j) \begin{cases} = 0, & j = 0,\ldots,\sigma_k - 2 \\ \neq 0, & j = \sigma_k - 1 \end{cases} \tag{3.6}$$

Meanwhile, there exists a PPD $\phi_c(k)$, and then system (3.1) can be transformed into the following CFDL data model:

$$y(k+1) - y(k - \sigma_k + 1) = \phi_c(k)\big(u(k) - u(k - \sigma_k)\big), \tag{3.7}$$

with $|\phi_c(k)| \leq b$ for any time k.

Proof

First, we use the mathematics induction to prove the first part of the conclusion.

When $k = k_0$, there is $\Delta u(k_0) \neq 0$ according to the assumption, namely, that (3.6) holds in the case of $\sigma_k = \sigma_{k_0} = 1$.

Assume that $\Delta u(i-j)\begin{cases} = 0, & j = 0,\ldots,\sigma_i - 2 \\ \neq 0, & j = \sigma_i - 1 \end{cases}$ holds for $k = i > k_0$; now we need to prove that proposition (3.6) is satisfied as $k = i + 1$. Consider the problem in two possibilities. If $\Delta u(i+1) \neq 0$, then the conclusion can be obviously obtained since we select $\sigma_{i+1} = 1$. On the other hand, if $\Delta u(i+1) = 0$, then we have $\Delta u(i+1-j)\begin{cases} = 0, & j = 1,\ldots,\sigma_i - 1 \\ \neq 0, & j = \sigma_i \end{cases}$ according to the assumption of the induction at $k = i$. Thus, conclusion (3.6) can still be guaranteed when we select $\sigma_{i+1} = \sigma_i + 1$.

Next, analogous to the proof of Theorem 3.1, we can easily show the existence of σ_k, the validity of the dynamic linearization model (3.7) under condition (3.6), and the detailed construction approach of $\phi_c(k)$.

Finally, the boundedness of $\phi_c(k)$ can be derived directly on the basis of Assumption 3.2. ■

Remark 3.4

An extreme case is excluded in Theorem 3.2, namely, the control input sequence satisfies $\Delta u(k) = 0, \forall k \geq 1$. In such circumstance, linearization data model (3.7) still holds if the initial state of the system is in equilibrium. An alternative way to deal with this extreme case is to adopt the other two dynamic linearization methods, which will be presented in the following two sections, to describe the controlled plant.

When $\sigma_k = 1$ for any integer $k \geq 1$, condition (3.6) becomes $\Delta u(k) \neq 0, \forall k \geq 1$ in Theorem 3.1. Based on this observation, Theorem 3.2 can be considered as an extension of Theorem 3.1. Without loss of generality, we will discuss the case of $\Delta u(k) \neq 0, \forall k \geq 1$ in subsequent chapters.

3.2.2 Partial Form Dynamic Linearization

From the proof of Theorem 3.1, we can see that the CFDL method transforms a general discrete-time nonlinear system into a linear time-varying dynamic data model, which has only one scalar parameter $\phi_c(k)$. All possible complicated behavior characteristics, such as nonlinearities, time-varying parameters, time-varying structure, and so on, of the original dynamic nonlinear system are compressed and fused into a single time-varying scalar parameter $\phi_c(k)$. Therefore, the behavior of PPD $\phi_c(k)$ may be too complicated to be estimated.

On the other hand, the CFDL method in essence merely considers the time-varying dynamic relationship between the change of the system output at the next time instant and the change of the control input at the current time instant. In fact, the change of the system output at the next time instant may be also related to changes of the previous control inputs besides the change of the control input at the current time instant. Based on this observation, we propose a new linearization method called partial form dynamic linearization (PFDL), which considers all the influences on the system output increment at the next time instant imposed by the control input increments within a fixed length moving time window at the current time. Theoretically speaking, the possible complicated behavior of the original system may be well captured, dispersed and lowered by introducing multiple parameters in the PFDL method, rather than a single PPD in the CFDL data model, during the dynamic linearization transformation.

Denote $\boldsymbol{U}_L(k) \in R^L$ as a vector consisting of all control input signals within a moving time window $[k-L+1,k]$,

$$\boldsymbol{U}_L(k) = [u(k),\ldots,u(k-L+1)]^T, \tag{3.8}$$

with $\boldsymbol{U}_L(k) = \mathbf{0}_L$ for $k \leq 0$. The integer L is called the control input *linearization length constant* (LLC), and $\mathbf{0}_L$ is the zero vector of dimension L.

For SISO discrete-time-varying nonlinear system (3.1), similar to Assumptions 3.1 and 3.2, two assumptions are made in this section as follows.

Assumption 3.3

The partial derivatives of $f(\cdots)$ with respect to the variables from the (n_y+2)th to the (n_y+L+1)th, namely, $u(k)\ldots,u(k-L+1)$, are continuous.

Assumption 3.4

System (3.1) satisfies the generalized *Lipschitz* condition, that is,

$$\left|y(k_1+1) - y(k_2+1)\right| \leq b\left\|\boldsymbol{U}_L(k_1) - \boldsymbol{U}_L(k_2)\right\|,$$

for $\boldsymbol{U}_L(k_1) \neq \boldsymbol{U}_L(k_2)$ and any $k_1 \neq k_2$, $k_1,k_2 \geq 0$, where $y(k_i+1) = f(y(k_i),\ldots,y(k_i-n_y), u(k_i),\ldots,u(k_i-n_u))$, $i = 1,2$, and b is a positive constant.

Denote $\Delta\boldsymbol{U}_L(k) = \boldsymbol{U}_L(k) - \boldsymbol{U}_L(k-1)$. The following theorem presents the PFDL method of system (3.1) in detail.

Theorem 3.3

Consider nonlinear system (3.1) satisfying Assumptions 3.3 and 3.4. For any fixed L, if $\|\Delta \boldsymbol{U}_L(k)\| \neq 0$, then there exists a time-varying vector $\boldsymbol{\phi}_{p,L}(k) \in R^L$, called the *pseudo gradient* (PG), such that system (3.1) can be transformed into the following PFDL data model:

$$\Delta y(k+1) = \boldsymbol{\phi}_{p,L}^T(k)\Delta \boldsymbol{U}_L(k), \tag{3.9}$$

with bounded $\boldsymbol{\phi}_{p,L}(k) = [\phi_1(k),\ldots,\phi_L(k)]^T$ for any time k.

Proof

It follows from system (3.1) that

$$\begin{aligned}\Delta y(k+1) = {} & f\big(y(k),\ldots,y(k-n_y),u(k),u(k-1),\ldots,u(k-n_u)\big)\\ & - f\big(y(k),\ldots,y(k-n_y),u(k-1),u(k-1),\ldots,u(k-n_u)\big)\\ & + f\big(y(k),\ldots,y(k-n_y),u(k-1),u(k-1),\ldots,u(k-n_u)\big)\\ & - f\big(y(k-1),\ldots,y(k-n_y-1),u(k-1),u(k-2),\ldots,u(k-n_u-1)\big).\end{aligned} \tag{3.10}$$

By virtue of Assumption 3.3 and Cauchy's mean value theorem, (3.10) can be rewritten as

$$\begin{aligned}\Delta y(k+1) = {} & \frac{\partial f^*}{\partial u(k)}\Delta u(k)\\ & + f\big(y(k),\ldots,y(k-n_y),u(k-1),u(k-1),\ldots,u(k-n_u)\big)\\ & - f\big(y(k-1),\ldots,y(k-n_y-1),u(k-1),u(k-2),\ldots,u(k-n_u-1)\big),\end{aligned} \tag{3.11}$$

where $\partial f^*/\partial u(k)$ denotes the partial derivative of $f(\cdots)$ with respect to the (n_y+2)th variable at a certain point between

$$[y(k),\ldots,y(k-n_y),u(k-1),u(k-1),\ldots,u(k-n_u)]^T$$

and

$$[y(k),\ldots,y(k-n_y),u(k),u(k-1),\ldots,u(k-n_u)]^T$$

From system (3.1), we have

$$y(k) = f(y(k-1),\dots,y(k-n_y-1),u(k-1),\dots,u(k-n_u-1)). \tag{3.12}$$

Substituting (3.12) into (3.11), and defining

$$\begin{aligned}
&\psi_1\left(y(k-1),\dots,y(k-n_y-1),u(k-1),u(k-2),\dots,u(k-n_u-1)\right)\\
&\quad\triangleq f\Big(f\left(y(k-1),\dots,y(k-n_y-1),u(k-1),\dots,u(k-n_u-1)\right),\\
&\qquad y(k-1),\dots,y(k-n_y),u(k-1),u(k-1),\dots,u(k-n_u)\Big)\\
&\qquad - f\left(y(k-1),\dots,y(k-n_y-1),u(k-1),\dots,u(k-n_u-1)\right).
\end{aligned} \tag{3.13}$$

Equation (3.11) can be rewritten as

$$\begin{aligned}
\Delta y(k+1) &= \frac{\partial f^*}{\partial u(k)}\Delta u(k)\\
&\quad + \psi_1(y(k-1),\dots,y(k-n_y-1),u(k-1),u(k-2),\dots,u(k-n_u-1))\\
&\quad - \psi_1(y(k-1),\dots,y(k-n_y-1),u(k-2),u(k-2),\dots,u(k-n_u-1))\\
&\quad + \psi_1(y(k-1),\dots,y(k-n_y-1),u(k-2),u(k-2),\dots,u(k-n_u-1))\\
&= \frac{\partial f^*}{\partial u(k)}\Delta u(k) + \frac{\partial \psi_1^*}{\partial u(k-1)}\Delta u(k-1)\\
&\quad + \psi_2(y(k-2),\dots,y(k-n_y-2),u(k-2),u(k-3),\dots,u(k-n_u-2)),
\end{aligned} \tag{3.14}$$

where $(\partial\psi_1^*)/(\partial u(k-1))$ denotes the partial derivative of $\psi_1(\cdots)$ with respect to the (n_y+2)th variable at a certain point between

$$[y(k-1),\dots,y(k-n_y-1),u(k-2),u(k-2),\dots,u(k-n_u-1)]^T$$

and

$$[y(k-1),\dots,y(k-n_y-1),u(k-1),u(k-2),\dots,u(k-n_u-1)]^T,$$

and

$$\begin{aligned}
&\psi_2\left(y(k-2),\dots,y(k-n_y-2),u(k-2),u(k-3),\dots,u(k-n_u-2)\right)\\
&\quad\triangleq \psi_1\Big(f\left(y(k-2),\dots,y(k-n_y-2),u(k-2),u(k-3),\dots,u(k-n_u-2)\right),\\
&\qquad y(k-2),\dots,y(k-n_y-1),u(k-2),u(k-2),\dots,u(k-n_u-1)\Big).
\end{aligned}$$

Analogously, from (3.14) with a fixed L, we have

$$\Delta y(k+1)=\frac{\partial f^*}{\partial u(k)}\Delta u(k)+\frac{\partial \psi_1^*}{\partial u(k-1)}\Delta u(k-1)+\cdots+\frac{\partial \psi_{L-1}^*}{\partial u(k-L+1)}\Delta u(k-L+1)$$
$$+\psi_L\left(y(k-L),\ldots,y(k-n_y-L),u(k-L),\ldots,u(k-n_u-L)\right), \tag{3.15}$$

where

$$\psi_i\left(y(k-i),\ldots,y(k-n_y-i),u(k-i),u(k-i-1),\ldots,u(k-n_u-i)\right)$$
$$\triangleq \psi_{i-1}\left(f\left(y(k-i),\ldots,y(k-n_y-i),u(k-i),u(k-i-1),\ldots,u(k-n_u-i)\right),\right.$$
$$\left. y(k-i),\ldots,y(k-n_y-i+1),u(k-i),u(k-i),\ldots,u(k-n_u-i+1)\right),$$

for $i=2,\ldots,L$.

For every fixed k, consider the following equation with a variable $\boldsymbol{\eta}(k)$

$$\psi_L\left(y(k-L),\ldots,y(k-n_y-L),u(k-L),\ldots,u(k-n_u-L)\right)$$
$$=\boldsymbol{\eta}^T(k)[\Delta u(k)\cdots\Delta u(k-L+1)]^T=\boldsymbol{\eta}^T(k)\Delta \boldsymbol{U}_L(k). \tag{3.16}$$

Since $\|\Delta \boldsymbol{U}_L(k)\|\neq 0$, there at least exists a solution $\boldsymbol{\eta}^*(k)$ to Equation (3.16). In fact, there are many finite solutions.

Let

$$\boldsymbol{\phi}_{p,L}(k)=\boldsymbol{\eta}^*(k)+\left[\frac{\partial f^*}{\partial u(k)},\frac{\partial \psi_1^*}{\partial u(k-1)},\ldots,\frac{\partial \psi_{L-1}^*}{\partial u(k-L+1)}\right]^T.$$

Equation (3.15) can be written as the PFDL data model (3.9).

Finally, using the PFDL data model (3.9) and Assumption 3.4, we have that

$$\left|\Delta y(k+1)\right|=\left|\boldsymbol{\phi}_{p,L}^T(k)\Delta \boldsymbol{U}_L(k)\right|\leq b\left\|\Delta \boldsymbol{U}_L(k)\right\|$$

holds for any k and $\|\Delta \boldsymbol{U}_L(k)\|\neq 0$. From the above inequality, we can see that if any element of $\boldsymbol{\phi}_{p,L}(k)$ is unbounded, it would violate the inequality, so the boundedness of $\boldsymbol{\phi}_{p,L}(k)$ for any k is guaranteed. ■

Remark 3.5

Analogous to PPD $\phi_c(k)$, PG $\boldsymbol{\phi}_{p,L}(k)$ is related to I/O signals of the controlled system until time k; it can also be regarded as a set of differential signals in a certain sense.

In particular, the PFDL data model (3.9) becomes the CFDL data model (3.2) when $L = 1$. Obviously, compared with the CFDL data model, the PFDL method synthetically takes all the relationships between the system output change at the next time instant $k + 1$ and the control input changes within a fixed length moving time window $[k - L + 1, k]$ into consideration, rather than imprudently compressing or fusing these factors into the single time-varying scalar parameter $\phi_c(k)$ in the last section. Although the dimension of PG in the PFDL data model increases compared with the CFDL model, the complexity of the dynamics of PG in the PFDL data model is reduced to that of PPD in the CFDL data model. Consequently, $\boldsymbol{\phi}_{p,L}(k)$ is easily estimated with the time-varying parameter estimation algorithm using the closed-loop input–output data when the PFDL data model is used in control system design.

Remark 3.6

From the proof of Theorem 3.3, we can see that the PG in the PFDL data model (3.9) of system (3.1) is not unique when L is fixed, and at least one bounded PG $\boldsymbol{\phi}_{p,L}(k)$ exists such that this dynamic linearization data model holds at any time instant. Further, we can obtain different PFDL data models by selecting different control input LLC L. Thus, the flexibility of the dynamic linearization approach of equivalently describing original nonlinear system can be enhanced by using different PFDL data models via a proper selection of PG and L.

Remark 3.7

For an LTI model

$$A(q^{-1})y(k) = B(q^{-1})u(k), \tag{3.17}$$

where $A(q^{-1}) = 1 + a_1q^{-1} + \cdots + a_{n_a}q^{-n_a}$ and $B(q^{-1}) = b_1q^{-1} + \cdots + b_{n_b}q^{-n_b}$ are polynomials with respect to unit delay operator q^{-1}, $a_1, \ldots, a_{n_a}$ and $b_1, \ldots, b_{n_b}$ are constant coefficients.

If all roots of $A(q^{-1})$ lie within the unit circle, then (3.17) can be written as

$$y(k+1) = \frac{B(q^{-1})}{A(q^{-1})q^{-1}}u(k) \doteq H(q^{-1})u(k), \tag{3.18}$$

where $H(q^{-1}) = h_0 + h_1q^{-1} + \cdots + h_{n_h}q^{-n_h}$ is the finite impulse response polynomial of the plant [128,129]. Model (3.18) is a good approximation of the true model (3.17) when n_h is sufficiently large.

Rewrite (3.18) in the following form:

$$\Delta y(k+1) = H(q^{-1})\Delta u(k) = \boldsymbol{\phi}_{n_h}^T \Delta \boldsymbol{U}_{n_h}(k), \tag{3.19}$$

where $\boldsymbol{\phi}_{n_h} \triangleq [h_0, h_1, \ldots, h_{n_h-1}]^T \in R^{n_h}$ is a constant vector, and $\Delta \boldsymbol{U}_{n_h}(k) \triangleq [\Delta u(k), \Delta u(k-1), \ldots, \Delta u(k-n_h+1)]^T \in R^{n_h}$.

Comparing (3.19) with the PFDL data model (3.9), it can be seen that, when $L \geq n_h$, the PG in the PFDL model is a time-invariant vector $\boldsymbol{\phi}_{p,L}(k) = [h_0, \ldots, h_{n_h-1}, 0, \ldots, 0]^T$; when $L < n_h$, even if system (3.1) is time-invariant, the PG is obviously time-varying, since the finite impulse response coefficients $h_0, \ldots, h_{n_h-1}$ would boil down to the L entries $\phi_1(k), \ldots, \phi_L(k)$ in PG.

There may exist explicit relationships between $h_0, \ldots, h_{n_h-1}$ and $\phi_1(k), \ldots, \phi_L(k)$, in the case that (3.18) well approximates the true model (3.17) with high precision when n_h is sufficiently large. In other words, the finite impulse response coefficients $h_0, \ldots, h_{n_h-1}$ are good approximations to $\phi_1(k), \ldots, \phi_L(k)$ of the PFDL data model for a stable LTI system. Conversely, the element of PG could be regarded as an extension of the finite impulse response coefficients from the LTI system to the nonlinear discrete-time system.

3.2.3 Full Form Dynamic Linearization

The PFDL method merely considers time-varying dynamic relationships between the change of the system output at the next time instant and the changes of the control inputs within the fixed-length moving time window at the current time. In fact, the change of the system output at the next time instant may be also related to the present and previous changes of system outputs. Based on this observation, we propose a new linearization method called the full form dynamic linearization (FFDL), which fully considers all the influences on the system output increment at the next time instant imposed by both the control input increments and the system output increments within input-related/output-related fixed length moving time windows at the current time instant, respectively. Using the FFDL method, the possible complicated behavior of the original system may be better captured and dispersed by introducing more parameters than a scalar PPD in the CFDL data model and a vector PG in the PFDL data model during the dynamic linearization transformation.

Denote $\boldsymbol{H}_{L_y,L_u}(k) \in R^{L_y+L_u}$ as a vector consisting of all control input signals within an input-related moving time window $[k-L_u+1, k]$ and all system output signals within a output-related moving time window $[k-L_y+1, k]$,

$$\boldsymbol{H}_{L_y,L_u}(k) = [y(k), \ldots, y(k-L_y+1), u(k), \ldots, u(k-L_u+1)]^T, \tag{3.20}$$

with $\boldsymbol{H}_{L_y,L_u}(k) = \mathbf{0}_{L_y+L_u}$ for $k \leq 0$. Two integers L_y and L_u ($0 \leq L_y \leq n_y$, $1 \leq L_u \leq n_u$) are called *pseudo orders* of the system, or *controlled output* LLC and *control input* LLC, respectively, similar to the definition of constant L in PFDL.

For SISO discrete-time nonlinear system (3.1), similar to Assumptions 3.1 and 3.2, two assumptions are made in this section as follows.

Assumption 3.5

The partial derivatives of $f(\cdots)$ with respect to all variables are continuous.

Assumption 3.6

System (3.1) satisfies the generalized *Lipschitz* condition, that is,

$$\left|y(k_1+1)-y(k_2+1)\right| \le b\left\|\boldsymbol{H}_{L_y,L_u}(k_1)-\boldsymbol{H}_{L_y,L_u}(k_2)\right\|,$$

for $\boldsymbol{H}_{L_y,L_u}(k_1) \neq \boldsymbol{H}_{L_y,L_u}(k_2)$ and any $k_1 \neq k_2$, $k_1,k_2 \ge 0$, where $y(k_i+1) = f(y(k_i),\ldots, y(k_i-n_y),u(k_i),\ldots,u(k_i-n_u))$, $i=1,2$, and b is a positive constant.

Denote $\Delta\boldsymbol{H}_{L_y,L_u}(k) = \boldsymbol{H}_{L_y,L_u}(k) - \boldsymbol{H}_{L_y,L_u}(k-1)$. The following theorem presents the FFDL method of system (3.1) in detail.

Theorem 3.4

Consider nonlinear system (3.1) satisfying Assumptions 3.5 and 3.6. For any fixed $0 \le L_y \le n_y$ and $1 \le L_u \le n_u$ if $\|\Delta\boldsymbol{H}_{L_y,L_u}(k)\| \neq 0$, then there exists a time-varying vector $\boldsymbol{\phi}_{f,L_y,L_u}(k) \in R^{L_y+L_u}$, called the *pseudo gradient* (PG), such that system (3.1) can be transformed into the following FFDL data model:

$$\Delta y(k+1) = \boldsymbol{\phi}^T_{f,L_y,L_u}(k)\Delta\boldsymbol{H}_{L_y,L_u}(k), \tag{3.21}$$

with bounded $\boldsymbol{\phi}_{f,L_y,L_u}(k) = [\phi_1(k),\ldots,\phi_{L_y}(k),\phi_{L_y+1}(k),\ldots,\phi_{L_y+L_u}(k)]^T$ for any time k.

Proof

It follows from system (3.1) that

$$\begin{aligned}\Delta y(k+1) &= f\left(y(k),\ldots,y(k-n_y),u(k),\ldots,u(k-n_u)\right)\\ &\quad - f\left(y(k-1),\ldots,y(k-n_y-1),u(k-1),\ldots,u(k-n_u-1)\right)\\ &= f\left(y(k),\ldots,y(k-L_y+1),y(k-L_y),\ldots,y(k-n_y),\right.\end{aligned}$$

$$
\begin{aligned}
&u(k),\ldots,u(k-L_u+1),u(k-L_u),\ldots,u(k-n_u)\big)\\
&-f\big(y(k-1),\ldots,y(k-L_y),y(k-L_y),\ldots,y(k-n_y),\\
&u(k-1),\ldots,u(k-L_u),u(k-L_u),\ldots,u(k-n_u)\big)\\
&+f\big(y(k-1),\ldots,y(k-L_y),y(k-L_y),\ldots,y(k-n_y),\\
&u(k-1),\ldots,u(k-L_u),u(k-L_u),\ldots,u(k-n_u)\big)\\
&-f\big(y(k-1),\ldots,y(k-L_y),y(k-L_y-1),\ldots,y(k-n_y-1),\\
&u(k-1),\ldots,u(k-L_u),u(k-L_u-1),\ldots,u(k-n_u-1)\big). \qquad (3.22)
\end{aligned}
$$

Denote

$$
\begin{aligned}
\psi(k) \triangleq\ & f\big(y(k-1),y(k-2),\ldots,y(k-L_y),y(k-L_y),\ldots,y(k-n_y),\\
&u(k-1),u(k-2),\ldots,u(k-L_u),u(k-L_u),\ldots,u(k-n_u)\big)\\
&-f\big(y(k-1),y(k-2),\ldots,y(k-L_y),y(k-L_y-1),\ldots,y(k-n_y-1),\\
&u(k-1),u(k-2),\ldots,u(k-L_u),u(k-L_u-1),\ldots,u(k-n_u-1)\big).
\end{aligned}
$$

By virtue of Assumption 3.5 and Cauchy's mean value theorem, (3.22) can be rewritten as

$$
\begin{aligned}
\Delta y(k+1) =\ & \frac{\partial f^*}{\partial y(k)}\Delta y(k)+\cdots+\frac{\partial f^*}{\partial y(k-L_y)}\Delta y(k-L_y+1)\\
&+\frac{\partial f^*}{\partial u(k)}\Delta u(k)+\cdots+\frac{\partial f^*}{\partial u(k-L_u)}\Delta u(k-L_u+1)+\psi(k), \qquad (3.23)
\end{aligned}
$$

where $\partial f^*/\partial y(k-i)$, $0 \le i \le L_y-1$ and $\partial f^*/\partial y(k-j)$, $0 \le j \le L_u-1$ denote the partial derivatives of $f(\cdots)$ with respect to the $(i+1)$th variable and the (n_y+2+j)th variable at a certain point between

$$
\begin{aligned}
&[y(k),y(k-1),\ldots,y(k-L_y+1),y(k-L_y),\ldots,y(k-n_y),\\
&u(k),u(k-1),\ldots,u(k-L_u+1),u(k-L_u),\ldots,u(k-n_u)]^T
\end{aligned}
$$

and

$$
\begin{aligned}
&[y(k-1),y(k-2),\ldots,y(k-L_y),y(k-L_y),\ldots,y(k-n_y),\\
&u(k-1),u(k-2),\ldots,u(k-L_u),u(k-L_u),\ldots,u(k-n_u)]^T.
\end{aligned}
$$

For every fixed time k, consider the following equation with a variable $\boldsymbol{\eta}(k)$:

$$\begin{aligned}\psi(k) &= \boldsymbol{\eta}^T(k)[\Delta y(k),\ldots,\Delta y(k-L_y+1),\Delta u(k),\ldots,\Delta u(k-L_u+1)]^T \\ &= \boldsymbol{\eta}^T(k)\Delta \boldsymbol{H}_{L_y,L_u}(k).\end{aligned} \tag{3.24}$$

Since $\left\|\Delta \boldsymbol{H}_{L_y,L_u}(k)\right\| \neq 0$, there at least exists a solution $\boldsymbol{\eta}^*(k)$ to (3.24). In fact, there are many finite solutions.

Let

$$\boldsymbol{\phi}_{f,L_y,L_u}(k) = \boldsymbol{\eta}^*(k) + \left[\frac{\partial f^*}{\partial y(k)},\ldots,\frac{\partial f^*}{\partial y(k-L_y)},\frac{\partial f^*}{\partial u(k)},\ldots,\frac{\partial f^*}{\partial u(k-L_u)}\right]^T.$$

Equation 3.23 can be written as the FFDL data model (3.21).

Finally, using the FFDL data model (3.21) and Assumption 3.6, we have that

$$\left|\Delta y(k+1)\right| = \left|\boldsymbol{\phi}_{f,L_y,L_u}^T(k)\Delta \boldsymbol{H}_{L_y,L_u}(k)\right| \leq b\left\|\Delta \boldsymbol{H}_{L_y,L_u}(k)\right\|$$

holds for any k and $\left\|\Delta \boldsymbol{H}_{L_y,L_u}(k)\right\| \neq 0$. From the above inequality, we can see that if any element of $\boldsymbol{\phi}_{f,L_y,L_u}(k)$ is unbounded, it would violate the inequality, so the boundedness of $\boldsymbol{\phi}_{f,L_y,L_u}(k)$ for any k is guaranteed. ■

Remark 3.8

In particular, the FFDL data model (3.21) becomes the PFDL data model (3.9) when $L_y = 0$ and $L_u = L$, and the FFDL data model (3.21) is transformed into the CFDL data model (3.2) when $L_y = 0$ and $L_u = 1$. Therefore, the FFDL proposed in this section can be regarded as the most general dynamic linearization method in this book. In practice, we can obtain different dynamic linearization models by selecting different pseudo orders. Generally, the FFDL with a higher pseudo order is needed for a complicated system, and the FFDL with a lower pseudo order is suitable to a simple system. Obviously, compared with the PFDL data model, the FFDL method has taken into consideration that all the relationships between the system output change at time instant $k+1$, and both the control input change within an input-related fixed length moving time window $[k-L_u+1,k]$ and the system output change within an output-related fixed-length moving time window $[k-L_y+1,k]$. Although the dimension of PG in the FFDL data model is the highest, the FFDL method enhances the applicability of the dynamic linearization method, and reduces the complexity of dynamics of PG element in the dynamic linearization data model. But it leads to a heavier computational burden for I/O data based controller design as discussed in the following chapters.

Remark 3.9

The pseudo orders L_y and L_u are adjustable. If the orders of the system n_y and n_u are known, it is reasonable to choose pseudo orders satisfying $L_y = n_y$ and $L_u = n_u$. In practical applications, the values of n_y and n_u are generally unknown *a priori*, and sometimes n_y and n_u are time-varying, so they are difficult to be determined. In such circumstance, we may select pseudo orders of the FFDL data model to be approximate values of the true orders of the system. On the other hand, when the orders of the plant are large, it is necessary to select pseudo orders in proper size in order to derive a dynamic linearization data model with lower orders, which will help to reduce the computational burden when it is used in practice. As we know, a high-order nonlinear model must lead to a high-order nonlinear controller, which definitely renders unavoidable difficulties in the design and application of the controller. Thus, with a proper selection of the pseudo orders, the simplification procedure for a complex system model or for a complex controller can be avoided, which is an inevitable procedure in model-based control theory.

Remark 3.10

For the LTI system (3.17), it can be rewritten as

$$y(k+1) = \bar{A}(q^{-1})y(k) + B(q^{-1})u(k), \tag{3.25}$$

where $\bar{A}(q^{-1}) = -a_1 q^{-1} - \cdots - a_{n_a} q^{-n_a}$.

From (3.25), we have

$$\Delta y(k+1) = \bar{A}(q^{-1})\Delta y(k) + B(q^{-1})\Delta u(k) = \boldsymbol{\phi}_{n_a,n_b}^T \Delta \boldsymbol{H}_{n_a,n_b}(k), \tag{3.26}$$

where $\boldsymbol{\phi}_{n_a,n_b} \triangleq [a_1,\ldots,a_{n_a},b_1\ldots,b_{n_b}]^T \in R^{n_a+n_b}$ is a constant vector, $\Delta \boldsymbol{H}_{n_a,n_b}(k) \triangleq [\Delta y(k),\ldots,\Delta y(k-n_a),\Delta u(k),\ldots,\Delta u(k-n_b)]^T \in R^{n_a+n_b}$. Comparing (3.26) with FFDL data model (3.21), we can see that when $L_y \geq n_a$ and $L_u \geq n_b$, PG in the FFDL model can be chosen as a time-invariant vector $\boldsymbol{\phi}_{f,L_y,L_u}(k) = [a_1,\ldots,a_{n_a}, 0,\ldots,0,b_1\ldots,b_{n_b},0,\ldots,0]^T$; when $L_y < n_a$ or $L_u < n_b$, PG is obviously time-varying even if system (3.1) is time-invariant.

3.3 MIMO Discrete-Time Nonlinear Systems

3.3.1 Compact Form Dynamic Linearization

In this section, the dynamic linearization method is extended to a class of MIMO discrete-time nonlinear systems as follows:

$$y(k+1) = f(y(k),\ldots,y(k-n_y),u(k),\ldots,u(k-n_u)), \tag{3.27}$$

where $\boldsymbol{u}(k) \in R^m$ and $\boldsymbol{y}(k) \in R^m$ are the control input and the system output at time instant k, respectively. n_y and n_u are two unknown positive integers, and $f(\cdots) = (f_1(\cdots),\cdots,f_m(\cdots))^T \in \prod_{n_u+n_y+2} R^m \mapsto R^m$ is an unknown nonlinear function vector.

Assumption 3.7

The partial derivatives of $f_i(\cdots)$, $i = 1,\ldots,m$, with respect to every entry of the $(n_y + 2)$th variable $\boldsymbol{u}(k)$ are continuous.

Assumption 3.8

System (3.27) satisfies the generalized *Lipschitz* condition, that is,

$$\left\| y(k_1+1) - y(k_2+1) \right\| \leq b \left\| u(k_1) - u(k_2) \right\|,$$

for $\boldsymbol{u}(k_1) \neq \boldsymbol{u}(k_2)$ and any $k_1 \neq k_2$, $k_1,k_2 \geq 0$, where $y(k_i+1) = f(y(k_i),\ldots, y(k_i - n_y),u(k_i),\ldots,u(k_i - n_u))$, $i = 1,2$, and b is a positive constant.

Theorem 3.5

Consider nonlinear system (3.27) satisfying Assumptions 3.7 and 3.8. If $||\Delta \boldsymbol{u}(k)|| \neq 0$, then there exists a time-varying matrix $\boldsymbol{\Phi}_c(k) \in R^{m\times m}$, called the *pseudo Jacobian matrix* (PJM), such that system (3.27) can be transformed into the following CFDL data model:

$$\Delta y(k+1) = \boldsymbol{\Phi}_c(k)\Delta \boldsymbol{u}(k), \tag{3.28}$$

with bounded $\boldsymbol{\Phi}_c(k)$ for any time k.

The proof is omitted here since it is analogous to Theorem 3.1.

Remark 3.11

It is worth pointing out that PJM $\mathbf{\Phi}_c(k)$ and the corresponding dynamic linearization data model (3.28) are not unique. It is different from the CFDL data model in the SISO case.

The dynamic linearization proposed in this section is an extension of the method in Section 3.2.1 to the MIMO system, hence, we may give all remarks similar to the ones in Section 3.2.1.

For the MISO case, that is, $\boldsymbol{u}(k) \in R^m$, $y(k) \in R$, the plant (3.27) becomes

$$y(k+1) = f(y(k),\ldots,y(k-n_y),\boldsymbol{u}(k),\ldots,\boldsymbol{u}(k-n_u)), \tag{3.29}$$

where $f(\cdots) \in \prod_{n_y+1} R \times \prod_{n_u+1} R^m \mapsto R$ is an unknown nonlinear function. In this case, the conditions of Theorem 3.5 must be revised as follows.

Assumption 3.7′

The partial derivatives of $f(\cdots)$ with respect to every entry of the (n_y+2)th variable $\boldsymbol{u}(k)$ are continuous.

Assumption 3.8′

System (3.29) satisfies the generalized *Lipschitz* condition, that is,

$$\left|y(k_1+1) - y(k_2+1)\right| \le b\left\|\boldsymbol{u}(k_1) - \boldsymbol{u}(k_2)\right\|,$$

for $\boldsymbol{u}(k_1) \neq \boldsymbol{u}(k_2)$ and any $k_1 \neq k_2$, $k_1,k_2 \ge 0$, where $y(k_i+1) = f(y(k_i),\ldots,y(k_i-n_y),$ $\boldsymbol{u}(k_i),\ldots,\boldsymbol{u}(k_i-n_u))$, $i = 1,2$, and b is a positive constant.

Corollary 3.1

Consider nonlinear system (3.29) satisfying Assumptions 3.7′ and 3.8′. If $\|\Delta\boldsymbol{u}(k)\| \neq 0$, then there exists a time-varying vector $\boldsymbol{\phi}_c(k) \in R^m$, called the *pseudo gradient* (PG), such that system (3.29) can be transformed into the following CFDL data model:

$$\Delta y(k+1) = \boldsymbol{\phi}_c^T(k)\Delta\boldsymbol{u}(k), \tag{3.30}$$

with bounded $\boldsymbol{\phi}_c(k)$ for any time k.

Remark 3.12

Obviously, PG $\boldsymbol{\phi}_c(k)$ of MISO linearization data model (3.30) proposed here is a special case of PJM $\boldsymbol{\Phi}_c(k)$ of MIMO linearization data model (3.28) in Theorem 3.5. Meanwhile, PG $\boldsymbol{\phi}_c(k)$ may also be treated as an extension of PPD $\phi_c(k)$ of SISO linearization data model (3.2) in Theorem 3.1. In other words, if the system input is of multiple dimension, the PPDs corresponding to all entries $\Delta u_1,\ldots,\Delta u_m$ of $\Delta\boldsymbol{u}(k)$ should be taken into consideration when the CFDL method is applied, and then synthesized into the PG $\boldsymbol{\phi}_c(k)$ of data model (3.30).

Remark 3.13

Dynamic linearization data model (3.30) is similar to data model (3.9) in form, and the two models are derived by using the same idea in theoretical proof. But it is worth pointing out that the two models are different not only in the linearization method used, but also in the controlled plant. The data model (3.30) is derived by applying the CFDL method to the MISO system (3.29) for input increment $\Delta\boldsymbol{u}(k)$ at time k. In contrast, the data model (3.9) is derived by applying the PFDL method to SISO system (3.1) for input increment $\Delta\boldsymbol{U}_L(k)$ within a moving time window $[k-L+1,k]$ with length L. Furthermore, we should note that PG $\boldsymbol{\phi}_c(k)$ of the data model (3.30) is different from PG $\boldsymbol{\phi}_{p,L}(k)$ of the data model (3.9) in essence. The dimension of the former depends on the dimension of the input signal $\boldsymbol{u}(k)$ in MISO system (3.29). It is determined by the system's intrinsic characteristic, and cannot be altered. In contrast, the dimension of the latter depends on the control input LLC L, that is, the length of the moving time window, which can be chosen accordingly.

3.3.2 Partial Form Dynamic Linearization

The PFDL data model for system (3.27) is formulated in this section.

Denote $\bar{U}_L(k) \in R^{mL}$ as a vector consisting of all control input vectors within a moving time window $[k-L+1,k]$,

$$\bar{U}_L(k) = [\boldsymbol{u}^T(k),\ldots,\boldsymbol{u}^T(k-L+1)]^T, \tag{3.31}$$

with $\bar{U}_L(k) = \mathbf{0}_{mL}$ for $k \le 0$. The integer L is the control input LLC.

For MIMO discrete-time-varying nonlinear system (3.27), similar to Assumptions 3.7 and 3.8, two assumptions are made in this section as follows.

Assumption 3.9

The partial derivatives of $f_i(\cdots)$, $i = 1,\ldots,m$, with respect to all entries of the variable from the $(n_y + 2)$th to the $(n_y + L + 1)$th, namely, $\boldsymbol{u}(k),\ldots,\boldsymbol{u}(k - L + 1)$, are continuous.

Assumption 3.10

System (3.27) satisfies the generalized *Lipschitz* condition, that is,

$$\left\|\boldsymbol{y}(k_1 + 1) - \boldsymbol{y}(k_2 + 1)\right\| \leq b\left\|\bar{\boldsymbol{U}}_L(k_1) - \bar{\boldsymbol{U}}_L(k_2)\right\|,$$

for $\bar{\boldsymbol{U}}_L(k_1) \neq \bar{\boldsymbol{U}}_L(k_2)$ and any $k_1 \neq k_2$, $k_1,k_2 \geq 0$, where $\boldsymbol{y}(k_i + 1) = \boldsymbol{f}(\boldsymbol{y}(k_i),\ldots,$ $\boldsymbol{y}(k_i - n_y),\boldsymbol{u}(k_i),\ldots,\boldsymbol{u}(k_i - n_u))$, $i = 1,2$, and b is a positive constant.

Denote $\Delta\bar{\boldsymbol{U}}_L(k) = \bar{\boldsymbol{U}}_L(k) - \bar{\boldsymbol{U}}_L(k - 1)$. The following theorem presents the PFDL method of system (3.27) in detail.

Theorem 3.6

Consider nonlinear system (3.27) satisfying Assumptions 3.9 and 3.10. For any fixed L, if $\left\|\Delta\bar{\boldsymbol{U}}_L(k)\right\| \neq 0$, then there exists a time-varying matrix $\boldsymbol{\Phi}_{p,L}(k) \in R^{m\times mL}$, called the *pseudo partitioned Jacobian matrix* (PPJM), such that system (3.27) can be transformed into the following PFDL data model:

$$\Delta\boldsymbol{y}(k + 1) = \boldsymbol{\Phi}_{p,L}(k)\Delta\bar{\boldsymbol{U}}_L(k), \tag{3.32}$$

with bounded $\boldsymbol{\Phi}_{p,L}(k) = [\boldsymbol{\Phi}_1(k) \ \cdots \ \boldsymbol{\Phi}_L(k)]$ for any time k, where $\boldsymbol{\Phi}_i(k) \in R^{m\times m}$, $i = 1,\ldots,L$.

The proof is omitted here since it is analogous to Theorem 3.3.

The dynamic linearization proposed in this section is an extension of the method in Section 3.2.2 to the MIMO system. Hence, we can give all remarks similar to the ones in Section 3.2.2.

Remark 3.14

$\boldsymbol{\Phi}_{p,L}(k)$ is an extension of PJM $\boldsymbol{\Phi}_c(k)$ in the last section. Actually, each subblock $\boldsymbol{\Phi}_i(k) \in R^{m\times m}$, $i = 1,\ldots,L$ of $\boldsymbol{\Phi}_{p,L}(k)$ corresponds to each entry $\Delta\boldsymbol{u}(k - i + 1)$, $i = 1,\ldots,$

L of $\Delta\bar{U}_L(k)$ one by one. That is the reason why we call it the partitioned Jacobian matrix.

Following the above derivation, we can extend the result in Theorem 3.6 to the MISO nonlinear system (3.29). In this case, the conditions of Theorem 3.6 could be revised accordingly as follows.

Assumption 3.9′

The partial derivatives of $f(\cdots)$ with respect to all entries of the variable from $(n_y + 2)$th to the $(n_y + L + 1)$th, namely, $\boldsymbol{u}(k),\ldots,\boldsymbol{u}(k-L+1)$, are continuous.

Assumption 3.10′

System (3.29) satisfies the generalized *Lipschitz* condition, that is,

$$\left|y(k_1+1) - y(k_2+1)\right| \le b\left\|\bar{U}_L(k_1) - \bar{U}_L(k_2)\right\|,$$

for $\bar{U}_L(k_1) \neq \bar{U}_L(k_2)$ and any $k_1 \neq k_2$, $k_1,k_2 \ge 0$, where $y(k_i+1) = f(y(k_i),\ldots, y(k_i - n_y),\boldsymbol{u}(k_i),\ldots,\boldsymbol{u}(k_i - n_u))$, $i = 1,2$, and b is a positive constant.

Corollary 3.2

Consider nonlinear system (3.29) satisfying Assumptions 3.9′ and 3.10′. For any fixed L, if $\left\|\Delta\bar{U}_L(k)\right\| \neq 0$, then there exists a time-varying vector $\bar{\boldsymbol{\phi}}_{p,L}(k) \in R^{mL}$, called the *pseudo partitioned gradient* (PPG), such that system (3.29) can be transformed into the following PFDL data model:

$$\Delta y(k+1) = \bar{\boldsymbol{\phi}}_{p,L}^T(k)\Delta\bar{U}_L(k), \tag{3.33}$$

with bounded $\bar{\boldsymbol{\phi}}_{p,L}(k) = \begin{bmatrix}\bar{\boldsymbol{\phi}}_1^T & \cdots & \bar{\boldsymbol{\phi}}_L^T\end{bmatrix}^T$ for any time k, where $\bar{\boldsymbol{\phi}}_i \in R^m, i = 1,\ldots,L$.

Remark 3.15

Obviously, PPG $\bar{\boldsymbol{\phi}}_{p,L}(k)$ of MISO linearization data model (3.33) proposed here is a special case of PPJM $\boldsymbol{\Phi}_{p,L}(k)$ of MIMO linearization data model (3.32) in Theorem 3.6.

Meanwhile, PPG $\bar{\boldsymbol{\phi}}_{p,L}(k)$ may also be treated as an extension of PG $\boldsymbol{\phi}_{p,L}(k)$ of SISO linearization data model (3.9) in Theorem 3.3. In other words, if the system input is of multiple dimension, the PGs corresponding to all entries $\Delta \boldsymbol{u}(k), \ldots, \Delta \boldsymbol{u}(k-L+1)$ of $\Delta \bar{\boldsymbol{U}}_L(k)$ should be taken into consideration when the PFDL method is applied and then synthesized into the PPG $\bar{\boldsymbol{\phi}}_{p,L}(k)$ of data model (3.33).

3.3.3 Full Form Dynamic Linearization

Following the same lines as those in Section 3.2.3, the FFDL data model for system (3.27) is formulated in this section.

Denote $\bar{\boldsymbol{H}}_{L_y,L_u}(k) \in R^{mL_y+mL_u}$ as a vector consisting of all control input vectors within an input-related moving time window $[k-L_u+1,k]$ and all system output vectors within an output-related moving time window $[k-L_y+1,k]$

$$\bar{\boldsymbol{H}}_{L_y,L_u}(k) = [\boldsymbol{y}^T(k),\ldots,\boldsymbol{y}^T(k-L_y+1),\boldsymbol{u}^T(k),\ldots,\boldsymbol{u}^T(k-L_u+1)]^T \quad (3.34)$$

with $\bar{\boldsymbol{H}}_{L_y,L_u}(k) = \boldsymbol{0}_{mL_y+mL_u}$ for $k \le 0$. Two integers L_y and L_u $(0 \le L_y \le n_y, 0 \le L_u \le n_u)$ are called *pseudo orders* of the system, or *control output* LLC and *control input* LLC, respectively.

For MIMO discrete-time-varying nonlinear system (3.27), similar to Assumptions 3.7 and 3.8, two assumptions are made in this section as follows.

Assumption 3.11

The partial derivatives of $f_i(\cdots)$, $i = 1,\ldots,m$, with respect to all entries of every variable are continuous.

Assumption 3.12

System (3.27) satisfies the generalized *Lipschitz* condition, that is,

$$\left\|\boldsymbol{y}(k_1+1) - \boldsymbol{y}(k_2+1)\right\| \le b\left\|\bar{\boldsymbol{H}}_{L_y,L_u}(k_1) - \bar{\boldsymbol{H}}_{L_y,L_u}(k_2)\right\|,$$

for $\bar{\boldsymbol{H}}_{L_y,L_u}(k_1) \ne \bar{\boldsymbol{H}}_{L_y,L_u}(k_2)$ and any $k_1 \ne k_2$, $k_1,k_2 \ge 0$, where $\boldsymbol{y}(k_i+1) = f(\boldsymbol{y}(k_i),\ldots,\boldsymbol{y}(k_i-n_y),\boldsymbol{u}(k_i),\ldots,\boldsymbol{u}(k_i-n_u))$, $i = 1,2$, and b is a positive constant.

Denote $\Delta\bar{\boldsymbol{H}}_{L_y,L_u}(k) = \bar{\boldsymbol{H}}_{L_y,L_u}(k) - \bar{\boldsymbol{H}}_{L_y,L_u}(k-1)$. The following theorem presents the FFDL method of system (3.27) in detail.

Theorem 3.7

Consider nonlinear system (3.27) satisfying Assumptions 3.11 and 3.12. For any fixed $0 \le L_y \le n_y$ and $1 \le L_u \le n_u$, if $\|\Delta\bar{\boldsymbol{H}}_{L_y,L_u}(k)\| \ne 0$, then there exists a time-varying matrix $\boldsymbol{\Phi}_{f,L_y,L_u}(k) \in R^{m\times(mL_y+mL_u)}$, called the *pseudo partitioned Jacobian matrix*, such that system (3.27) can be transformed into the following FFDL data model:

$$\Delta\boldsymbol{y}(k+1) = \boldsymbol{\Phi}_{f,L_y,L_u}(k)\Delta\bar{\boldsymbol{H}}_{L_y,L_u}(k), \tag{3.35}$$

with bounded $\boldsymbol{\Phi}_{f,L_y,L_u}(k) = [\boldsymbol{\Phi}_1(k) \quad \cdots \quad \boldsymbol{\Phi}_{L_y+L_u}(k)]$ for any time k, where $\boldsymbol{\Phi}_i(k) \in R^{m\times m}$, $i = 1,\ldots,L_y + L_u$.

The proof is omitted here since it is analogous to Theorem 3.4.

Remark 3.16

$\boldsymbol{\Phi}_{f,L_y,L_u}(k)$ is an extension of PPJM $\boldsymbol{\Phi}_{p,L}(k)$ in the last section. Actually, each subblock $\boldsymbol{\Phi}_i(k) \in R^{m\times m}$, $i = 1,\ldots,L_y + L_u$ of $\boldsymbol{\Phi}_{f,L_y,L_u}(k)$ corresponds to each entry $\Delta\boldsymbol{y}(k-i-1)$, $i = 1,\ldots,L_y$ and $\Delta\boldsymbol{u}(k-i-1)$, $i = 1,\ldots,L_u$ in $\Delta\bar{\boldsymbol{H}}_{L_y,L_u}$ one by one.

Following the above derivation, we may extend the result in Theorem 3.7 to the MISO system (3.29). In this case, denote $\breve{\boldsymbol{H}}_{L_y,L_u}(k) \in R^{L_y+mL_u}$ as a new vector consisting of all control input signal vectors within an input-related moving time window $[k - L_u + 1,k]$ and all system output signals within an output-related moving time window $[k - L_y + 1,k]$,

$$\breve{\boldsymbol{H}}_{L_y,L_u}(k) = [y^T(k),\ldots,y^T(k-L_y+1),\boldsymbol{u}^T(k),\ldots,\boldsymbol{u}^T(k-L_u+1)]^T, \tag{3.36}$$

with $\breve{\boldsymbol{H}}_{L_y,L_u}(k) = \boldsymbol{0}_{L_y+mL_u}$ for $k \le 0$.

The conditions of Theorem 3.7 are modified as follows.

Assumption 3.11′

The partial derivatives of $f(\cdots)$ with respect to all entries of every variable are continuous.

Assumption 3.12′

System (3.29) satisfies the generalized *Lipschitz* condition, that is,

$$\left|y(k_1+1) - y(k_2+1)\right| \le b\left\|\breve{\boldsymbol{H}}_{L_y,L_u}(k_1) - \breve{\boldsymbol{H}}_{L_y,L_u}(k_2)\right\|,$$

for $\breve{\boldsymbol{H}}_{L_y,L_u}(k_1) \neq \breve{\boldsymbol{H}}_{L_y,L_u}(k_2)$ and any $k_1 \neq k_2$, $k_1,k_2 \geq 0$, where $y(k_i+1) = f(y(k_i),\ldots, y(k_i-n_y),\boldsymbol{u}(k_i),\ldots,\boldsymbol{u}(k_i-n_u))$, $i = 1,2$, and b is a positive constant.

Corollary 3.3

Consider nonlinear system (3.29) satisfying Assumptions 3.11′ and 3.12′. For any fixed $0 \leq L_y \leq n_y$ and $1 \leq L_u \leq n_u$, if $\left\|\Delta\breve{\boldsymbol{H}}_{L_y,L_u}(k)\right\| \neq 0$, then there exists a time-varying vector $\bar{\boldsymbol{\phi}}_{f,L_y,L_u}(k) \in R^{L_y+mL_u}$, called the *pseudo partitioned gradient*, such that system (3.29) can be transformed into the following FFDL data model:

$$\Delta y(k+1) = \bar{\boldsymbol{\phi}}^T_{f,L_y,L_u}(k)\Delta\breve{\boldsymbol{H}}_{L_y,L_u}(k), \tag{3.37}$$

with bounded $\bar{\boldsymbol{\phi}}_{f,L_y,L_u}(k) = [\phi_1(k),\ldots,\phi_{L_y}(k),\bar{\boldsymbol{\phi}}^T_{L_y+1}(k),\ldots,\bar{\boldsymbol{\phi}}^T_{L_y+L_u}(k),]^T$ for any time k, where $\phi_i(k) \in R$, $\bar{\boldsymbol{\phi}}_{L_y+j}(k) \in R^m$, $i = 1,\ldots,L_y$, $j = 1,\ldots,L_u$.

Remark 3.17

Obviously, PPG $\bar{\boldsymbol{\phi}}_{f,L_y,L_u}(k)$ of MISO linearization data model (3.37) proposed here is a special case of PPJM $\boldsymbol{\Phi}_{f,L_y,L_u}(k)$ of MIMO linearization data model (3.35) in Theorem 3.7. Meanwhile, PPG $\bar{\boldsymbol{\phi}}_{f,L_y,L_u}(k)$ may also be treated as an extension of PG $\boldsymbol{\phi}_{f,L_y,L_u}(k)$ of SISO linearization data model (3.21) in Theorem 3.4. In other words, if system input is of multiple dimensions, the PPDs corresponding to all entries $\Delta y(k),\ldots,\Delta y(k-L_y+1)$ of $\Delta\breve{\boldsymbol{H}}_{L_y,L_u}(k)$ and the PGs corresponding to all entries $\Delta\boldsymbol{u}(k),\ldots,\Delta\boldsymbol{u}(k-L+1)$ of $\Delta\breve{\boldsymbol{H}}_{L_y,L_u}(k)$ should be taken into consideration when the FFDL method is applied and then synthesized into the PPG $\bar{\boldsymbol{\phi}}_{f,L_y,L_u}(k)$ in the data model (3.37).

3.4 Conclusions

In this chapter, a novel dynamic linearization approach for discrete-time nonlinear systems is proposed in detail, and three kinds of dynamic linearization data models are presented for SISO, MISO, and MIMO nonlinear systems, respectively. Compared with existing linearization methods, the dynamic linearization approach proposed in the chapter has the following merits:

1. The process of dynamic linearization only depends on the measured I/O data of the controlled plant, without any *a priori* knowledge of the system's dynamic model and structure information. Therefore, the dynamic linearization approach developed in this chapter can be applied to a very general class

of dynamic systems, no matter whether the system parameters and model structure are time-varying or not.

2. The linearization data model is in the form of I/O increment. It brings much convenience for us to develop data-driven estimation algorithms for the parameters PPD, PG, PJM, and so on and data-driven controller algorithms by utilizing the I/O data directly. It also facilitates the stability analysis for the control system, which will be shown in the next chapters.
3. The equivalent dynamic linearization data models use novel concepts such as PPD, PG, PJM, and so on to describe changes of the controlled system, whose existence is guaranteed by Cauchy's mean value in calculus and the solution of certain numerical equations. The proposed dynamic linearization approach is quite different from the other linearization methods, which are usually model-based ones. It is worth pointing out that the change of the PPD, PG, or PJM may not be sensitive to the system's uncertainties, such as time-varying parameters, time-varying structure, changes of orders, and so on. This property would bring revolutionary advantages for the coming model-free control system design based on aforementioned dynamic linearization data models.
4. The dynamic linearization approach provides an accurate, equivalent, I/O data based dynamic linearization data model in an incremental form to the original nonlinear system at each sampling point. Therefore, the unmodeled dynamics, which is an inevitable issue for the traditional model-based control system design, does not exist in this data-driven framework by using this dynamic linearization data model.
5. Although the description of dynamic linearization data models is very simple, it is valid for many complex systems with strong nonlinearity, nonminimum phase, time-varying delay, and so on. A smaller number of parameters are involved in these dynamic linearization data models. For SISO nonlinear systems, there is only one scalar PPD in the CFDL data model, and $(L_y + L_u)$ dimensional PG in the FFDL data model. Moreover, the amount of parameters can be selected *a priori*.
6. For a certain specific discrete-time nonlinear system, its dynamic linearization model can be expressed in different forms such as CFDL, PFDL, and FFDL. Thus, this could bring much flexibility to controller design. The CFDL data model has the simplest form, in which the dynamics of PPD may be very complicated for certain complex systems. In such a circumstance, it is difficult to design estimation algorithms for well capturing the dynamics. On the other hand, the FFDL data model has the most general form. The complicated behavior of the complex system is dispersed into the $L_y + L_u$ elements in PG. Therefore, it is convenient for us to design parameter estimation algorithms to evaluate its change, while we have to face the cost that higher order in the FFDL data model leads to a more complicated controller than the one based on the CFDL data model. This will bring more difficulties in

theoretical analysis and practical applications for the corresponding control systems.

7. It is worth pointing out that the dynamic linearization data model is control-design-oriented; that is, it is only applicable to controller design, but may not be good at diagnosis and monitoring for the controlled system and long-term prediction of the system output. The data models derived by the dynamic linearization approach are not based on physical laws; thus the physical meanings of the parameters PPD, PG, PJM, and PPJM are not as clear as the parameters in the first-principles-based models.

Chapter 4

Model-Free Adaptive Control of SISO Discrete-Time Nonlinear Systems

4.1 Introduction

Adaptive control theories and methodologies assume that the structure of the controlled plant is known but the parameters are unknown, slowly time-varying, or time-invariant. At present, adaptive control for linear systems has been well developed, many monographs [76,130–132] and thousands of papers have been published on this subject, and plenty of applications could be found in Refs. [133–135]. However, nonlinearity is a common characteristic of practical systems in the real world, such as aircraft, robots, industrial processes, traffic systems, electronic systems, and so on. Though research on the adaptive control of nonlinear systems has attracted increasing attention in recent years, compared with the adaptive control of linear systems, only a few special classes of nonlinear systems have been studied and only some typical control methods in this field have been recognized [136,137].

The NARMAX model [99], the Wiener model [97,138], the bilinear model [96], and the Hammerstein model [95] are typical nonlinear systems, and dead-zone, input saturation, hysteresis, and so on are the typical nonlinear elements in practical systems. The classical adaptive control design methods for nonlinear systems include feedback linearization based adaptive control [139,140], the backstepping method [141–144], adaptive predictive control [145–148], the multiple model method [149,150], sliding mode adaptive control [151,152], and so on. All these methods are model-based control design methods since accurate plant dynamic models of the controlled system are required for control system design and analysis.

When knowledge of the system model is unavailable or the model dynamics is with large uncertainties, the aforementioned methods are hard to be applied and analyzed. As we know, building a state-space model or an input–output model of a practical plant is not an easy thing and is sometimes impossible. Even if the model of the controlled plant is established, unmodeled dynamics is also inevitable. Thus, the closed-loop control system design based on the system model with uncertainties under additional mathematical assumptions may cause unpredictable problems in practical applications or may even become unsafe [6]. Although a huge number of papers have been published to study the robustness of a control system, only a few robust control methods can be used successfully to address practical control problems. Thus, it is of great significance to develop the data-driven MFAC for nonlinear systems, not only in theory, but also in practical applications [14,123].

Although fuzzy adaptive control [153–155] and neural-network-based adaptive control [156–160] can realize adaptive control for nonlinear systems without the need of an accurate model of plant dynamics, they need either a comprehensive understanding of the controlled system in order to establish fuzzy rules or massive operation data of the system to train neural networks. Generally speaking, their controller design depends on fuzzy rules and the neural network model; thus, the fundamental problems of model-based control still exist there.

MFAC was first proposed for a class of nonlinear systems in 1994 [23]. Undergoing incessant development and improvement in the last two decades [24–26,124–126], a systematic framework of MFAC methods has been established. The basic idea of the MFAC design is implemented by building an equivalent dynamic linearization data model of the nonlinear system at each operation point first, then estimating the system's PPD online by using I/O data of the controlled plant, and designing the controller according to some weighted one-step-ahead cost functions.

In this chapter, a series of MFAC schemes and analysis results are presented based on dynamic linearization data models given in Chapter 3 for a class of SISO discrete-time nonlinear systems. Compared with conventional adaptive control, MFAC has the following distinguished features. First, only the measurement I/O data of the closed-loop controlled system, rather than the information about the system model, are required for the controller design. The traditional unmodeled dynamics does not exist in this data-driven framework, and this makes MFAC suitable for industrial systems. Second, MFAC is a low-cost control method with a low computational burden and simple structure, since no mathematics model, experimental signal, test signal, or training process is needed. Finally, adaptive control for a nonlinear system with time-varying parameters and time-varying structure can be uniformly realized by applying the MFAC method.

This chapter is organized as follows. In Section 4.2, the CFDL data model based model-free adaptive control (CFDL–MFAC) scheme, the input–output stability, the monotonic convergence of the tracking error, and simulation results are presented for a class of SISO discrete-time nonlinear systems. The corresponding results on the PFDL data model based model-free adaptive control (PFDL–MFAC) and FFDL data model based model-free adaptive control (FFDL–MFAC) schemes are proposed in Sections 4.3 and 4.4, respectively. Section 4.5 presents the conclusions.

4.2 CFDL Data Model Based MFAC

4.2.1 Control System Design

Consider a class of SISO discrete-time nonlinear systems as follows:

$$y(k+1) = f(y(k),\ldots,y(k-n_y),u(k),\ldots,u(k-n_u)), \tag{4.1}$$

where $y(k) \in R$ and $u(k) \in R$ are the system output and the control input at time k, respectively. n_y and n_u are two unknown integers, and $f(\cdots) : R^{n_u+n_y+2} \mapsto R$ is an unknown nonlinear function.

From Theorem 3.1, we can see that system (4.1), satisfying Assumptions 3.1 and 3.2 with $|\Delta u(k)| \neq 0$ for all k, can be transformed into the following CFDL data model:

$$y(k+1) = y(k) + \phi_c(k)\Delta u(k), \tag{4.2}$$

where $\phi_c(k) \in R$ is the bounded PPD of system (4.1).

Equation (4.2) is a virtual equivalent dynamic linearization description of nonlinear system (4.1). It is a controller-design-oriented linear time-varying data model with a scalar parameter and in a simple incremental form. In this sense, it is totally different from the other linearized models derived from the first principles and the other linearization methods.

The control system scheme will be designed based on this virtual dynamic linearization data model (4.2) in what follows. It is worth pointing out that the control scheme design in this book is given only for the case $\sigma_k = 1$, that is, $\Delta u(k) \neq 0$. For the case $\sigma_k > 1$, all the discussions are similar.

4.2.1.1 Controller Algorithm

For a discrete-time system, the controller algorithm obtained from one-step-ahead prediction error cost function may yield excessive control effort, which might damage the control system, whereas the controller algorithm obtained from weighted one-step-ahead prediction error cost function may lead to steady tracking error. Thus, we use the following cost function of controller input to design the controller algorithm,

$$J(u(k)) = \left|y^*(k+1) - y(k+1)\right|^2 + \lambda\left|u(k) - u(k-1)\right|^2, \tag{4.3}$$

where $\lambda > 0$ is a weighting factor introduced to restrict the changing rate of the control input. $y^*(k+1)$ is the desired reference output signal.

Substituting CFDL data model (4.2) into cost function (4.3), differentiating cost function (4.3) with respect to $u(k)$ and setting it to zero, yields the following control law:

$$u(k) = u(k-1) + \frac{\rho\phi_c(k)}{\lambda + \left|\phi_c(k)\right|^2}\left(y^*(k+1) - y(k)\right), \tag{4.4}$$

where the step factor $\rho \in (0,1]$ is added to make the controller algorithm more general.

Remark 4.1

It is obvious that λ in controller algorithm (4.4) is a penalty factor on the change of the control input $\Delta u(k)$, which is often used in control system design to ensure the smoothness of the control input signal. In fact, λ is an important adjustable parameter for MFAC system design. Theoretical analysis and numerical simulations will show that a proper selection of λ can guarantee the stability and good output tracking performance of the controlled system.

4.2.1.2 PPD Estimation Algorithm

Theorem 3.1 shows that nonlinear system (4.1), satisfying Assumptions 3.1 and 3.2, can be described by dynamic linearization data model (4.2) with a time-varying PPD parameter $\phi_c(k)$. Controller algorithm (4.4) can be used only if PPD is known. The accurate value of PPD, nevertheless, is difficult to obtain, since the mathematical model of the system is unknown and PPD is a time-varying parameter as discussed in Remark 3.1. Therefore, it is necessary to design a certain time-varying parameter estimation algorithm using the I/O data of the controlled plant to estimate PPD online.

The common cost function of the time-varying parameter estimation is to minimize the square of the error between real system output and model output. By applying the estimation algorithm derived from this kind of cost function, however, the estimated value of the parameter is often sensitive to some inexact sampling data, which may be caused by disturbance or faulty sensor. To overcome this drawback, a new cost function of PPD estimation is proposed as follows:

$$J(\phi_c(k)) = |y(k) - y(k-1) - \phi_c(k)\Delta u(k-1)|^2 + \mu|\phi_c(k) - \hat{\phi}_c(k-1)|^2, \tag{4.5}$$

where $\mu > 0$ is a weighting factor.

Minimizing cost function (4.5) with respect to $\phi_c(k)$ gives the following PPD estimation algorithm:

$$\hat{\phi}_c(k) = \hat{\phi}_c(k-1) + \frac{\eta\Delta u(k-1)}{\mu + \Delta u(k-1)^2}\left(\Delta y(k) - \hat{\phi}_c(k-1)\Delta u(k-1)\right), \tag{4.6}$$

where the step factor $\eta \in (0,2]$ is added to make algorithm (4.6) more general and more flexible, and $\hat{\phi}_c(k)$ denotes the estimation of $\phi_c(k)$.

Remark 4.2

The difference between algorithm (4.6) and the projection algorithm [76] lies in the fact that the introduction of constant μ into the denominator in the projection algorithm is to avoid division by zero. In contrast, μ in algorithm (4.6) is a weighting factor to punish the change rate of PPD estimation.

4.2.1.3 System Control Scheme

By integrating controller algorithm (4.4) and parameter estimation algorithm (4.6), the CFDL–MFAC scheme is constructed as follows:

$$\hat{\phi}_c(k) = \hat{\phi}_c(k-1) + \frac{\eta\Delta u(k-1)}{\mu + \Delta u(k-1)^2}\left(\Delta y(k) - \hat{\phi}_c(k-1)\Delta u(k-1)\right), \tag{4.7}$$

$$\hat{\phi}_c(k) = \hat{\phi}_c(1), \quad \text{if } |\hat{\phi}_c(k)| \le \varepsilon \text{ or } |\Delta u(k-1)| \le \varepsilon \text{ or } \operatorname{sign}(\hat{\phi}_c(k)) \ne \operatorname{sign}(\hat{\phi}_c(1)), \tag{4.8}$$

$$u(k) = u(k-1) + \frac{\rho\hat{\phi}_c(k)}{\lambda + |\hat{\phi}_c(k)|^2}\left(y^*(k+1) - y(k)\right), \tag{4.9}$$

where $\lambda > 0$, $\mu > 0$, $\rho \in (0,1]$, $\eta \in (0,2]$, ε is a small positive constant, and $\hat{\phi}_c(1)$ is the initial value of $\hat{\phi}_c(k)$.

Remark 4.3

The reset mechanism (4.8) introduced here is to endow the parameter estimation algorithm (4.7) with a strong ability to track time-varying parameter.

Remark 4.4

From CFDL–MFAC scheme (4.7)–(4.9), we can see that the MFAC scheme only utilizes the online measurement I/O data of the closed-loop controlled system, and does not explicitly or implicitly include any information on the system dynamic model. This is the reason why we call it model-free adaptive control. Since the PPD $\phi_c(k)$ is not sensitive to the time-varying parameters, structure, and delay, the CFDL–MFAC scheme has very strong adaptability and robustness, which are difficult to achieve in the framework of model-based adaptive control system design. However, it does not mean that MFAC is a universal adaptive control approach. A universal control method does not exist in the real world. In fact, the MFAC scheme is developed for a certain class of discrete-time nonlinear systems satisfying Assumptions 3.1 and 3.2 in Chapter 3 and Assumptions 4.1 and 4.2 in the next section.

4.2.2 Stability Analysis

For the rigorous analysis of stability, two assumptions are made about system (4.1) as follows.

Assumption 4.1

For a given bounded desired output signal $y^*(k+1)$, there exists a bounded control input $u^*(k)$, such that the system output driven by $u^*(k)$ is equal to $y^*(k+1)$.

Assumption 4.2

The sign of PPD is assumed unchanged for all k and $\Delta u(k) \neq 0$, that is, $\phi_c(k) > \underline{\varepsilon} > 0$ (or $\phi_c(k) < -\underline{\varepsilon}$) is satisfied, where $\underline{\varepsilon}$ is a small positive constant.

Without loss of generality, we only discuss the case of $\phi_c(k) > \underline{\varepsilon}$ in this book.

Remark 4.5

Assumption 4.1 implies that the control problem is solvable. In other words, system (4.1) is output controllable. Refer to Remark 3.4 for details. The physical meaning of Assumption 4.2 is clear: the system output does not decrease as the corresponding control input increases, which may be treated as a kind of linear-like characteristic. This assumption is similar to the assumption on the control direction in traditional model-based control methods [76], where the sign of the control direction is assumed known, or at least unchanged. It is not a severe assumption as many practical industrial systems, such as the temperature control system and the pressure control system, satisfy such a property.

Theorem 4.1

If nonlinear system (4.1), satisfying Assumptions 3.1, 3.2, 4.1, and 4.2, is controlled by the CFDL–MFAC scheme (4.7)–(4.9) for a regulation problem, that is, $y^*(k+1) = y^* = \text{const.}$, then there exists a constant $\lambda_{\min} > 0$ such that the following two properties hold for any $\lambda > \lambda_{\min}$:

a. System output tracking error converges monotonically, and $\lim_{k\to\infty} |y^* - y(k+1)| = 0$.
b. The closed-loop system is BIBO stable, namely, $\{y(k)\}$ and $\{u(k)\}$ are bounded.

Proof

If one of the conditions $|\hat{\phi}_c(k)| \le \varepsilon$, $|\Delta u(k-1)| \le \varepsilon$, and $\text{sign}(\hat{\phi}_c(k)) \ne \text{sign}(\hat{\phi}_c(1))$ is satisfied, then the boundedness of $\hat{\phi}_c(k)$ is straightforward.

In the other case, defining the PPD estimation error as $\tilde{\phi}_c(k) = \hat{\phi}_c(k) - \phi_c(k)$ and subtracting $\phi_c(k)$ from both sides of parameter estimation algorithm (4.7), we have

$$\tilde{\phi}_c(k) = \left(1 - \frac{\eta|\Delta u(k-1)|^2}{\mu + |\Delta u(k-1)|^2}\right)\tilde{\phi}_c(k-1) + \phi_c(k-1) - \phi_c(k). \tag{4.10}$$

Taking the absolute value on both sides of (4.10) yields

$$|\tilde{\phi}_c(k)| \le \left|\left(1 - \frac{\eta|\Delta u(k-1)|^2}{\mu + |\Delta u(k-1)|^2}\right)\right| |\tilde{\phi}_c(k-1)| + |\phi_c(k-1) - \phi_c(k)|. \tag{4.11}$$

Note that the term $\eta|\Delta u(k-1)|^2/(\mu + |\Delta u(k-1)|^2)$ is monotonically increasing with respect to $|\Delta u(k-1)|^2$ and its minimum value is $\eta\varepsilon^2/(\mu + \varepsilon^2)$. When $0 < \eta \le 2$ and $\mu > 0$, there must exist a constant d_1 such that

$$0 \le \left|\left(1 - \frac{\eta|\Delta u(k-1)|^2}{\mu + |\Delta u(k-1)|^2}\right)\right| \le 1 - \frac{\eta\varepsilon^2}{\mu + \varepsilon^2} = d_1 < 1. \tag{4.12}$$

Further, the property $|\phi_c(k)| \le \bar{b}$ in Theorem 3.1 leads to $|\phi_c(k-1) - \phi_c(k)| \le 2\bar{b}$. Thus, from (4.11) and (4.12), we have

$$\begin{aligned} |\tilde{\phi}_c(k)| &\le d_1 |\tilde{\phi}_c(k-1)| + 2\bar{b} \le d_1^2 |\tilde{\phi}_c(k-2)| + 2d_1\bar{b} + 2\bar{b} \\ &\le \cdots \le d_1^{k-1} |\tilde{\phi}_c(1)| + \frac{2\bar{b}(1 - d_1^{k-1})}{1 - d_1}, \end{aligned} \tag{4.13}$$

which implies that $\tilde{\phi}_c(k)$ is bounded. Then, the boundedness of $\hat{\phi}_c(k)$ is also guaranteed as $\phi_c(k)$ is bounded.

Define the tracking error as follows:

$$e(k+1) = y^* - y(k+1). \tag{4.14}$$

Substituting the CFDL data model (4.2) into (4.14) and taking the absolute value yield

$$\begin{aligned} |e(k+1)| &= |y^* - y(k+1)| = |y^* - y(k) - \phi_c(k)\Delta u(k)| \\ &\le \left| 1 - \frac{\rho \phi_c(k) \hat{\phi}_c(k)}{\lambda + |\hat{\phi}_c(k)|^2} \right| |e(k)|. \end{aligned} \tag{4.15}$$

From Assumption 4.2 and resetting algorithm (4.8), $\phi_c(k)\hat{\phi}_c(k) \ge 0$ is guaranteed. Let $\lambda_{\min} = \bar{b}^2/4$. Utilizing $\alpha^2 + \beta^2 \ge 2\alpha\beta$, and according to $\hat{\phi}_c(k) > \varepsilon$ from Assumption 4.2, boundedness of $\hat{\phi}_c(k)$ from the first step of this proof and $\hat{\phi}_c(k) > \varepsilon$ from resetting algorithm (4.8), there must exist a constant M_1 $(0 < M_1 < 1)$ such that

$$0 < M_1 \le \frac{\phi_c(k)\hat{\phi}_c(k)}{\lambda + |\hat{\phi}_c(k)|^2} \le \frac{\bar{b}\,\hat{\phi}_c(k)}{\lambda + |\hat{\phi}_c(k)|^2} \le \frac{\bar{b}\,\hat{\phi}_c(k)}{2\sqrt{\lambda}\,\hat{\phi}_c(k)} < \frac{\bar{b}}{2\sqrt{\lambda_{\min}}} = 1 \tag{4.16}$$

holds for $\lambda > \lambda_{\min}$, where $\bar{b}$ is the constant satisfying $|\phi_c(k)| \le \bar{b}$ in Theorem 3.1.

According to (4.16), $0 < \rho \le 1$ and $\lambda > \lambda_{\min}$, there exists a positive constant $d_2 < 1$ such that

$$\left| 1 - \frac{\rho\phi_c(k)\hat{\phi}_c(k)}{\lambda + |\hat{\phi}_c(k)|^2} \right| = 1 - \frac{\rho\phi_c(k)\hat{\phi}_c(k)}{\lambda + |\hat{\phi}_c(k)|^2} \le 1 - \rho M_1 = d_2 < 1. \tag{4.17}$$

Combining (4.15) and (4.17) yields

$$|e(k+1)| \leq d_2 |e(k)| \leq d_2^2 |e(k-1)| \leq \cdots \leq d_2^k |e(1)|. \tag{4.18}$$

This gives conclusion (a).

As $y^*(k)$ is constant, the convergence of $e(k)$ implies that $y(k)$ is also bounded.

Using $(\sqrt{\lambda})^2 + |\hat{\phi}_c(k)|^2 \geq 2\sqrt{\lambda}\,\hat{\phi}_c(k)$ and $\lambda > \lambda_{\min}$, the following inequality can be derived from (4.9):

$$\begin{aligned} |\Delta u(k)| &= \left| \frac{\rho \hat{\phi}_c(k)\left(y^* - y(k)\right)}{\lambda + |\hat{\phi}_c(k)|^2} \right| = \left| \frac{\rho \hat{\phi}_c(k)}{\lambda + |\hat{\phi}_c(k)|^2} \right| |e(k)| \\ &\leq \left| \frac{\rho \hat{\phi}_c(k)}{2\sqrt{\lambda}\,\hat{\phi}_c(k)} \right| |e(k)| \leq \left| \frac{\rho}{2\sqrt{\lambda_{\min}}} \right| |e(k)| = M_2 |e(k)|, \end{aligned} \tag{4.19}$$

where $M_2 = \rho/(2\sqrt{\lambda_{\min}})$ is a bounded constant.

From (4.18) and (4.19), we have

$$\begin{aligned} |u(k)| &\leq |u(k) - u(k-1)| + |u(k-1)| \\ &\leq |u(k) - u(k-1)| + |u(k-1) - u(k-2)| + |u(k-2)| \\ &\leq |\Delta u(k)| + |\Delta u(k-1)| + \cdots + |\Delta u(2)| + |u(1)| \\ &\leq M_2 \left(|e(k)| + |e(k-1)| + \cdots + |e(2)| \right) + |u(1)| \\ &\leq M_2 \left(d_2^{k-1} |e(1)| + d_2^{k-1} |e(1)| + \cdots + d_2 |e(1)| \right) + |u(1)| \\ &< M_2 \frac{d_2}{1 - d_2} |e(1)| + |u(1)|. \end{aligned} \tag{4.20}$$

Thus, conclusion (b) is the direct result of (4.20). ■

Remark 4.6

In Theorem 4.1, it has been proved that the MFAC scheme applying to the unknown nonlinear plant for the regulation problem can guarantee the stability and error convergence. Since the MFAC method is a pure data-driven control method and any plant model information is not involved in the control system design, the tracking control problem can also be easily addressed in a similar way. Consider a new controlled plant:

$$z(k+1) = y(k+1) - y^*(k+1),$$

where $y(k+1)$ is the output of the unknown plant (4.1), and the $y^*(k+1)$ is the known desired output signal tracked.

This new controlled plant originated from plant (4.1) is also a unknown nonlinear system. The regulation problem of this new controlled plant is equivalent to the tracking control problem of the original one. Thus, the tracking problem for MFAC has also been solved. All the results on the regulation problem in following chapters can be easily extended to the tracking case with the same discussion like this.

In Theorem 4.1, the stability of the CFDL–MFAC scheme is proved for the case $\sigma_k = 1$, that is, $\Delta u(k) \neq 0$. When $\sigma_k > 1$, the same conclusion can be drawn.

If $\sigma_k > 1$, then $\Delta u(k-j) = 0$, $j = 0, \ldots, \sigma_k - 2$, $\Delta u(k - \sigma_k + 1) \neq 0$ holds. From controller algorithm (4.9), we have $e(k) = e(k-1) = \cdots = e(k - \sigma_k + 2) = 0$. Thus, substituting dynamic linearization data model (3.7) and controller algorithm (4.9) into (4.14) yields

$$\begin{aligned}
|e(k+1)| &= |y^* - y(k+1)| \\
&= |y^* - y(k - \sigma_k + 1) - \phi_c(k)(u(k) - u(k - \sigma_k))| \\
&= |y^* - y(k - \sigma_k + 1) - \phi_c(k)\big(u(k - \sigma_k + 1) - u(k - \sigma_k)\big)| \\
&= \left|\left(1 - \frac{\rho\phi_c(k)\hat{\phi}_c(k - \sigma_k)}{\lambda + |\hat{\phi}_c(k - \sigma_k)|^2}\right) e(k - \sigma_k + 1)\right| \\
&\leq \left|1 - \frac{\rho\phi_c(k)\hat{\phi}_c(k - \sigma_k)}{\lambda + |\hat{\phi}_c(k - \sigma_k)|^2}\right| |e(k - \sigma_k + 1)|.
\end{aligned}$$

The above inequality implies that the tracking error converges to zero asymptotically.

Remark 4.7

The unknown parameter $\phi_c(k)$ is time-varying. Usually, it cannot be properly estimated by the conventional projection algorithm or least-squares algorithm, which is only suitable for time-invariant parameter estimation. From the above proof, we find that any time-varying parameter estimation algorithm, which can guarantee boundedness of the estimation, could be used to estimate the unknown PPD, such as the least-squares algorithm with covariance resetting, the least-squares algorithm with covariance modification, the time-varying parameter estimation algorithm with Kalman filter, the modified projection algorithm [23,24,124], the least-squares algorithm with time-varying forgetting factor [76,161], and so on.

Lemma 4.1

$|\tilde{\phi}_c(k)| \le |\phi_c(k)|$ is a sufficient condition for the inequality $\hat{\phi}_c(k)\phi_c(k) > 0$.

The proof of Lemma 4.1 is straightforward.

Lemma 4.2

Suppose $\sup_k|\Delta\phi_c(k)| \le \alpha|\phi_c(k)|$, $0 < \alpha < 1$ holds except for finite time instants. $\hat{\phi}_c(k)\phi_c(k) > 0$ is guaranteed if the following parameter estimation algorithm is applied with proper η and μ when $|\Delta\, u(k-1)| \neq 0$:

$$\hat{\phi}_c(k) = \hat{\phi}_c(k-1) + \frac{\eta\Delta u(k-1)}{\mu + \Delta u(k-1)^2}\left(\Delta y(k) - \hat{\phi}_c(k-1)\Delta u(k-1)\right).$$

Proof

From (4.11), we have

$$\left|\tilde{\phi}_c(k)\right| \le \left|\left(1 - \frac{\eta\Delta u(k-1)^2}{\mu + \left|\Delta u(k-1)\right|^2}\right)\right|\left|\tilde{\phi}_c(k-1)\right| + \left|\phi_c(k-1) - \phi_c(k)\right|.$$

Since $0 < \eta < 1$ and $\mu > 0$, there exists a constant d such that

$$0 < 1 - \frac{\eta\left|\Delta u(k-1)\right|^2}{\mu + \left|\Delta u(k-1)\right|^2} \le d < 1,$$

which implies

$$\begin{aligned}\left|\tilde{\phi}_c(k)\right| &\le d\left|\tilde{\phi}_c(k-1)\right| + \left|\Delta\phi_c(k)\right| \\ &\le d^2\left|\tilde{\phi}_c(k-2)\right| + d\left|\Delta\phi_c(k-1)\right| + \left|\Delta\phi_c(k)\right| \\ &\le \cdots \le d^{k-1}\left|\tilde{\phi}_c(1)\right| + \frac{\sup_k\left|\Delta\phi_c(k)\right|}{1-d}.\end{aligned}$$

By virtue of the condition $\sup_k|\Delta\phi_c(k)| \leq \alpha|\phi_c(k)|$, the above equation leads to

$$\left|\tilde{\phi}_c(k)\right| \leq d^{k-1}\left|\tilde{\phi}_c(1)\right| + \frac{\alpha\left|\phi_c(k)\right|}{1-d}.$$

Thus, if k is sufficiently large and $\alpha \leq 1-d$, then the result $\hat{\phi}_c(k)\phi_c(k) > 0$ is guaranteed by Lemma 4.1. ■

Theorem 4.2

If nonlinear system (4.1), satisfying Assumptions 3.1, 3.2, and 4.1 and the condition of Lemma 4.2, is controlled by the CFDL–MFAC scheme (4.7)–(4.9) for a regulation problem, that is, $y^*(k+1) = y^* = \text{const.}$, then there exists $\lambda_{\min} > 0$, such that the following two properties hold for any $\lambda > \lambda_{\min}$:

a. System output tracking error converges monotonically, and $\lim_{k\to\infty}\left|y^* - y(k+1)\right| = 0$.
b. The closed-loop system is BIBO stable, namely, $\{y(k)\}$ and $\{u(k)\}$ are bounded.

Proof

The result follows immediately by using Lemmas 4.1 and 4.2. ■

The significant difference between the model-based adaptive control and the CFDL–MFAC scheme proposed in this section lies in the following three main aspects:

1. The plants to be controlled are different. The MFAC scheme is proposed for a class of unknown nonlinear systems, whereas traditional model-based adaptive control is proposed for a time-invariant or slowly time-varying system with a known system structure and system order.
2. The design procedures are different. For the MFAC approach, an equivalent dynamic linearization incremental data model is built at each operation point of the closed-loop system with the help of the new concept of PPD, which can be estimated by only using the online I/O data of the controlled plant, and then it is used to design a weighted one-step-ahead adaptive controller algorithm. As a result, no structure information about the plant model is needed during the design of the controller and the estimator, and no unmodeled dynamics exists. For applying the traditional model-based adaptive control approach, the first step is to model the plant using the first principles or system identification approach, and then to develop a controller based on the mathematical model by using the certainty equivalence principle. In the sequel, the unmodeled dynamics always exists in the control

system, and all the conclusions on stability and tracking performance are dependent on the model accuracy.
3. The methods for stability analysis are also different. An I/O data-driven contraction mapping approach is used to analyze the stability of the MFAC system, whereas the key technical lemma or *Lyapunov* approach is used in traditional model-based adaptive control.

In addition, compared with robust adaptive control and neural-network-based adaptive control, MFAC also has its own advantages.

Robust adaptive control is designed for systems with uncertainties that can be parameterized or be with known bound, aiming at generating an adaptive controller that guarantees bounded input and bounded output in the presence of bounded disturbance and unmodeled dynamics. Thus, the robust adaptive controller is largely dependent upon the structure and accuracy of the nominal model of the controlled systems. The MFAC controller is constructed only based on a virtual equivalent time-varying linearization data model, in which neither the system model is utilized explicitly (or implicitly) nor the unmodeled dynamics is involved. The control performance of MFAC would not be influenced by traditional unmodeled dynamics. On the basis of this observation, we can say that the MFAC scheme should have stronger robustness compared with model-based control methods. It is also worth pointing out that the MFAC scheme not only has a simple and practical structure but also guarantees BIBO stability and monotonic convergence of the output tracking error for the controlled plant.

For unknown nonlinear systems, a neural network has been also applied successfully to design the adaptive controller, and some neural-network-based adaptive control methods can guarantee the stability of the controlled system. For those data-driven control methods that approximate the controller directly by the neural network, some prior knowledge of the controlled system, such as system order, is also needed, while an adequate training process and a heavy computational burden are the other two issues. Comparatively, the MFAC scheme neither requires any prior knowledge of the controlled system nor needs the training process, and has low computational burden.

4.2.3 Simulation Results

In this section, numerical simulations of three totally different SISO discrete-time nonlinear systems controlled by the same CFDL–MFAC scheme are given to verify the correctness and effectiveness of the CFDL–MFAC scheme. It is worth pointing out that the model dynamics, including system structure (linear or nonlinear), order, and relative degree, and so on, is not involved in CFDL–MFAC scheme, and the models in the following three examples are merely used to generate the I/O data of the controlled plant, but not for the MFAC design.

The initial conditions of the controlled systems are $u(1) = u(2) = 0$, $y(1) = -1$, $y(2) = 1$, and $\hat{\phi}_c(1) = 2$; the step factors are set as $\rho = 0.6$, $\eta = 1$; ε is set to 10^{-5}.

Example 4.1

Nonlinear system

$$y(k+1)=\begin{cases}\dfrac{y(k)}{1+y(k)^2}+u^3(k), & k\le 500\\[2mm] \dfrac{y(k)y(k-1)y(k-2)u(k-1)(y(k-2)-1)+a(k)u(k)}{1+y^2(k-1)+y^2(k-2)}, & k>500,\end{cases}$$

where $a(k) = \text{round}(k/500)$ is a time-varying parameter.

This nonlinear system consists of two subsystems in series. These two subsystems are both taken from Ref. [162], where they are controlled by a neural-network-based control approach separately, and there is no such parameter $a(k)$ in the second subsystem in Ref. [162]. Obviously, the controlled system studied here is with time-varying structure, time-varying parameter, and time-varying order.

The desired output signal is as follows:

$$y^*(k+1)=\begin{cases}0.5\times(-1)^{\text{round}(k/500)}, & k\le 300\\ 0.5\sin(k\pi/100)+0.3\cos(k\pi/50), & 300<k\le 700\\ 0.5\times(-1)^{\text{round}(k/500)}, & k>700\end{cases}$$

The simulation results are shown in Figure 4.1. From the simulation, one can find that the CFDL–MFAC scheme gives a satisfactory control performance if the weighting factors are set to $\lambda = 2$ and $\mu = 1$, and a little faster system response and a larger overshoot occur if the weighting factors are selected as $\lambda = 0.1$ and $\mu = 1$. Furthermore, the tracking performance and the PPD's estimation are not influenced by changes in the parameter and system structure. Finally, PPD is a bounded slowing time-varying parameter as shown in Figure 4.1c, and its dynamics behavior is related to the operation point of the closed-loop system, control input signal, and so on.

Example 4.2

Nonlinear system

$$y(k+1)=\begin{cases}\dfrac{5y(k)y(k-1)}{1+y^2(k)+y^2(k-1)+y^2(k-2)}+u(k)+1.1u(k-1), & k\le 500\\[2mm] \dfrac{2.5y(k)y(k-1)}{1+y^2(k)+y^2(k-1)}+1.2u(k)+1.4u(k-1)\\ \quad +0.7\sin(0.5(y(k)+y(k-1)))\cos(0.5(y(k)+y(k-1))), & k>500\end{cases}$$

The nonlinear system consists of two subsystems in series, which are taken from Refs. [162,163], respectively. The first subsystem controlled by the neural-network-based control method does not yield satisfactory performance due to the plant's inherent nonlinearity and nonminimum-phase characteristics [162]. The

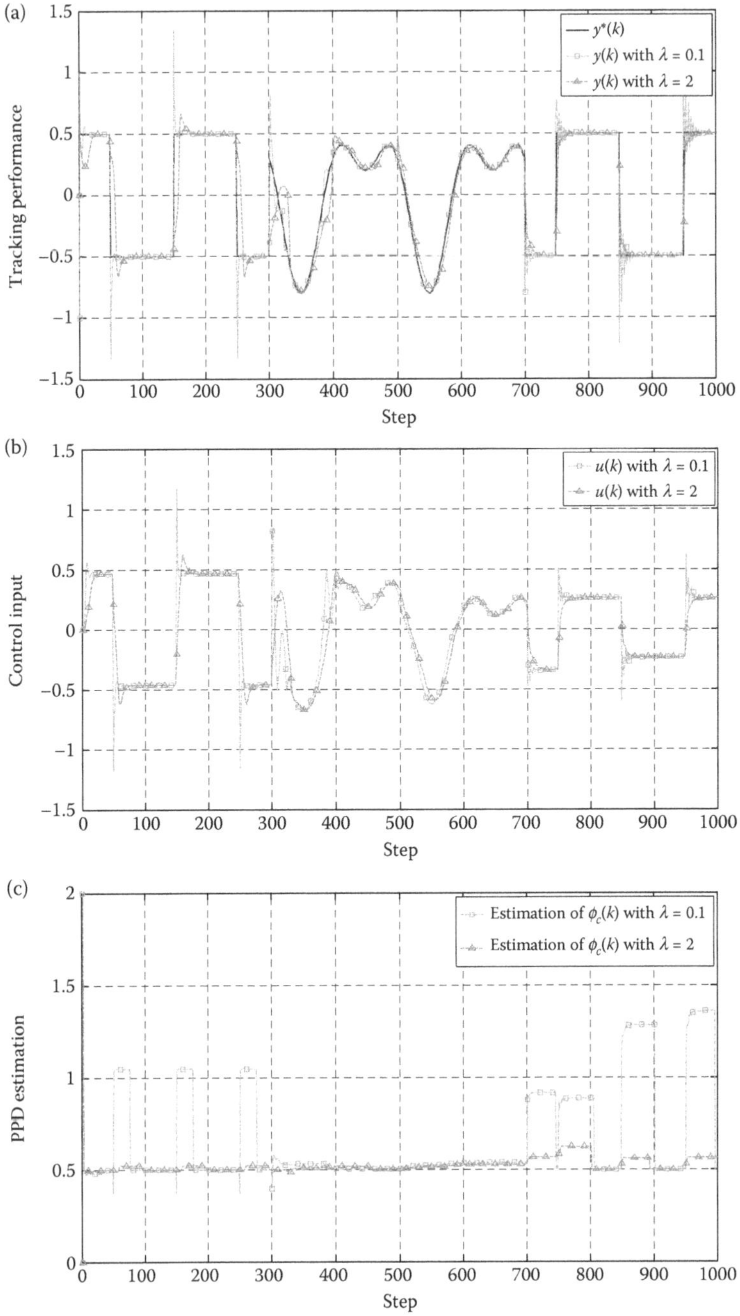

Figure 4.1 Simulation results in Example 4.1. (a) Tracking performance. (b) Control input. (c) Estimated PPD $\hat{\phi}_c(k)$.

second subsystem is also a nonminimum-phase nonlinear system due to the term $1.4u(k-1)$ added, which is different from the original system in Ref. [163]. Obviously, the system simulated here is a nonlinear nonminimum-phase system with time-varying structure and time-varying order.

The desired output signal is as follows:

$$y^*(k+1) = \begin{cases} 5\sin(k\pi/50) + 2\cos(k\pi/100), & k \leq 300 \\ 5(-1)^{\text{round}(k/100)}, & 300 < k \leq 700 \\ 5\sin(k\pi/50) + 2\cos(k\pi/100), & k > 700 \end{cases}$$

The simulation results using the proposed CFDL–MFAC scheme are shown in Figure 4.2, and the simulation results demonstrate a conclusion similar to that of Example 4.1. In addition, the small spike in the output tracking profile at the 500th time instant is caused by the change in the controlled system structure, and the minor changes could also be found in the corresponding control input and estimated PPD.

Example 4.3

Linear system

$$y(k+1) = 1.5y(k) - 0.7y(k-1)$$
$$+0.1 \times \begin{cases} u(k) + b(k)u(k-1), & 1 \leq k \leq 200 \\ u(k-2) + b(k)u(k-3), & 200 < k \leq 400 \\ u(k-4) + b(k)u(k-5), & 400 < k \leq 600 \\ u(k-6) + b(k)u(k-7), & 600 < k \leq 800 \\ u(k-8) + b(k)u(k-9), & 800 < k \leq 1000 \end{cases}$$

where $b(k) = 0.1 + 0.1\text{round}(k/100)$ is a time-varying parameter. Obviously, it is a linear system with time-varying structure and time-varying delays.

The desired output signal is

$$y^*(k+1) = 0.5 + 0.5 \times 0.1^{\text{round}(k/200)}$$

The simulation results using the proposed CFDL–MFAC scheme are shown in Figure 4.3.

From the simulation results illustrated in Figure 4.3, one can see that the CFDL–MFAC scheme gives satisfactory control performance if the weighting factors are set to $\lambda = 15$ and $\mu = 1$, while a faster system response and a larger overshoot occur if the weighting factors are selected as $\lambda = 2$ and $\mu = 1$. This numerical simulation clearly illustrates robustness of the CFDL–MFAC scheme against time-varying structure and large time-varying control input delays.

From the simulation results of the above three examples, one can see that the CFDL–MFAC scheme has the following merits.

First, the same CFDL–MFAC scheme with the same initial values can tackle the adaptive control problems for quite a large class of nonlinear systems with quite a good control performance, no matter whether the unknown system is

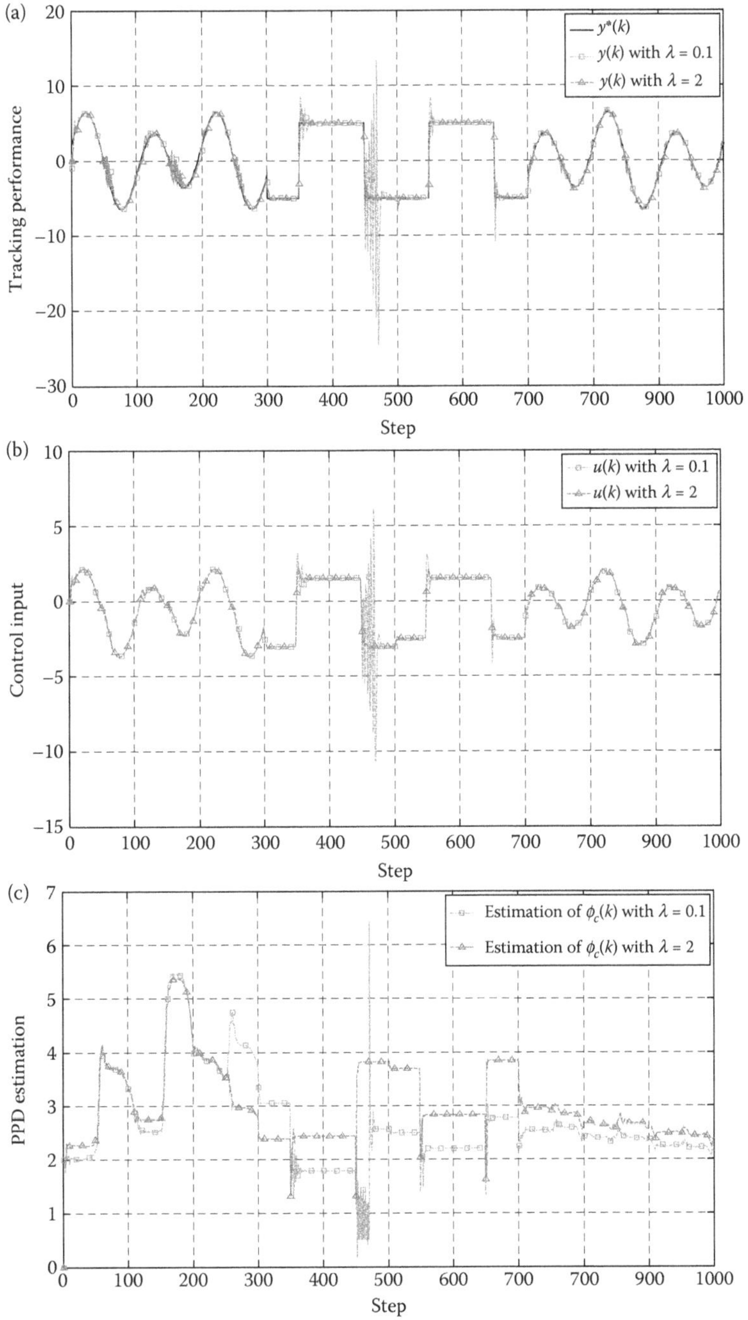

Figure 4.2 Simulation results in Example 4.2. (a) Tracking performance. (b) Control input. (c) Estimated PPD $\hat{\phi}_c(k)$.

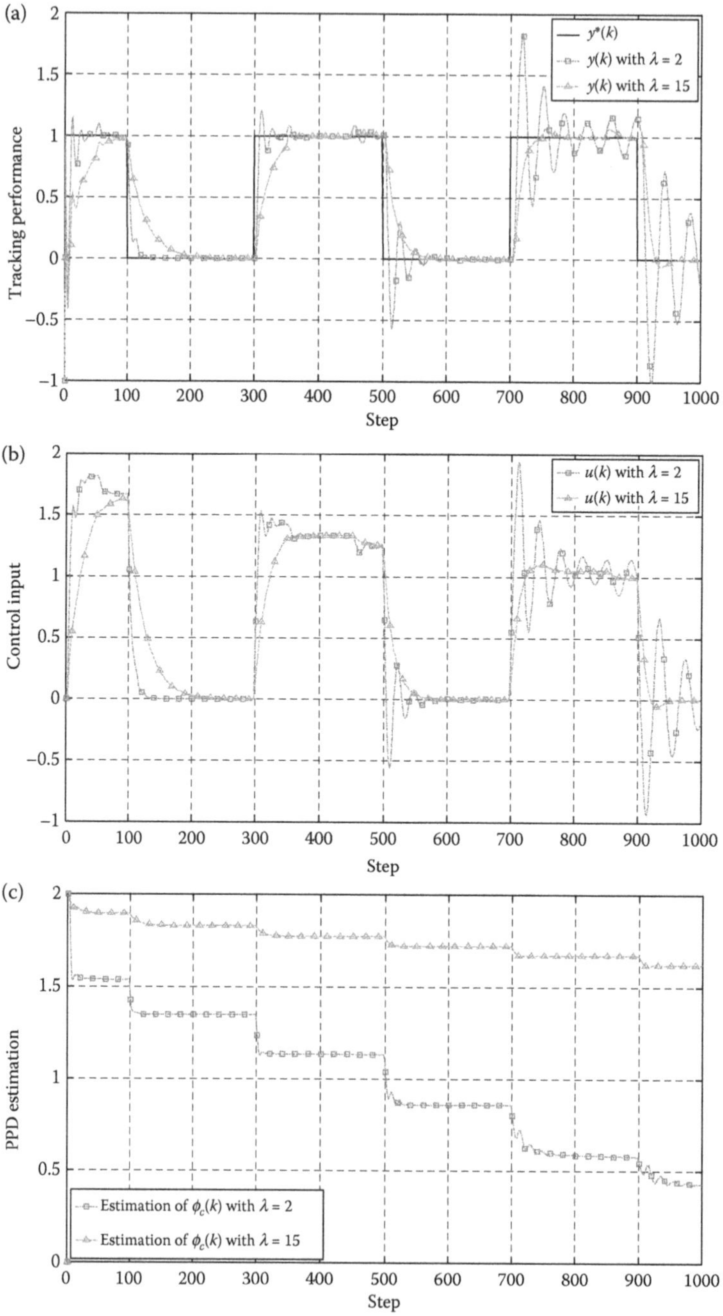

Figure 4.3 Simulation results in Example 4.3. (a) Tracking performance. (b) Control input. (c) Estimated PPD $\hat{\phi}_c(k)$.

time-varying in structure, parameter, delay, and phase, or not. It is difficult for traditional adaptive control to achieve this property even when the system models are exactly known. This implies that the CFDL–MFAC scheme has very strong adaptability and robustness against various kinds of changes in a controlled system.

Second, a trade-off between the response time and the overshoot can be realized by selecting appropriate weighting factors. Different selections of the weighting factor λ lead to different control performances. The smaller the weighting factor λ, the shorter the rising time, and the larger the overshoot, and vice versa. For $\lambda = 0$, oscillation or instability of the closed-loop system may occur in most cases.

Third, the PPD behavior of the system is relatively simple, and is a slowly time-varying scalar parameter. It has a relationship with the controlled system dynamics, closed-loop operation points, control input, and so on, but it is not sensitive to the time-varying structure, orders, delay, and parameter of the controlled system.

Finally, the CFDL–MFAC scheme has a very simple structure and is easy to implement with low computational burden, since there is only one scalar parameter to tune online.

4.3 PFDL Data Model Based MFAC

4.3.1 Control System Design

It is mentioned in Section 3.2.2 that the dynamics of PPD $\phi_c(k)$ in the CFDL data model might be too complicated when a complex nonlinear system is considered. In this circumstance, an efficient time-varying parameter estimation algorithm is required to achieve good control performance when the CFDL–MFAC scheme is applied to a practical plant. In fact, the change of the system output at time $k+1$ may be sensitive to the changes of certain control inputs within a fixed length moving time window at time k. The controller design without the consideration of these sensitive factors may make the closed-loop control system unstable. Taking this into account, the PFDL–MFAC scheme including stability analysis is presented on the basis of the PFDL data model given in Section 3.2.2. Compared with the CFDL–MFAC scheme, the PFDL–MFAC scheme has more adjustable degrees of freedom and more design flexibility.

From Theorem 3.3, nonlinear system (4.1), satisfying Assumptions 3.3 and 3.4 with $\|\Delta \boldsymbol{U}_L(k)\| \neq 0$ for all k, can be transformed into the following PFDL data model:

$$\Delta y(k+1) = \boldsymbol{\phi}_{p,L}^T(k)\Delta \boldsymbol{U}_L(k), \tag{4.21}$$

where $\boldsymbol{\phi}_{p,L}(k) = [\phi_1(k), \ldots, \phi_L(k)]^T \in R^L$ is the unknown but bounded PG, $\Delta \boldsymbol{U}_L(k) = [\Delta u(k), \ldots, \Delta u(k-L+1)]^T$, and L is the control input LLC.

4.3.1.1 Controller Algorithm

Consider a cost function of the control input as follows:

$$J(u(k)) = \left|y^*(k+1) - y(k+1)\right|^2 + \lambda\left|u(k) - u(k-1)\right|^2, \tag{4.22}$$

where $\lambda > 0$ is a weighting factor.

Substituting PFDL data model (4.21) into cost function (4.22) and minimizing this cost function with respect to $u(k)$ yield the following controller algorithm:

$$u(k) = u(k-1) + \frac{\rho_1\phi_1(k)\,(y^*(k+1) - y(k))}{\lambda + \left|\phi_1(k)\right|^2} - \frac{\phi_1(k)\sum_{i=2}^{L}\rho_i\phi_i(k)\Delta u(k-i+1)}{\lambda + \left|\phi_1(k)\right|^2}, \tag{4.23}$$

where the step factor $\rho_i \in (0,1]$, $i = 1,2,\ldots,L$ is added to make the controller algorithm more flexible.

4.3.1.2 PG Estimation Algorithm

Analogous to the analysis in Section 4.2.1, a cost function of PG estimation is put forward as follows:

$$J(\boldsymbol{\phi}_{p,L}(k)) = \left|y(k) - y(k-1) - \boldsymbol{\phi}_{p,L}^T(k)\Delta\boldsymbol{U}_L(k-1)\right|^2 + \mu\,\|\boldsymbol{\phi}_{p,L}(k) - \hat{\boldsymbol{\phi}}_{p,L}(k-1)\|^2, \tag{4.24}$$

where $\mu > 0$ is a weighting factor.

Lemma 4.3 [131] (Matrix Inversion Lemma)

Let A,B,C, and D be real matrices with appropriate dimensions. If the inversions of matrices A, C, and $DA^{-1}B + C^{-1}$ exist, it follows that

$$[A + BCD]^{-1} = A^{-1} - A^{-1}B[DA^{-1}B + C^{-1}]^{-1}DA^{-1}.$$

Minimizing (4.24) with respect to $\boldsymbol{\phi}_{p,L}(k)$ according to the optimality condition and using the above matrix inversion lemma give

$$\hat{\boldsymbol{\phi}}_{p,L}(k) = \hat{\boldsymbol{\phi}}_{p,L}(k-1) + \frac{\eta\Delta\boldsymbol{U}_L(k-1)\,(y(k) - y(k-1) - \hat{\boldsymbol{\phi}}_{p,L}^T(k-1)\Delta\boldsymbol{U}_L(k-1))}{\mu + \left\|\Delta\boldsymbol{U}_L(k-1)\right\|^2}, \tag{4.25}$$

where the step factor $\eta \in (0,2]$ is added to make the estimation algorithm more generic, and $\hat{\boldsymbol{\phi}}_{p,L}(k)$ is the estimation of unknown PG $\boldsymbol{\phi}_{p,L}(k)$.

The unknown parameter $\boldsymbol{\phi}_{p,L}(k)$ is a time-varying vector with dimension L, rather than the time-varying scalar parameter $\phi_c(k)$ in the CFDL data model. Thus, its behavior can be dispersed and easily captured by a time-varying parameter estimation algorithm. In fact, many time-varying parameter estimation algorithms could be applied to estimate the unknown parameter vector PG $\boldsymbol{\phi}_{p,L}(k)$, such as a modified projection algorithm [23,24,124], least-squares algorithm with a time-varying forgetting factor [76,161], and so on.

4.3.1.3 System Control Scheme

By synthesizing PG estimation algorithm (4.25) and controller algorithm (4.23), the PFDL–MFAC scheme is presented as follows:

$$\hat{\boldsymbol{\phi}}_{p,L}(k) = \hat{\boldsymbol{\phi}}_{p,L}(k-1) + \frac{\eta \Delta \boldsymbol{U}_L(k-1)(y(k) - y(k-1) - \hat{\boldsymbol{\phi}}^T_{p,L}(k-1)\Delta \boldsymbol{U}_L(k-1))}{\mu + \|\Delta \boldsymbol{U}_L(k-1)\|^2}, \tag{4.26}$$

$$\hat{\boldsymbol{\phi}}_{p,L}(k) = \hat{\boldsymbol{\phi}}_{p,L}(1) \quad \text{if } \|\hat{\boldsymbol{\phi}}_{p,L}(k)\| \le \varepsilon \text{ or } \|\Delta \boldsymbol{U}_L(k-1)\| \le \varepsilon \text{ or}$$
$$\text{sign}(\hat{\phi}_1(k)) \ne \text{sign}(\hat{\phi}_1(1)), \tag{4.27}$$

$$u(k) = u(k-1) + \frac{\rho_1 \hat{\phi}_1(k)\left(y^*(k+1) - y(k)\right)}{\lambda + |\hat{\phi}_1(k)|^2} - \frac{\hat{\phi}_1(k)\sum_{i=2}^{L} \rho_i \hat{\phi}_i(k)\Delta u(k-i+1)}{\lambda + |\hat{\phi}_1(k)|^2}, \tag{4.28}$$

where $\lambda > 0$, $\mu > 0$, $\eta \in (0,2]$, $\rho_i \in (0,1]$, $i = 1,2,\ldots,L$, ε is a small positive constant, and $\hat{\boldsymbol{\phi}}_{p,L}(1)$ is the initial value of $\hat{\boldsymbol{\phi}}_{p,L}(k)$. The reset mechanism (4.8) added here is to endow parameter estimation algorithm (4.27) with a strong ability to track the time-varying parameter.

Remark 4.8

There is an L-dimensional vector $\hat{\boldsymbol{\phi}}_{p,L}(k)$, namely, estimated PG, to be tuned online in the PFDL–MFAC scheme, while the control input length linearization constant L is also adjustable. Extensive numerical simulations and experiments indicate that L should be set to be an integer within an interval between 1 and the sum of the approximated orders of the plant if n_y and n_u are unknown. From the practical point of view, L could be selected to be 1 for a simple system, and larger value as the complexity of the system increases. When $L = 1$, PFDL–MFAC scheme (4.26)–(4.28) becomes CFDL–MFAC scheme (4.7)–(4.9). Compared with the CFDL–MFAC scheme, the PFDL–MFAC scheme has more adjustable degrees of freedom and more design flexibility due to the introduction of step factors $\rho_1, \rho_2, \ldots, \rho_L$.

4.3.2 Stability Analysis

For the rigorous analysis of the stability, another assumption and a lemma are given as follows.

Assumption 4.3

The sign of the first element in PG $\boldsymbol{\phi}_{p,L}(k)$ is assumed unchanged for all k and $\|\Delta \mathbf{U}_L(k)\| \neq 0$, that is, $\phi_1(k) > \underline{\varepsilon} > 0$ (or $\phi_1(k) < -\underline{\varepsilon}$) is satisfied, where $\underline{\varepsilon}$ is a small positive constant.

Without loss of generality, we only discuss the case of $\phi_1(k) > \underline{\varepsilon}$ in this book.

Lemma 4.4 [164]

Let

$$A = \begin{bmatrix} a_1 & a_2 & \cdots & a_{L-1} & a_L \\ 1 & 0 & \cdots & 0 & 0 \\ 0 & 1 & \cdots & 0 & 0 \\ \vdots & \vdots & \vdots & \vdots & \vdots \\ 0 & 0 & \cdots & 1 & 0 \end{bmatrix},$$

if $\sum_{i=1}^{L} |a_i| < 1$, then $s(A) < 1$, where $s(\cdot)$ is the spectral radius.

Theorem 4.3

If nonlinear system (4.1), satisfying Assumptions 3.1, 3.2, 4.1, and 4.3, is controlled by the PFDL–MFAC scheme (4.26)–(4.28) for a regulation problem $y^*(k+1) = y^* = \text{const.}$, then there exists $\lambda_{\min} > 0$ such that the following two properties hold for any $\lambda > \lambda_{\min}$:

a. System output tracking error converges asymptotically, and $\lim_{k\to\infty}\left|y^* - y(k+1)\right| = 0$.
b. The closed-loop system is BIBO stable, namely, $\{y(k)\}$ and $\{u(k)\}$ are bounded for all k.

Proof

This proof consists of two steps. The first step is to prove the boundedness of PG estimation. The second step is to prove the convergence of the tracking error and BIBO stability.

Step 1 If one of the conditions $|\hat{\boldsymbol{\phi}}_{p,L}(k)| \le \varepsilon$ or $\|\Delta \boldsymbol{U}_L(k-1)\| \le \varepsilon$ or $\text{sign}(\hat{\phi}_1(k)) \ne \text{sign}(\hat{\phi}_1(1))$ is satisfied, then the boundedness of $\hat{\boldsymbol{\phi}}_{p,L}(k)$ is straightforward.

In the other case, define the PG estimation error as $\tilde{\boldsymbol{\phi}}_{p,L}(k) = \hat{\boldsymbol{\phi}}_{p,L}(k) - \boldsymbol{\phi}_{p,L}(k)$. Subtracting $\boldsymbol{\phi}_{p,L}(k)$ from both sides of (4.26) yields

$$\tilde{\boldsymbol{\phi}}_{p,L}(k) = \left(\mathbf{I} - \frac{\eta \Delta \boldsymbol{U}_L(k-1)\Delta \boldsymbol{U}_L^T(k-1)}{\mu + \|\Delta \boldsymbol{U}_L(k-1)\|^2}\right)\tilde{\boldsymbol{\phi}}_{p,L}(k-1) + \boldsymbol{\phi}_{p,L}(k-1) - \boldsymbol{\phi}_{p,L}(k), \tag{4.29}$$

where $\mathbf{I}$ is an identity matrix with the corresponding dimension.

In Theorem 3.3, the boundedness of $\|\boldsymbol{\phi}_{p,L}(k)\|$ is proven. Assume its bound as a constant $\bar{b}$. Taking norms on both sides of (4.29) leads to

$$\begin{aligned}\left\|\tilde{\boldsymbol{\phi}}_{p,L}(k)\right\| &\le \left\|\left(\mathbf{I} - \frac{\eta \Delta \boldsymbol{U}_L(k-1)\Delta \boldsymbol{U}_L^T(k-1)}{\mu + \|\Delta \boldsymbol{U}_L(k-1)\|^2}\right)\tilde{\boldsymbol{\phi}}_{p,L}(k-1)\right\| + \left\|\boldsymbol{\phi}_{p,L}(k-1) - \boldsymbol{\phi}_{p,L}(k)\right\| \\ &\le \left\|\left(\mathbf{I} - \frac{\eta \Delta \boldsymbol{U}_L(k-1)\Delta \boldsymbol{U}_L^T(k-1)}{\mu + \|\Delta \boldsymbol{U}_L(k-1)\|^2}\right)\tilde{\boldsymbol{\phi}}_{p,L}(k-1)\right\| + 2\bar{b}.\end{aligned} \tag{4.30}$$

Squaring the first term on the right side of (4.30) gives

$$\begin{aligned}&\left\|\left(\mathbf{I} - \frac{\eta \Delta \boldsymbol{U}_L(k-1)\Delta \boldsymbol{U}_L^T(k-1)}{\mu + \|\Delta \boldsymbol{U}_L(k-1)\|^2}\right)\tilde{\boldsymbol{\phi}}_{p,L}(k-1)\right\|^2 < \left\|\tilde{\boldsymbol{\phi}}_{p,L}(k-1)\right\|^2 \\ &\quad + \left(-2 + \frac{\eta\|\Delta \boldsymbol{U}_L(k-1)\|^2}{\mu + \|\Delta \boldsymbol{U}_L(k-1)\|^2}\right)\frac{\eta\left(\tilde{\boldsymbol{\phi}}_{p,L}^T(k-1)\Delta \boldsymbol{U}_L(k-1)\right)^2}{\mu + \|\Delta \boldsymbol{U}_L(k-1)\|^2}.\end{aligned} \tag{4.31}$$

For $0 < \eta \leq 2$ and $\mu > 0$, the following inequality holds:

$$-2 + \frac{\eta \left\| \Delta \boldsymbol{U}_L(k-1) \right\|^2}{\mu + \left\| \Delta \boldsymbol{U}_L(k-1) \right\|^2} < 0. \tag{4.32}$$

Combining inequalities (4.31) and (4.32) yields

$$\left\| \left(\mathbf{I} - \frac{\eta \Delta \boldsymbol{U}_L(k-1) \Delta \boldsymbol{U}_L^T(k-1)}{\mu + \left\| \Delta \boldsymbol{U}_L(k-1) \right\|^2} \right) \tilde{\boldsymbol{\phi}}_{p,L}(k-1) \right\|^2 < \left\| \tilde{\boldsymbol{\phi}}_{p,L}(k-1) \right\|^2, \tag{4.33}$$

which implies that there exists a constant $0 < d_1 < 1$ such that

$$\left\| \left(\mathbf{I} - \frac{\eta \Delta \boldsymbol{U}_L(k-1) \Delta \boldsymbol{U}_L^T(k-1)}{\mu + \left\| \Delta \boldsymbol{U}_L(k-1) \right\|^2} \right) \tilde{\boldsymbol{\phi}}_{p,L}(k-1) \right\| \leq d_1 \left\| \tilde{\boldsymbol{\phi}}_{p,L}(k-1) \right\|. \tag{4.34}$$

Note that we only need to show the existence of the constant d_1 instead of giving its exact value.

Substituting (4.34) into (4.30) yields

$$\begin{aligned} \left\| \tilde{\boldsymbol{\phi}}_{p,L}(k) \right\| &\leq d_1 \left\| \tilde{\boldsymbol{\phi}}_{p,L}(k-1) \right\| + 2\bar{b} \\ &\leq d_1^2 \left\| \tilde{\boldsymbol{\phi}}_{p,L}(k-2) \right\| + 2d_1\bar{b} + 2\bar{b} \\ &\leq \cdots \leq d_1^{k-1} \left\| \tilde{\boldsymbol{\phi}}_{p,L}(1) \right\| + \frac{2\bar{b}(1-d_1^{k-1})}{1-d_1}, \end{aligned} \tag{4.35}$$

which means that $\hat{\boldsymbol{\phi}}_{p,L}(k)$ is bounded, since the boundedness of $\tilde{\boldsymbol{\phi}}_{p,L}(k)$ and $\boldsymbol{\phi}_{p,L}(k)$ is guaranteed from (4.35) and Theorem 3.3, respectively.

Step 2 Since $\boldsymbol{\phi}_{p,L}(k)$ and $\hat{\boldsymbol{\phi}}_{p,L}(k)$ are bounded, there exist bounded constants M_2, M_3, M_4, and $\lambda_{\min} > 0$, such that the following inequalities hold when $\lambda > \lambda_{\min}$:

$$\left| \frac{\hat{\phi}_1(k)}{\lambda + |\hat{\phi}_1(k)|^2} \right| \leq \left| \frac{\hat{\phi}_1(k)}{2\sqrt{\lambda}|\hat{\phi}_1(k)|} \right| < \frac{1}{2\sqrt{\lambda_{\min}}} \triangleq M_1 < \frac{0.5}{\bar{b}}, \tag{4.36}$$

$$0 < M_2 \leq \left| \frac{\hat{\phi}_1(k)\hat{\phi}_i(k)}{\lambda + |\hat{\phi}_1(k)|^2} \right| \leq \bar{b} \left| \frac{\hat{\phi}_1(k)}{2\sqrt{\lambda}|\hat{\phi}_1(k)|} \right| < \frac{\bar{b}}{2\sqrt{\lambda_{\min}}} < 0.5, \tag{4.37}$$

$$M_1 \left\| \boldsymbol{\phi}_{p,L}(k) \right\|_v \leq M_3 < 0.5, \tag{4.38}$$

$$M_2 + M_3 < 1, \tag{4.39}$$

$$\left(\sum_{i=2}^{L} \left| \frac{\hat{\phi}_1(k)\hat{\phi}_i(k)}{\lambda + |\hat{\phi}_1(k)|^2} \right| \right)^{1/(L-1)} \leq M_4. \tag{4.40}$$

Select $\max\limits_{i=2,\ldots,L} \rho_i$ such that

$$\sum_{i=2}^{L} \rho_i \left| \frac{\hat{\phi}_1(k)\hat{\phi}_i(k)}{\lambda + |\hat{\phi}_1(k)|^2} \right| \leq \left(\max_{i=2,\ldots,L} \rho_i \right) \sum_{i=2}^{L} \left| \frac{\hat{\phi}_1(k)\hat{\phi}_i(k)}{\lambda + |\hat{\phi}_1(k)|^2} \right| \leq \left(\max_{i=2,\ldots,L} \rho_i \right) M_4^{L-1} \triangleq M_5 < 1. \tag{4.41}$$

Define the tracking error as

$$e(k) = y^* - y(k). \tag{4.42}$$

Let

$$A(k) = \begin{bmatrix} -\dfrac{\rho_2 \hat{\phi}_1(k)\hat{\phi}_2(k)}{\lambda + |\hat{\phi}_1(k)|^2} & -\dfrac{\rho_3 \hat{\phi}_1(k)\hat{\phi}_3(k)}{\lambda + |\hat{\phi}_1(k)|^2} & \cdots & -\dfrac{\rho_L \hat{\phi}_1(k)\hat{\phi}_L(k)}{\lambda + |\hat{\phi}_1(k)|^2} & 0 \\ 1 & 0 & \cdots & 0 & 0 \\ 0 & 1 & \cdots & 0 & 0 \\ \vdots & \vdots & \vdots & \vdots & \vdots \\ 0 & 0 & \cdots & 1 & 0 \end{bmatrix}_{L \times L},$$

$$\Delta \boldsymbol{U}_L(k) = \left[\Delta u(k) \ \cdots \ \Delta u(k-L+1) \right]^T, \ C = \left[1 \ 0 \ \cdots \ 0 \right]^T \in R^L.$$

Then, controller algorithm (4.28) can be rewritten as

$$\begin{aligned}\Delta \boldsymbol{U}_L(k) &= \left[\Delta u(k) \quad \cdots \quad \Delta u(k-L+1)\right]^T \\ &= A(k)\left[\Delta u(k-1) \quad \cdots \quad \Delta u(k-L)\right]^T + \frac{\rho_1 \hat{\phi}_1(k)}{\lambda + |\hat{\phi}_1(k)|^2} Ce(k) \\ &= A(k)\Delta \boldsymbol{U}_L(k-1) + \frac{\rho_1 \hat{\phi}_1(k)}{\lambda + |\hat{\phi}_1(k)|^2} Ce(k). \end{aligned} \tag{4.43}$$

The characteristic equation of $A(k)$ is

$$z^L + \frac{\rho_2 \hat{\phi}_1(k)\hat{\phi}_2(k)}{\lambda + |\hat{\phi}_1(k)|^2} z^{L-1} + \cdots + \frac{\rho_L \hat{\phi}_1(k)\hat{\phi}_L(k)}{\lambda + |\hat{\phi}_1(k)|^2} z = 0. \tag{4.44}$$

From (4.41) and Lemma 4.4, we have $|z| < 1$. Thus, the following inequality holds:

$$|z|^{L-1} \le \sum_{i=2}^{L} \rho_i \left| \frac{\hat{\phi}_1(k)\hat{\phi}_i(k)}{\lambda + |\hat{\phi}_1(k)|^2} \right| |z|^{L-i} \le \sum_{i=2}^{L} \rho_i \left| \frac{\hat{\phi}_1(k)\hat{\phi}_i(k)}{\lambda + |\hat{\phi}_1(k)|^2} \right| \le \left(\max_{i=2,\ldots,L} \rho_i \right) M_4^{L-1} < 1, \tag{4.45}$$

which implies $|z| \le \left(\max_{i=2,\ldots,L} \rho_i \right)^{1/(L-1)} M_4 < 1$. Furthermore, there exists an arbitrary small positive constant ε_1 such that

$$\|A(k)\|_v \le s(A(k)) + \varepsilon_1 \le \left(\max_{i=2,\ldots,L} \rho_i \right)^{1/(L-1)} M_4 + \varepsilon_1 < 1, \tag{4.46}$$

where $\|A(k)\|_v$ is the consistent matrix norm of $A(k)$.

Let $d_2 = \left(\max_{i=2,\ldots,L} \rho_i \right)^{1/(L-1)} M_4 + \varepsilon_1$. From the definition of $\boldsymbol{U}_L(k)$, $k \le 0$, we have $\|\Delta \boldsymbol{U}_L(0)\|_v = 0$. Taking the norm on both sides of (4.43) yields

$$\begin{aligned}\|\Delta \boldsymbol{U}_L(k)\|_v &\le \|A(k)\|_v\,\|\Delta \boldsymbol{U}_L(k-1)\|_v + \rho_1\left|\frac{\hat{\phi}_1(k)}{\lambda+|\hat{\phi}_1(k)|^2}\right||e(k)|\\ &< d_2\,\|\Delta \boldsymbol{U}(k-1)\|_v + \rho_1 M_1\,|e(k)|\\ &\ \ \vdots\\ &= \rho_1 M_1\sum_{i=1}^{k} d_2^{k-i}\,|e(i)|. \end{aligned} \tag{4.47}$$

Substituting PFDL data model (4.21) and controller algorithm (4.43) into (4.42) yields

$$\begin{aligned} e(k+1) &= y^* - y(k+1) = y^* - y(k) - \boldsymbol{\phi}_{p,L}^T(k)\Delta \boldsymbol{U}_L(k)\\ &= e(k) - \boldsymbol{\phi}_{p,L}^T(k)\left(A(k)\Delta \boldsymbol{U}_L(k-1) + \rho_1\frac{\hat{\phi}_1(k)}{\lambda+|\hat{\phi}_1(k)|^2}Ce(k)\right)\\ &= \left(1-\frac{\rho_1\hat{\phi}_1(k)\phi_1(k)}{\lambda+|\hat{\phi}_1(k)|^2}\right)e(k) - \boldsymbol{\phi}_{p,L}^T(k)A(k)\Delta \boldsymbol{U}_L(k-1). \end{aligned} \tag{4.48}$$

From (4.37), one can choose $0<\rho_1\le 1$ such that

$$\left|1-\frac{\rho_1\hat{\phi}_1(k)\phi_1(k)}{\lambda+|\hat{\phi}_1(k)|^2}\right| = 1-\left|\frac{\rho_1\hat{\phi}_1(k)\phi_1(k)}{\lambda+|\hat{\phi}_1(k)|^2}\right| \le 1-\rho_1 M_2 < 1.$$

Let $d_3 = 1-\rho_1 M_2$. Taking norms on both sides of (4.48) yields

$$\begin{aligned} |e(k+1)| &\le \left|\left(1-\frac{\rho_1\hat{\phi}_1(k)\phi_1(k)}{\lambda+|\hat{\phi}_1(k)|^2}\right)\right||e(k)| + \|\phi_{p,L}(k)\|_v\,\|A(k)\|_v\,\|\Delta \boldsymbol{U}_L(k-1)\|_v\\ &< d_3\,|e(k)| + d_2\,\|\phi_{p,L}(k)\|_v\,\|\Delta \boldsymbol{U}_L(k-1)\|_v \end{aligned}$$

$$
\begin{aligned}
&< \cdots < d_3^k |e(1)| + d_2 \sum_{i=1}^{k-1} d_3^{k-1-i} \left\|\boldsymbol{\phi}_{p,L}(i+1)\right\|_v \left\|\Delta \boldsymbol{U}_L(i)\right\|_v \\
&< d_3^k |e(1)| + d_2 \sum_{i=1}^{k-1} d_3^{k-1-i} \left\|\boldsymbol{\phi}_{p,L}(i+1)\right\|_v \left\|\Delta \boldsymbol{U}_L(i)\right\|_v \\
&< d_3^k |e(1)| + d_2 \sum_{i=1}^{k-1} d_3^{k-1-i} \left\|\boldsymbol{\phi}_{p,L}(i+1)\right\|_v \rho_1 M_1 \sum_{j=1}^{i} d_2^{i-j} |e(j)|. \qquad (4.49)
\end{aligned}
$$

Let $d_4 = \rho_1 M_3$. Combining (4.38) and (4.49) gives rise to

$$
|e(k+1)| < d_3^k |e(1)| + d_2 d_4 \sum_{i=1}^{k-1} d_3^{k-1-i} \sum_{j=1}^{i} d_2^{i-j} |e(j)|. \qquad (4.50)
$$

Denote

$$
g(k+1) = d_3^k |e(1)| + d_2 d_4 \sum_{i=1}^{k-1} d_3^{k-1-i} \sum_{j=1}^{i} d_2^{i-j} |e(j)|. \qquad (4.51)
$$

Then, inequality (4.50) can be rewritten as

$$
|e(k+1)| < g(k+1), \quad \forall k = 1,2,\ldots \qquad (4.52)
$$

with $g(2) = d_3|e(1)|$.

Obviously, if $g(k+1)$ converges monotonically to zero, $e(k+1)$ also converges to zero. Computing $g(k+2)$ gives

$$
\begin{aligned}
g(k+2) &= d_3^{k+1} |e(1)| + d_2 d_4 \sum_{i=1}^{k} d_3^{k-i} \sum_{j=1}^{i} d_2^{i-j} |e(j)| \\
&= d_3 g(k+1) + d_4 d_2^k |e(1)| + \cdots + d_4 d_2^2 |e(k-1)| + d_4 d_2 |e(k)| \\
&< d_3 g(k+1) + d_4 d_2^k |e(1)| + \cdots + d_4 d_2^2 |e(k-1)| + d_4 d_2 |g(k)| \\
&= d_3 g(k+1) + h(k), \qquad (4.53)
\end{aligned}
$$

where $h(k) \triangleq d_4 d_2^k |e(1)| + \cdots + d_4 d_2^2 |e(k-1)| + d_4 d_2 |g(k)|$.

From (4.39), we have $d_3 = 1 - \rho_1 M_2 > \rho_1 (M_2 + M_3) - \rho_1 M_2 = \rho_1 M_3 = d_4$. Thus, the following inequality holds:

$$
\begin{aligned}
h(k) &< d_4 d_2^k |e(1)| + \cdots + d_4 d_2^2 |e(k-1)| + d_3 d_2 |g(k)| \\
&< d_4 d_2^k |e(1)| + \cdots + d_4 d_2^2 |e(k-1)| \\
&\quad + d_3 d_2 \left(d_3^{k-1} |e(1)| + d_2 d_4 \sum_{i=1}^{k-2} d_3^{k-2-i} \sum_{j=1}^{i} d_2^{i-j} |e(j)| \right) \\
&= d_2 \left(d_3^k |e(1)| + d_2 d_4 \sum_{i=1}^{k-1} d_3^{k-1-i} \sum_{j=1}^{i} d_2^{i-j} |e(j)| \right) \\
&= d_2 g(k+1).
\end{aligned} \tag{4.54}
$$

Substituting (4.54) into (4.53) yields

$$
g(k+2) < d_3 g(k+1) + h(k) < (d_3 + d_2) g(k+1). \tag{4.55}
$$

Select $0 < \rho_1 \leq 1, \ldots, 0 < \rho_L \leq 1$ such that $0 < \max_{i=2,\ldots,L} \{\rho_i\}^{1/(L-1)} M_4 < \rho_1 M_2 < 1$, and

$$
0 < 1 - \rho_1 M_2 + \max_{i=2,\ldots,L} \{\rho_i\}^{\frac{1}{L-1}} M_4 < 1. \tag{4.56}
$$

Since ε_1 is an arbitrary small positive constant, the following inequality holds:

$$
d_3 + d_2 = 1 - \rho_1 M_2 + \max_{i \in \{2,L\}} \{\rho_i\}^{\frac{1}{L-1}} M_4 + \varepsilon_1 < 1. \tag{4.57}
$$

Substituting (4.57) into (4.55) yields

$$
\lim_{k \to \infty} g(k+2) < \lim_{k \to \infty} (d_3 + d_2) g(k+1) < \cdots < \lim_{k \to \infty} (d_3 + d_2)^k g(2) = 0. \tag{4.58}
$$

Thus, conclusion (a) is a direct result of (4.52) and (4.58).

In virtue of the boundedness of y^* and $e(k)$, $y(k)$ is also bounded. From inequalities (4.47), (4.52), and (4.58), we have

$$
\begin{aligned}
\|\boldsymbol{U}_L(k)\|_v &\leq \sum_{i=1}^{k} \|\Delta \boldsymbol{U}_L(i)\|_v < \rho_1 M_1 \sum_{i=1}^{k} \sum_{j=1}^{i} d_2^{i-j} |e(j)| \\
&< \frac{\rho_1 M_1}{1 - d_2} \left(|e(1)| + \cdots + |e(k)| \right) < \frac{\rho_1 M_1}{1 - d_2} \left(e(1) + g(2) + \cdots + g(k) \right) \\
&< \frac{\rho_1 M_1}{1 - d_2} \left(e(1) + \frac{g(2)}{1 - d_2 - d_3} \right).
\end{aligned} \tag{4.59}
$$

Thus, conclusion (b) is obtained. ■

4.3.3 Simulation Results

In this section, numerical simulations for three totally different SISO discrete-time nonlinear systems controlled by the same PFDL–MFAC control scheme are given to verify the correctness and effectiveness of the proposed control method. It is noted that the model dynamics, including system structure (linear or nonlinear), orders, and relative degree, is not involved in PFDL–MFAC scheme. The models in the following three examples are only used to generate the I/O data of the controlled plant.

The initial conditions are $u(1) = u(2) = \cdots = u(5) = 0$, $y(1) = y(2) = y(3) = 0$, $y(4) = 1$, and $y(5) = y(6) = 0$. The control input linearization length constant, the step factors, the weighting factors, and the initial value of the PG estimation in the PFDL–MFAC scheme are set to $L = 3$, $\rho_1 = \rho_2 = \rho_3 = 0.5$, $\eta = 0.5$, $\lambda = 0.01$, $\mu = 1$, $\hat{\boldsymbol{\phi}}_{p,L}(1) = [1\ \ 0\ \ 0]^T$, respectively; ε is set to be 10^{-5}.

Example 4.4

Nonlinear system

$$y(k+1) = \begin{cases} 2.5y(k)y(k-1)/(1+y^2(k)+y^2(k-1)) \\ \quad +0.7\sin(0.5(y(k)+y(k-1))) \\ \quad +1.4u(k-1)+1.2u(k), & k \le 200, \\ \quad -0.1y(k)-0.2y(k-1)-0.3y(k-2) \\ \quad +0.1u(k)+0.02u(k-1)+0.03u(k-2), & k > 200, \end{cases}$$

where the two subsystems are nonminimum-phase nonlinear system and minimum-phase linear system, respectively. Obviously, the controlled system is with time-varying structure, time-varying order, and time-varying phase.

The desired output signal is $y^*(k+1) = 5 \times (-1)^{\mathrm{round}(k/80)}$.

For comparison, both the classical PID method and the PFDL–MFAC scheme are simulated in this example. The PID controller is as follows:

$$u(k) = K_p\left[e(k) + \sum_{j=0}^{k} e(j)/T_I + T_d(e(k) - e(k-1))\right].$$

After several trial and error runs, the best PID parameters are set to $K_p = 0.15, T_I = 0.5$, and $T_D = 0$.

The simulation comparison results are shown in Figure 4.4. From the simulation results, one can see that the PFDL–MFAC scheme has a faster system response, almost the same overshoot, and a shorter settling time, compared with the PID method. Moreover, there are fewer parameters in the PFDL–MFAC scheme than in the PID method. This makes it convenient to adjust the parameters in the PFDL–MFAC scheme, which is also with a simple structure and is easy to implement.

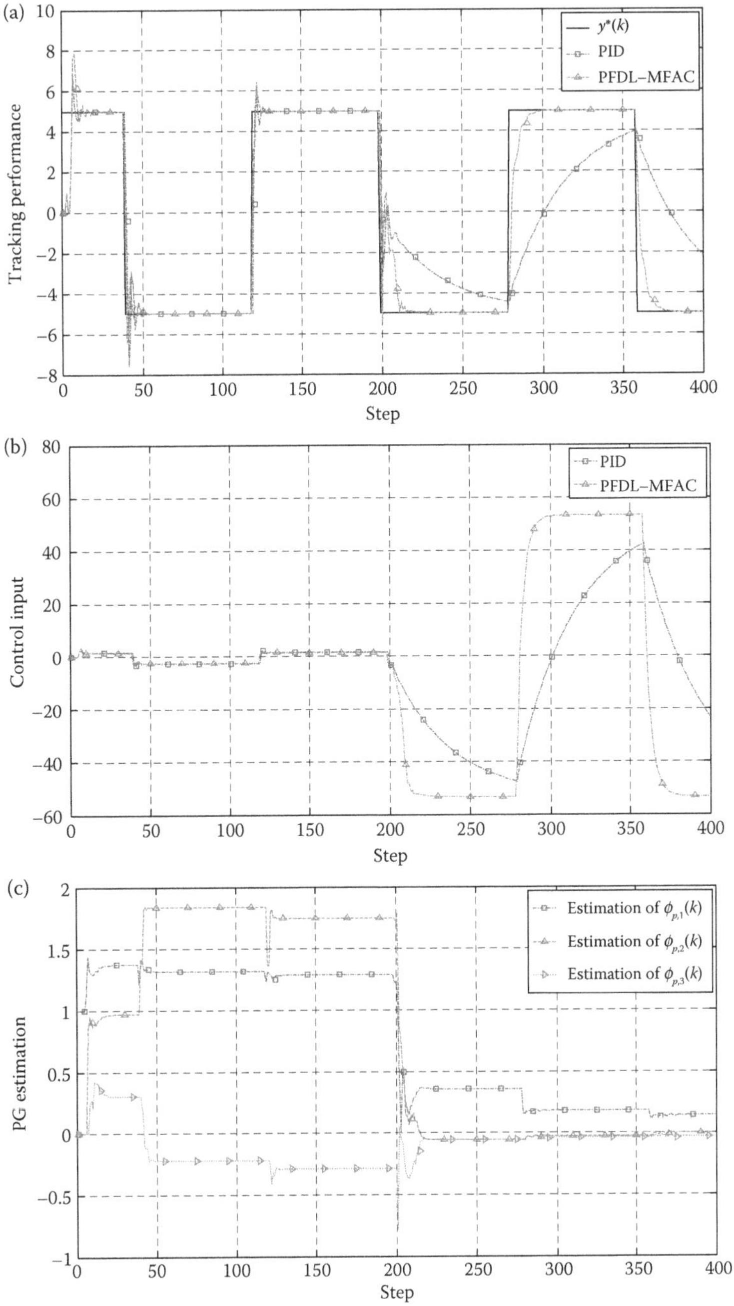

Figure 4.4 Performance comparison using PFDL–MFAC and PID in Example 4.4. (a) Tracking performance. (b) Control input. (c) Estimated PG $\hat{\phi}_{p,L}(k)$.

Example 4.5

Nonlinear system

$$y(k+1)=\frac{y(k)y(k-1)y(k-2)u(k-1)(y(k)-1)+(1+a(k))u(k)}{1+y^2(k)+y^2(k-1)+y^2(k-2)},$$

where $a(k)$ is a time-varying parameter.

The desired output signal is given as

$$y^*(k+1)=0.5\times(-1)^{\mathrm{round}(k/50)}.$$

The simulation results with $a(k)=1$ are shown in Figure 4.5. Figure 4.5a gives the tracking performance. Figure 4.5b demonstrates the profile of the control input.

In Ref. [165], a multilayered recurrent neural network with a dynamic back propagation (DBP) learning algorithm was used to approximate this nonlinear

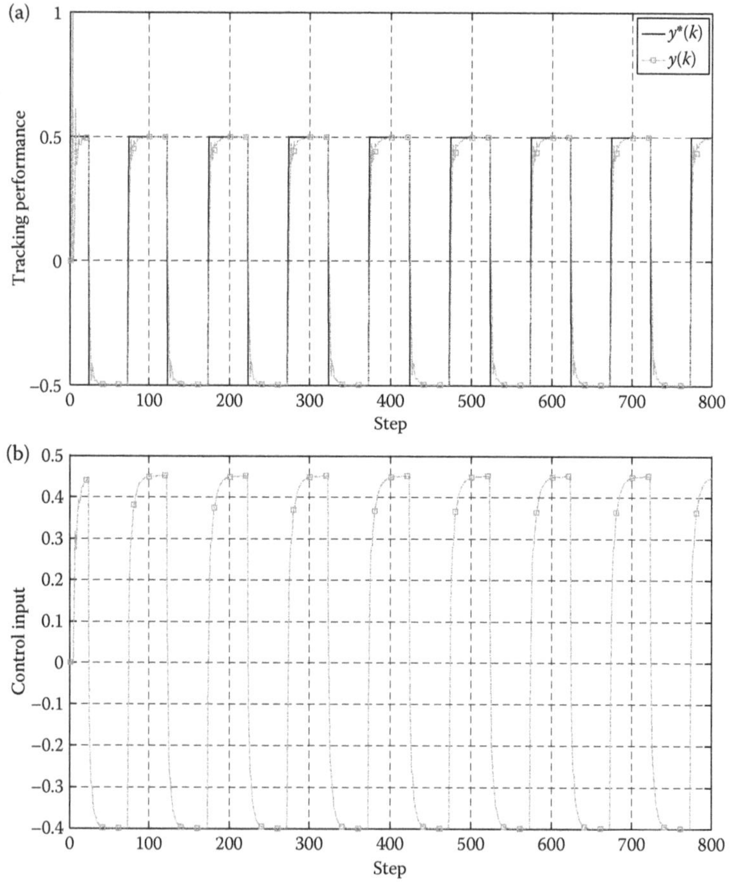

Figure 4.5 Simulation results in Example 4.5 with $a(k)=1$. (a) Tracking performance. (b) Control input.

system, and the control performance using neural-network-based control algorithm is acceptable. For the details see Ref. [165].

From Figure 4.5a, one can see that the PFDL–MFAC scheme gives a satisfactory performance, which is much better than that obtained using the neural-network-based control method, and the controller's computational burden is much lower in comparison with the neural-network-based control.

In the case of

$$a(k) = \begin{cases} 0, & k \le 100; \\ 1, & 100 < k \le 300; \\ 2, & 300 < k \le 500; \\ 3, & 500 < k \le 800, \end{cases}$$

the simulation results are shown in Figure 4.6. The control performance of the PFDL–MFAC scheme is still quite good, even when abrupt changes of the

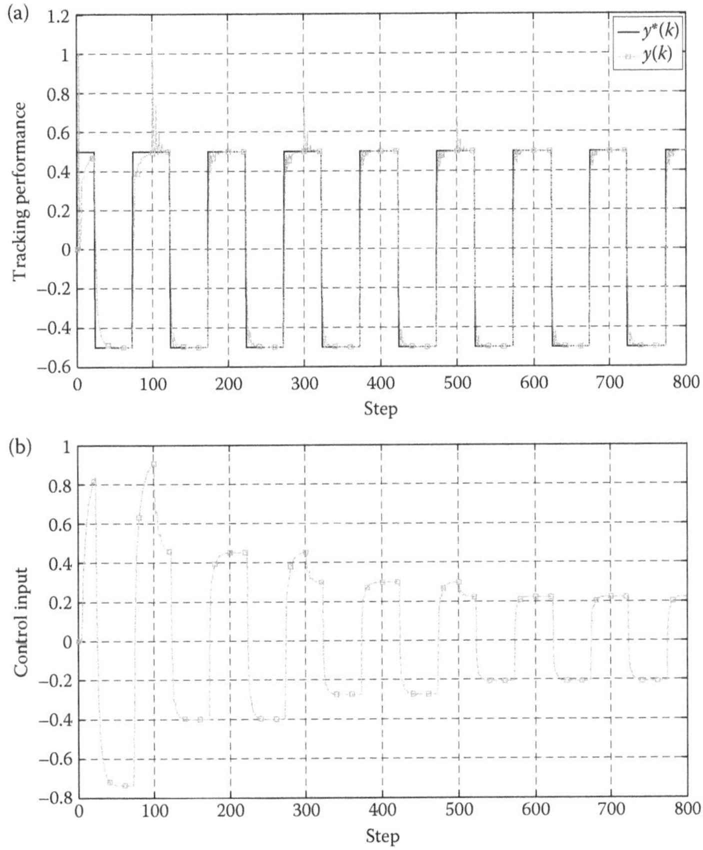

Figure 4.6 Simulation results in Example 4.5 with time-varying parameter *a*(*k*). (a) Tracking performance. (b) Control input.

controller gain caused by the time-varying parameter $a(k)$ come forth at the 100th, 300th, and 500th steps.

Example 4.6

Nonlinear system

$$y(k+1)=\begin{cases}\dfrac{2.5y(k)y(k-1)}{1+y^2(k)+y^2(k-1)}+1.2u(k)+0.09u(k)u(k-1)+1.6u(k-2)\\ \quad+0.7\sin(0.5(y(k)+y(k-1)))\cos(0.5(y(k)+y(k-1))), & k\le 400;\\ \dfrac{5y(k)y(k-1)}{1+y^2(k)+y^2(k-1)+y^2(k-2)}+u(k)+1.1u(k-1), & k>400,\end{cases}$$

where both subsystems are nonlinear and nonminimum-phase. Obviously, the system is with time-varying structure, and time-varying orders.

The desired output signal is given as

$$y^*(k+1) = 5\sin(k\pi/50)+2\cos(k\pi/20).$$

The comparison between CFDL–MFAC scheme (4.7)–(4.9) and PFDL–MFAC scheme (4.26)–(4.28) is given in this simulation. For the CFDL–MFAC scheme, the initial value of the PPD estimation $\hat{\phi}_c(1)$ is set to be 1; step factors and weighting factors are set as $\rho=0.5$, $\eta=0.5$, $\lambda=0.01$, and $\mu=1$, respectively; ε is set to be 10^{-5}.

From the simulation results shown in Figure 4.7, one can see that the dynamics of $\hat{\phi}_c(k)$ in the CFDL–MFAC scheme is too complex to track for the proposed projection estimation algorithm, which makes its control performance poor. The dynamics of PG $\hat{\boldsymbol{\phi}}_{p,L}(k)$ of the PFDL–MFAC scheme, however, is relatively simple; thus, the tracking performance of the PFDL–MFAC scheme is better.

From the simulation results in the above three examples, one can see that quite a good control performance can be obtained by applying the same PFDL–MFAC scheme with the same initial setting to totally different unknown nonlinear systems, even with a time-varying structure, a time-varying parameter, and a time-varying delay, whereas it is difficult for traditional adaptive control methods to deal with such nonlinear systems, even when the models are completely known. The complicated behavior of the PPD in the CFDL–MFAC scheme is dispersed in the PFDL–MFAC scheme due to the introduction of more parameters in the PG. Therefore, a better control performance is achieved when the PFDL–MFAC scheme is applied to complex nonlinear systems.

4.4 FFDL Data Model Based MFAC

4.4.1 Control System Design

The PFDL data model, in essence, just focuses on dynamic relationships between the change $\Delta y(k+1)$ of the system output at time $k+1$ and the changes of

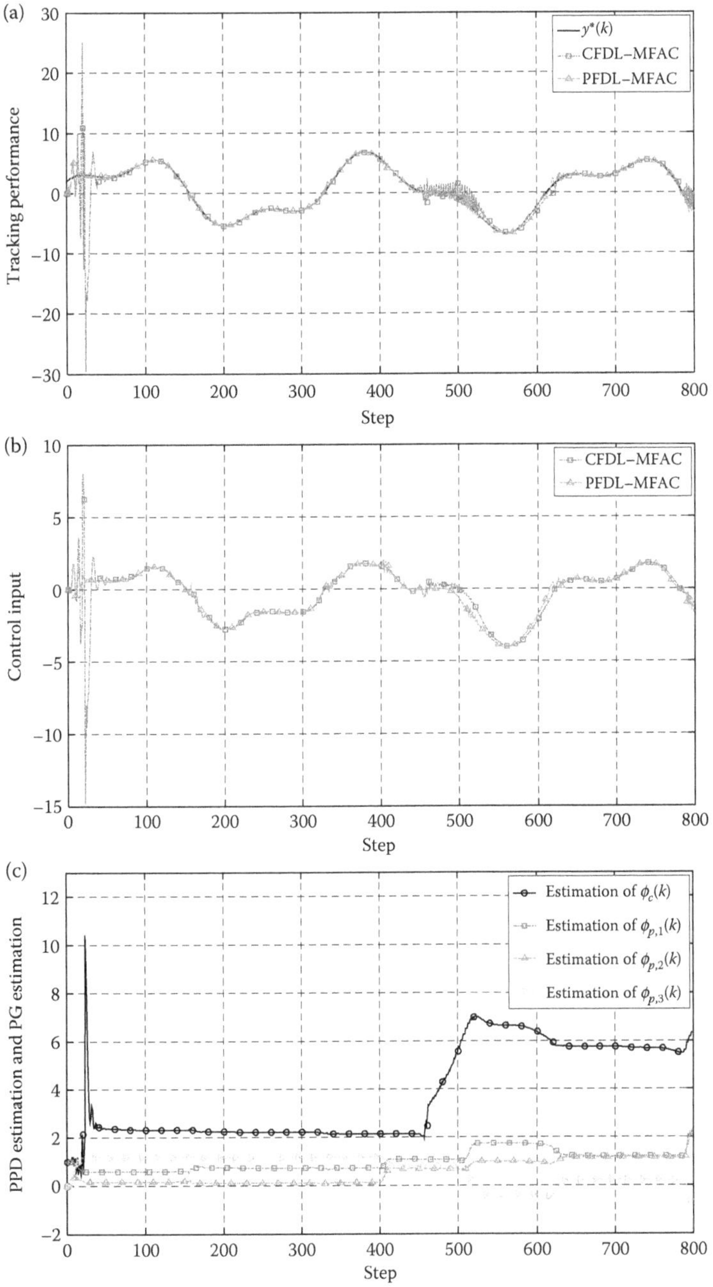

Figure 4.7 Simulation results in Example 4.6. (a) Tracking performance. (b) Control input. (c) Estimated PPD $\hat{\phi}_c(k)$ in CFDL–MFAC scheme and estimated PG $\hat{\phi}_{p,L}(k)$ in PFDL–MFAC scheme.

control inputs within an input-related fixed-length moving time window at time k. However, the change $\Delta y(k+1)$ of the system output at time $k+1$ may be sensitive not only to the changes of the control inputs within the input-related moving time window at time k, but also to the changes of certain system outputs within an output-related fixed-length moving time window at time k. In this circumstance, the controller design should take these changes into consideration. Otherwise, it might lead to poor performance or even instability of the closed-loop control system. On the basis of this observation, the FFDL–MFAC scheme design is presented in virtue of the FFDL data model presented in Section 3.2.3. Compared with the PFDL–MFAC scheme, the FFDL–MFAC scheme has more adjustable degrees of freedom and more design flexibility.

From Theorem 3.4, nonlinear system (4.1) satisfying Assumptions 3.5 and 3.6 with $\|\Delta \boldsymbol{H}_{L_y,L_u}(k)\| \neq 0$ for all k can be transformed into the following FFDL data model:

$$\Delta y(k+1) = \boldsymbol{\phi}^T_{f,L_y,L_u}(k)\Delta \boldsymbol{H}_{L_y,L_u}(k), \tag{4.60}$$

where $\boldsymbol{\phi}_{f,L_y,L_u}(k) = [\phi_1(k),\ldots,\phi_{L_y}(k),\phi_{L_y+1}(k),\ldots,\phi_{L_y+L_u}(k)]^T \in R^{L_y+L_u}$ is the unknown bounded PG, $\Delta \boldsymbol{H}_{L_y,L_u}(k) = [\Delta y(k),\ldots,\Delta y(k-L_y+1),\Delta u(k),\ldots,\Delta u(k-L_u+1)]^T$, and L_y and L_u are called *pseudo orders.*

4.4.1.1 Controller Algorithm

A cost function of control input is constructed as

$$J(u(k)) = \left|y^*(k+1) - y(k+1)\right|^2 + \lambda\left|u(k) - u(k-1)\right|^2, \tag{4.61}$$

where $\lambda > 0$ is a weighting factor.

Substituting FFDL data model (4.60) into cost function (4.61) and minimizing it with respect to $u(k)$ yield the controller algorithm as follows:

$$u(k) = u(k-1) + \frac{\rho_{L_y+1}\phi_{L_y+1}(k)(y^*(k+1) - y(k))}{\lambda + \left|\phi_{f,L_y+1}(k)\right|^2}$$
$$- \frac{\phi_{L_y+1}(k)\sum_{i=1}^{L_y}\rho_i\phi_i(k)\Delta y(k-i+1)}{\lambda + \left|\phi_{L_y+1}(k)\right|^2} - \frac{\phi_{L_y+1}(k)\sum_{i=L_y+2}^{L_y+L_u}\rho_i\phi_i(k)\Delta u(k+L_y-i+1)}{\lambda + \left|\phi_{L_y+1}(k)\right|^2}, \tag{4.62}$$

where the step factor $\rho_i \in (0,1]$, $i = 1,2,\ldots, L_y + L_u$ is added to make the controller algorithm more flexible.

4.4.1.2 PG Estimation Algorithm

A cost function of PG estimation is given as follows:

$$J(\boldsymbol{\phi}_{f,L_y,L_u}(k)) = \left| y(k) - y(k-1) - \boldsymbol{\phi}^T_{f,L_y,L_u}(k)\Delta \boldsymbol{H}_{L_y,L_u}(k-1) \right|^2 + \mu \left\| \boldsymbol{\phi}_{f,L_y,L_u}(k) - \hat{\boldsymbol{\phi}}_{f,L_y,L_u}(k-1) \right\|^2, \tag{4.63}$$

where $\mu > 0$ is a weighting factor.

Minimizing cost function (4.63) with respect to $\boldsymbol{\phi}_{f,L_y,L_u}(k)$ according to the optimality condition and using the matrix inversion lemma give

$$\hat{\boldsymbol{\phi}}_{f,L_y,L_u}(k) = \hat{\boldsymbol{\phi}}_{f,L_y,L_u}(k-1) + \frac{\eta \Delta \boldsymbol{H}_{L_y,L_u}(k-1)\left(y(k) - y(k-1) - \hat{\boldsymbol{\phi}}^T_{f,L_y,L_u}(k-1)\Delta \boldsymbol{H}_{L_y,L_u}(k-1) \right)}{\mu + \left\| \Delta \boldsymbol{H}_{L_y,L_u}(k-1) \right\|^2}, \tag{4.64}$$

where the step factor $\eta \in (0,2]$ is added to make the estimation algorithm more generic, and $\hat{\boldsymbol{\phi}}_{f,L_y,L_u}(k)$ is the estimation of PG $\boldsymbol{\phi}_{f,L_y,L_u}(k)$.

4.4.1.3 System Control Scheme

By synthesizing PG estimation algorithm (4.64) and controller algorithm (4.62), the FFDL–MFAC scheme is presented as follows:

$$\hat{\boldsymbol{\phi}}_{f,L_y,L_u}(k) = \hat{\boldsymbol{\phi}}_{f,L_y,L_u}(k-1) + \frac{\eta \Delta \boldsymbol{H}_{L_y,L_u}(k-1)\left(y(k) - y(k-1) - \hat{\boldsymbol{\phi}}^T_{f,L_y,L_u}(k-1)\Delta \boldsymbol{H}_{L_y,l_u}(k-1) \right)}{\mu + \left\| \Delta \boldsymbol{H}_{L_y,L_u}(k-1) \right\|^2}, \tag{4.65}$$

$$\hat{\boldsymbol{\phi}}_{f,L_y,L_u}(k) = \hat{\boldsymbol{\phi}}_{f,L_y,L_u}(1) \quad \text{if } \|\hat{\boldsymbol{\phi}}_{f,L_y,L_u}(k)\| \le \varepsilon \text{ or } \|\Delta \boldsymbol{H}_{L_y,L_u}(k-1)\| \le \varepsilon \text{ or } \operatorname{sign}(\hat{\phi}_{L_y+1}(k)) \ne \operatorname{sign}(\hat{\phi}_{L_y+1}(1)), \tag{4.66}$$

$$u(k)=u(k-1)+\frac{\rho_{L_y+1}\hat{\phi}_{L_y+1}(k)(y^*(k+1)-y(k))}{\lambda+\left|\hat{\phi}_{L_y+1}(k)\right|^2}$$

$$-\frac{\hat{\phi}_{L_y+1}(k)\sum_{i=1}^{L_y}\rho_i\hat{\phi}_i(k)\Delta y(k-i+1)}{\lambda+\left|\hat{\phi}_{L_y+1}(k)\right|^2}-\frac{\hat{\phi}_{L_y+1}(k)\sum_{i=L_y+2}^{L_y+L_u}\rho_i\hat{\phi}_i(k)\Delta u(k+L_y-i+1)}{\lambda+\left|\hat{\phi}_{L_y+1}(k)\right|^2},$$

(4.67)

where $\lambda > 0$, $\mu > 0$, $\eta \in (0,2]$, $\rho_i \in (0,1]$, $i = 1,2,\ldots,L_y + L_u$, ε is a small positive constant, and $\hat{\boldsymbol{\phi}}_{f,L_y,L_u}(1)$ is the initial value of $\hat{\boldsymbol{\phi}}_{f,L_y,L_u}(k)$.

Remark 4.9

There is an $(L_y + L_u)$-dimensional vector $\hat{\boldsymbol{\phi}}_{f,L_y.L_u}(k)$, namely, estimated PG, to tune online in the FFDL–MFAC scheme, while the *pseudo orders* L_y and L_u are also adjustable. When $L_y = 0$ and $L_u = L$, the FFDL–MFAC scheme (4.65)–(4.67) becomes the PFDL–MFAC scheme (4.26)–(4.28), and when $L_y = 0$ and $L_u = 1$, the FFDL–MFAC scheme (4.65)–(4.67) becomes the CFDL–MFAC scheme (4.7)–(4.9). Compared with the CFDL–MFAC scheme and the PFDL–MFAC scheme, the FFDL–MFAC scheme has more adjustable degrees of freedom and more design flexibility due to the introduction of more step factors as $\rho_1, \rho_2, \ldots, \rho_{L_y+L_u}$.

Remark 4.10

For the LTI systems with a known model structure, the proposed FFDL–MFAC scheme becomes the traditional adaptive control scheme if the weighting factor and *pseudo orders* are set to $\lambda = 0$, $L_y = n_y$, and $L_u = n_u$ [76]. For unknown systems, the *pseudo orders* L_y and L_u can be chosen as the approximations of n_y and n_u, or just as small as possible to get a simple controller with low computational burden. The control scheme in Ref. [166] is a special case of the FFDL–MFAC scheme with $L_y = 1$ and $L_u = 1$. By properly selecting the *pseudo orders* of the proposed control scheme, order-reduction design of traditional model-based adaptive control systems can be avoided without involving any unmodeled dynamics.

Remark 4.11

Data model (4.60) can be interpreted as a linear time-varying system. It is well known that stability analysis for the adaptive control of linear time-varying systems is very challenging, and there are very few results in the existing literature except for some special cases where very strong assumptions are made on the change rate of the time-varying coefficients. Therefore, the complete stability analysis of the FFDL–MFAC scheme is an open problem that needs further investigation in the future.

4.4.2 Simulation Results

In this section, numerical simulations for three totally different SISO discrete-time nonlinear systems are given to verify the correctness and the effectiveness of the FFDL–MFAC scheme.

In the following three examples, the same initial conditions and the same FFDL–MFAC scheme are used. The initial conditions are $u(1) = u(2) = \cdots = u(4) = 0$, $u(5) = 0.5$, $y(1) = y(2) = y(3) = 0$, $y(4) = 1$, $y(5) = 0.2$, and $y(6) = 0$. The *pseudo orders* of the FFDL–MFAC scheme are selected to be $L_y = 1$ and $L_u = 2$; the step factors and the weighting factors are set to $\rho_1 = \rho_2 = \rho_3 = 0.7$, $\eta = 0.2$, $\lambda = 7$, and $\mu = 1$, respectively; ε is set to 10^{-5}.

Example 4.7

Nonlinear system

$$y(k+1) = \frac{-0.9y(k) + (a(k)+1)u(k)}{1 + y^2(k)},$$

where $a(k) = 4 \times \text{round}(k/100) + \sin(k/100)$ is a time-varying parameter. The above system with $a(k) = 0$ was given in Ref. [167], where the neural network based control approach was applied.

The desired output signal is as follows:

$$y^*(k+1) = \begin{cases} 0.4^{\text{round}(k/50)}, & k \le 490 \\ 0.1 + 0.1 \times (-1)^{\text{round}(k/50)}, & k > 490. \end{cases}$$

The initial value of PG is set to $\hat{\boldsymbol{\phi}}_{f,L_y,L_u}(1) = [-2 \quad 0.5 \quad 0.9]^T$. The simulation results are illustrated in Figure 4.8, which shows that the FFDL–MFAC gives a satisfactory control performance. The spikes in Figure 4.8a are caused by the abrupt changes of the parameter $a(k)$. From the profile of the system output, a better tracking performance is obtained than that obtained using the neural network-based control method; see Ref. [167] for details.

Example 4.8

Nonlinear system

$$y(k+1) = 1.6y(k) - 0.63y(k-1) + u'(k) - 0.5u'(k-1),$$

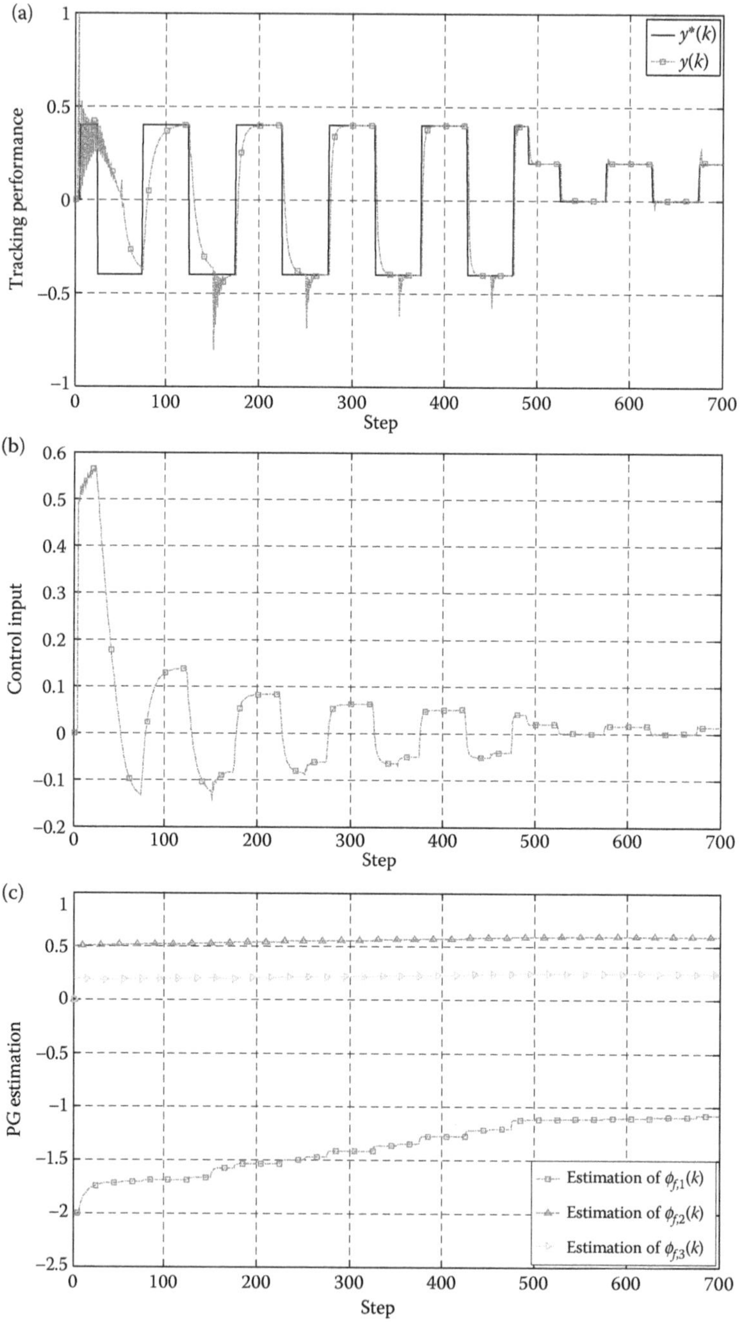

Figure 4.8 Simulation results in Example 4.7. (a) Tracking performance. (b) Control input. (c) Estimated PG $\hat{\phi}_{f,L_y,L_u}(k)$.

where

$$u'(k) = \begin{cases} 0, & \text{if } |u(k)| \le 2 \\ u(k) - 0.2 \times \text{sign}(u(k)), & \text{if } |u(k)| > 2, \end{cases}$$

and sign(·) is the sign function. The system is taken from Ref. [167], where the neural-network-based control approach was applied.

The desired trajectory is as follows:

$$y^*(k+1) = \begin{cases} 5^{\text{round}(k/50)}, & k \le 490 \\ 3.5 + 0.5^{\text{round}(k/100)}, & k > 490. \end{cases}$$

The initial value of the PG is set to be $\hat{\boldsymbol{\phi}}_{f,L_y,L_u}(1) = \begin{bmatrix} 2 & 0.5 & 0.2 \end{bmatrix}^T$. The other settings are the same as Example 4.7. The simulation results illustrated in Figure 4.9 show that the FFDL–MFAC scheme gives a satisfactory control performance with a small overshoot, which is better than that obtained using the neural-network-based control method; see Ref. [167] for details.

Example 4.9

Linear system

$$y(k+1) = \begin{cases} 0.55y(k) + 0.46y(k-1) + 0.07y(k-2) \\ \quad + 0.1u(k) + 0.02u(k-1) + 0.03u(k-2), & \text{if } k \le 400; \\ -0.1y(k) - 0.2y(k-1) - 0.3y(k-2) \\ \quad + 0.1u(k) + 0.02u(k-1) + 0.03u(k-2), & \text{if } k > 400. \end{cases}$$

The system is taken from Ref. [168], where the neural-network-based control approach was applied. Obviously, the first linear subsystem is open-loop unstable, and the whole system is with a time-varying structure.

The desired trajectory is

$$y^*(k+1) = 2\sin(k/50) + \cos(k/20).$$

The initial value of PG is set to $\hat{\boldsymbol{\phi}}_{f,L_y,L_u}(1) = [2 \quad 0.5 \quad 0]^T$, and the weighting factor is $\lambda = 0.001$. The other settings are the same as in Example 4.7. The simulation results illustrated in Figure 4.10 show that the FFDL–MFAC scheme gives a satisfactory control performance, which is better than that obtained using the neural-network-based control method [168]. The small oscillation at the 400th step is caused by a change of the system structure.

From the simulation results in the above three examples, one can see that quite a good control performance can be obtained by applying the same FFDL–MFAC scheme with the same initial setting to totally different unknown nonlinear systems. In addition, PG is a slowly time-varying parameter, and is insensitive to the time-variant in system structure, order, and parameter.

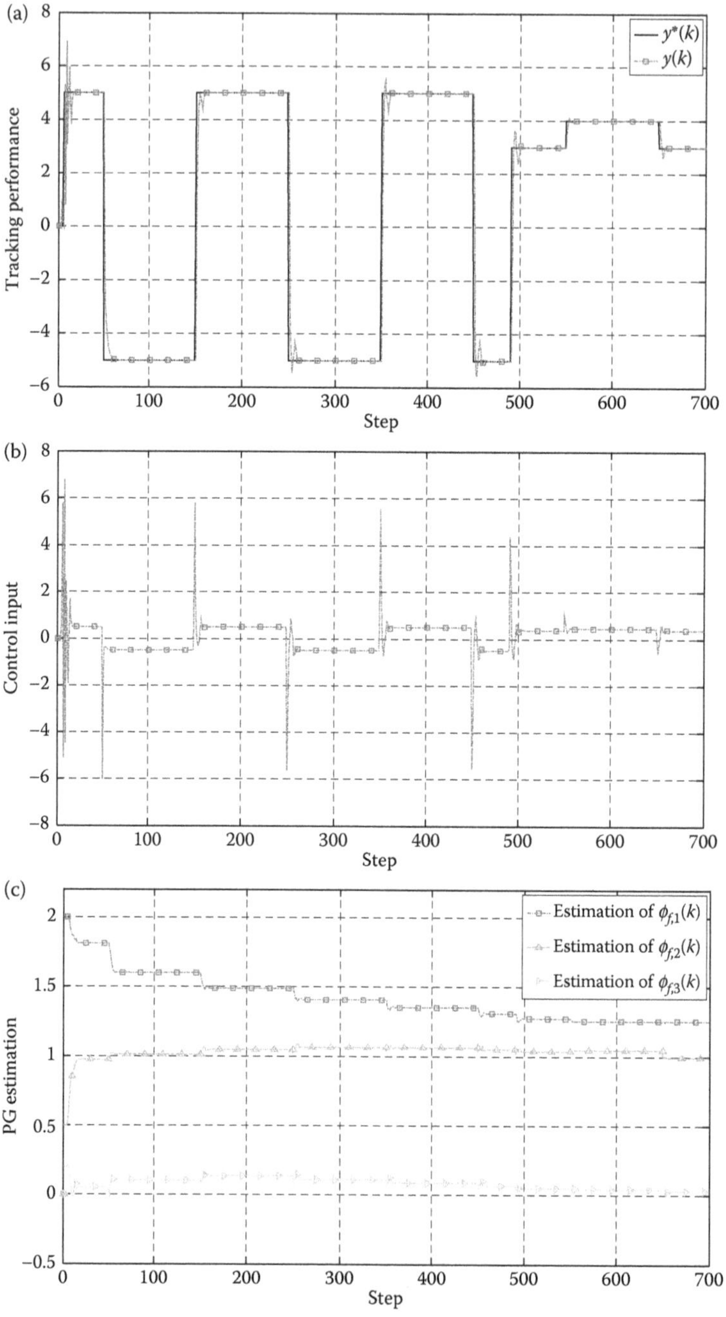

Figure 4.9 Simulation results in Example 4.8. (a) Tracking performance. (b) Control input. (c) Estimated PG $\hat{\boldsymbol{\phi}}_{f,L_y,L_u}(k)$.

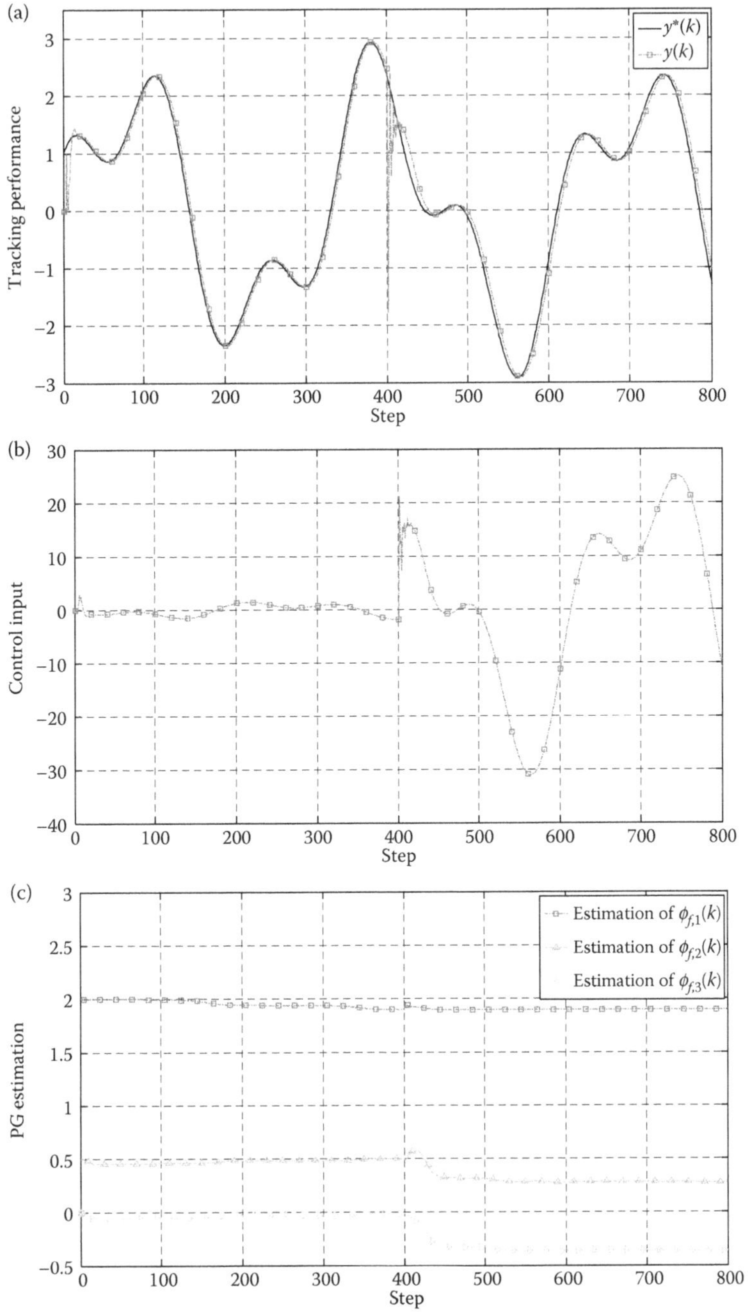

Figure 4.10 Simulation results in Example 4.9. (a) Tracking performance. (b) Control input. (c) Estimated PG $\hat{\phi}_{f,L_y,L_u}(k)$.

4.5 Conclusions

On the basis of the dynamic linearization approach presented in Chapter 3, three kinds of MFAC schemes for the unknown SISO discrete-time nonlinear system are studied in this chapter, including CFDL–MFAC, PFDL–MFAC, and FFDL–MFAC. The proposed MFAC approach merely utilizes online I/O data of the controlled plant, and no model information of the plant is involved. Thus, it is a data-driven MFAC method. Moreover, since the controller in the MFAC scheme is of a simple incremental form, and has only a few parameters to be updated online, it is a low-cost controller. Next, the MFAC method shows very strong robustness, as it can uniformly deal with control problems for nonlinear systems with time-varying structure, time-varying parameter, and time-varying delay. Finally, the BIBO stability and monotonic convergence of the tracking error in the first two schemes are proven based on the contraction mapping method. That is different from the traditional *Lyapunov* method, which cannot guarantee monotonic convergence. The effectiveness is verified via extensive numerical simulations on different nonlinear systems.

It is worth pointing out that there are three possible MFAC schemes for a given discrete-time nonlinear system (4.1). They are the CFDL–MFAC scheme, the PFDL–MFAC scheme, and the FFDL–MFAC scheme. The first one is designed for simple SISO discrete-time nonlinear systems and has the simplest form of a controller with only one scalar PPD updated online. The second includes the first one as a special case, and the third in turn includes the second. The last two schemes have an L-dimensional and an $(L_y + L_u)$-dimensional vector PG needed to be updated online, respectively. Further, the length of the PG can be designed by the researcher. Compared with the CFDL–MFAC scheme, the PFDL–MFAC and the FFDL–MFAC schemes enjoy more adjustable degrees of freedom and more design flexibility by introducing different kinds of PGs, which consider the change of system output possibly caused by the previous changes of the control inputs in a specified length moving time window at current time instant k, or the change of system output caused by the previous changes of both the inputs and outputs in a specified length moving time window at current time instant k. The trade-off between the computation burden and the simplicity of the control system design among these three MFAC schemes could be achieved through careful selection of the different kinds of control schemes of the MFAC methods, and the vital criterion in the choice of the controller depends on the control system performance as well as the cost one could afford.

Chapter 5

Model-Free Adaptive Control of MIMO Discrete-Time Nonlinear Systems

5.1 Introduction

Most practical industrial processes are nonlinear and multivariable in nature. For MIMO nonlinear systems, the controller design is very difficult due to the couplings among different inputs and outputs [169]. Typical adaptive control methods for MIMO systems are the model reference adaptive control (MRAC) [76,170,171], adaptive backstepping control [172,173], adaptive dynamic surface control (DSC) [174,175,176], and so on. The all aforementioned adaptive control methods, proposed for linear systems or nonlinear systems, depend on the known structure information of the controlled system, especially on coupling relationships between inputs and outputs. However, it is more difficult to construct an accurate first-principles or the identified model for a practical MIMO system due to its more complex dynamics compared with SISO systems. Furthermore, it is much hard to cancel certain nonlinear elements accurately by using Lyapunov based adaptive control methods of MIMO nonlinear systems when there exist uncertainties and disturbances. Thus, it is difficult to obtain desired control performance by applying model based adaptive control methods in practical applications.

It is worth pointing out that, to get out of the dependence on the accurate plant model for the control system design, the fuzzy control and neural network control, are proposed [158,177]. However, the fuzzy rules are difficult to establish and update, and the neural network is hard to train for a practical control problem.

In this chapter, three kinds of virtual equivalent dynamic linearization data models, that is, the CFDL data model, the PFDL data model, and the FFDL data model, are built along the dynamic operating points for a class of unknown MIMO discrete-time nonlinear and nonaffine systems, and then the PJM in data models are estimated merely using the I/O data of the controlled plant. Using these data models and PJM estimation algorithms, the MFAC schemes are presented with a rigorous analysis of BIBO stability [23,26,178–180]. The main features of the proposed MFAC approaches are: the controller design depends merely upon the real-time measured I/O data of the controlled system without involving any plant model information, and the stability and convergence of the closed-loop system are guaranteed under some practical assumptions. Finally, the proposed MFAC approaches are easy to design and implement with a small computational burden.

This chapter is organized as follows. In Sections 5.2 and 5.3, the CFDL–MFAC scheme and the PFDL–MFAC scheme for a class of unknown discrete-time MIMO nonlinear nonaffine systems are presented with stability analysis and numerical simulation results, respectively. The FFDL–MFAC scheme and simulation results are given in Section 5.4. Some conclusions are presented in Section 5.5.

5.2 CFDL Data Model Based MFAC

5.2.1 Control System Design

Consider a MIMO nonlinear discrete-time system represented by the following equation:

$$\boldsymbol{y}(k+1) = \boldsymbol{f}(\boldsymbol{y}(k),\ldots,\boldsymbol{y}(k-n_y),\boldsymbol{u}(k),\ldots,\boldsymbol{u}(k-n_u)), \tag{5.1}$$

where $\boldsymbol{u}(k) \in R^m$ and $\boldsymbol{y}(k) \in R^m$ are the control input and system output at time k, respectively, n_y and n_u are two unknown integers, and $\boldsymbol{f}(\cdots) = (f_1(\cdots),\ldots, f_m(\cdots))^T \in \prod_{n_u+n_y+2} R^m \mapsto R^m$ is an unknown nonlinear vector-valued function.

From Theorem 3.5, one can see that system (5.1), satisfying Assumptions 3.7 and 3.8 with $\|\Delta\boldsymbol{u}(k)\| \neq 0$ for all k, can be transformed into the following CFDL data model:

$$\Delta\boldsymbol{y}(k+1) = \boldsymbol{\Phi}_c(k)\Delta\boldsymbol{u}(k), \tag{5.2}$$

where

$$\mathbf{\Phi}_c(k) = \begin{bmatrix} \phi_{11}(k) & \phi_{12}(k) & \cdots & \phi_{1m}(k) \\ \phi_{21}(k) & \phi_{22}(k) & \cdots & \phi_{2m}(k) \\ \vdots & \vdots & \vdots & \vdots \\ \phi_{m1}(k) & \phi_{m2}(k) & \cdots & \phi_{mm}(k) \end{bmatrix} \in R^{m\times m}$$

is the unknown bounded PJM of system (5.1).

For rigorous analysis of stability, another assumption is given.

Assumption 5.1

The PJM $\mathbf{\Phi}_c(k)$ is a diagonally dominant matrix in the following sense: $|\phi_{ij}(k)| \le b_1$, $b_2 \le |\phi_{ii}(k)| \le \alpha b_2$, $\alpha \ge 1$, $b_2 > b_1(2\alpha+1)(m-1)$, $i = 1,\ldots,m$, $j = 1,\ldots,m$, $i \ne j$, and the signs of all the elements of $\mathbf{\Phi}_c(k)$ are fixed.

Remark 5.1

Assumption 5.1 describes the relationship between input and output data of the closed-loop system, and it is similar to the assumption in Ref. [181]. For the discrete-time SISO nonlinear system, this assumption is the same as Assumption 4.2. Since the plant model is unavailable, and only the system operation data can be used, the diagonal dominance condition of the system I/O data relationship may be the last choice to describe system coupling. The condition on the fixed sign of all the elements of $\mathbf{\Phi}_c(k)$ is similar to the assumption on the control direction in the model-based control methods. For instance, in most of the existing adaptive control methods, the control gain is usually assumed to be constant with a known sign. Roughly speaking, this condition could be verified if the I/O data of the controlled system is rich and accurate enough, so it is a reasonable assumption.

5.2.1.1 Controller Algorithm

Consider the following cost function of the control input:

$$J(\boldsymbol{u}(k)) = \left\|\boldsymbol{y}^*(k+1) - \boldsymbol{y}(k+1)\right\|^2 + \lambda\left\|\boldsymbol{u}(k) - \boldsymbol{u}(k-1)\right\|^2, \tag{5.3}$$

where the weighting factor $\lambda > 0$ is introduced to restrain the control input changes, and $\boldsymbol{y}^*(k+1)$ is the desired output signal.

Substituting CFDL data model (5.2) into cost function (5.3), differentiating cost function (5.3) with respect to $\boldsymbol{u}(k)$ and setting it to be zero, yields the following control law:

$$\boldsymbol{u}(k) = \boldsymbol{u}(k-1) + \left(\lambda \mathbf{I} + \boldsymbol{\Phi}_c^T(k)\boldsymbol{\Phi}_c(k)\right)^{-1} \boldsymbol{\Phi}_c^T(k)\left(\boldsymbol{y}^*(k+1) - \boldsymbol{y}(k)\right). \quad (5.4)$$

It is noted that controller algorithm (5.4) requires calculating the matrix inversion, which is very difficult when the dimensions of the system input and output are large; hence, controller algorithm (5.4) is not suitable for practical applications. To solve this problem, a simplified controller algorithm is given as follows, similar to that of Section 4.2:

$$\boldsymbol{u}(k) = \boldsymbol{u}(k-1) + \frac{\rho \boldsymbol{\Phi}_c^T(k)\left(\boldsymbol{y}^*(k+1) - \boldsymbol{y}(k)\right)}{\lambda + \left\|\boldsymbol{\Phi}_c(k)\right\|^2}, \quad (5.5)$$

where the step factor $\rho \in (0,1]$ is introduced to make the controller algorithm more general, and it will be used in the stability analysis.

5.2.1.2 PJM Estimation Algorithm

Analogous to the analysis of the estimation algorithm in the CFDL–MFAC scheme for discrete-time SISO nonlinear systems, a cost function of PJM estimation for MIMO systems is given as follows, according to the CFDL data model (5.2):

$$J(\boldsymbol{\Phi}_c(k)) = \left\|\Delta \boldsymbol{y}(k) - \boldsymbol{\Phi}_c(k)\Delta \boldsymbol{u}(k-1)\right\|^2 + \mu \left\|\boldsymbol{\Phi}_c(k) - \hat{\boldsymbol{\Phi}}_c(k-1)\right\|^2, \quad (5.6)$$

where the weighting factor $\mu > 0$ is used to restrain the change of the PJM estimation.

Minimizing cost function (5.6), we have the following projection algorithm:

$$\begin{aligned} \hat{\boldsymbol{\Phi}}_c(k) = {} & \hat{\boldsymbol{\Phi}}_c(k-1) + (\Delta \boldsymbol{y}(k) - \hat{\boldsymbol{\Phi}}_c(k-1)\Delta \boldsymbol{u}(k-1))\Delta \boldsymbol{u}^T(k-1) \\ & \times (\mu I + \Delta \boldsymbol{u}(k-1)\Delta \boldsymbol{u}^T(k-1))^{-1}. \end{aligned} \quad (5.7)$$

Since the estimation algorithm (5.7) also contains matrix inversion computation, the following simplified PJM estimation algorithm is used:

$$\hat{\boldsymbol{\Phi}}_c(k) = \hat{\boldsymbol{\Phi}}_c(k-1) + \frac{\eta(\Delta \boldsymbol{y}(k) - \hat{\boldsymbol{\Phi}}_c(k-1)\Delta \boldsymbol{u}(k-1))\Delta \boldsymbol{u}^T(k-1)}{\mu + \left\|\Delta \boldsymbol{u}(k-1)\right\|^2}, \quad (5.8)$$

where $\eta \in (0,2]$ is a weighting factor, and

$$\hat{\mathbf{\Phi}}_c(k) = \begin{bmatrix} \hat{\phi}_{11}(k) & \hat{\phi}_{12}(k) & \cdots & \hat{\phi}_{1m}(k) \\ \hat{\phi}_{21}(k) & \hat{\phi}_{22}(k) & \cdots & \hat{\phi}_{2m}(k) \\ \vdots & \vdots & \vdots & \vdots \\ \hat{\phi}_{m1}(k) & \hat{\phi}_{m2}(k) & \cdots & \hat{\phi}_{mm}(k) \end{bmatrix} \in R^{m\times m}$$

is the estimation of the unknown PJM $\mathbf{\Phi}_c(k)$.

5.2.1.3 System Control Scheme

By combining PJM estimation algorithm (5.8) and controller algorithm (5.5), the CFDL–MFAC scheme of discrete-time MIMO nonlinear systems is constructed as follows:

$$\hat{\mathbf{\Phi}}_c(k) = \hat{\mathbf{\Phi}}_c(k-1) + \frac{\eta(\Delta \boldsymbol{y}(k) - \hat{\mathbf{\Phi}}_c(k-1)\Delta \boldsymbol{u}(k-1))\Delta \boldsymbol{u}^T(k-1)}{\mu + \|\Delta \boldsymbol{u}(k-1)\|^2}, \tag{5.9}$$

$$\begin{aligned} &\hat{\phi}_{ii}(k) = \hat{\phi}_{ii}(1), \text{ if } |\hat{\phi}_{ii}(k)| < b_2 \text{ or } |\hat{\phi}_{ii}(k)| > \alpha b_2 \text{ or} \\ &\quad \operatorname{sign}(\hat{\phi}_{ii}(k)) \neq \operatorname{sign}(\hat{\phi}_{ii}(1)), i = 1,\ldots,m \end{aligned} \tag{5.10}$$

$$\begin{aligned} &\hat{\phi}_{ij}(k) = \hat{\phi}_{ij}(1), \text{ if } |\hat{\phi}_{ij}(k)| > b_1 \text{ or } \operatorname{sign}(\hat{\phi}_{ij}(k)) \neq \operatorname{sign}(\hat{\phi}_{ij}(1)), \\ &\quad i,j = 1,\ldots,m, i \neq j \end{aligned} \tag{5.11}$$

$$\boldsymbol{u}(k) = \boldsymbol{u}(k-1) + \frac{\rho \hat{\mathbf{\Phi}}_c^T(k)\left(\boldsymbol{y}^*(k+1) - \boldsymbol{y}(k)\right)}{\lambda + \|\hat{\mathbf{\Phi}}_c(k)\|^2}, \tag{5.12}$$

where $\hat{\phi}_{ij}(1)$ is the initial value of $\hat{\phi}_{ij}(k)$, $i = 1,\ldots,m$, $j = 1,\ldots,m$, $\lambda > 0$, $\mu > 0$, $\eta \in (0,2]$, and $\rho \in (0,1]$.

Remark 5.2

Similar discussions as Remarks 4.1–4.5 could be listed here.

Remark 5.3

When $\boldsymbol{u}(k) \in R^m$ and $y(k) \in R$, system (5.1) becomes the following discrete-time MISO nonlinear system:

$$y(k+1) = f(y(k),\ldots,y(k-n_y),\boldsymbol{u}(k),\ldots,\boldsymbol{u}(k-n_u)), \tag{5.13}$$

where $f(\cdots) \in \prod_{n_y+1} R \times \prod_{n_u+1} R^m \mapsto R$ is an unknown nonlinear function.

Similarly, using the CFDL data model $y(k+1) = y(k) + \boldsymbol{\phi}_c^T(k)\Delta\boldsymbol{u}(k)$ of the discrete-time MISO nonlinear system, the CFDL–MFAC scheme of system (5.13) is given as follows:

$$\hat{\boldsymbol{\phi}}_c(k) = \hat{\boldsymbol{\phi}}_c(k-1) + \frac{\eta\Delta\boldsymbol{u}(k-1)(\Delta y(k) - \hat{\boldsymbol{\phi}}_c^T(k-1)\Delta\boldsymbol{u}(k-1))}{\mu + \|\Delta\boldsymbol{u}(k-1)\|^2}, \tag{5.14}$$

$$\hat{\phi}_i(k) = \hat{\phi}_i(1), \quad \text{if } |\hat{\phi}_i(k)| < b_2 \text{ or } \mathrm{sign}(\hat{\phi}_i(k)) \neq \mathrm{sign}(\hat{\phi}_i(1)), i = 1,\ldots,m, \tag{5.15}$$

$$\boldsymbol{u}(k) = \boldsymbol{u}(k-1) + \frac{\rho\hat{\boldsymbol{\phi}}_c(k)\left(y^*(k+1) - y(k)\right)}{\lambda + \|\hat{\boldsymbol{\phi}}_c(k)\|^2}, \tag{5.16}$$

where $\hat{\boldsymbol{\phi}}_c(k) = [\hat{\phi}_1(k),\hat{\phi}_2(k),\ldots,\hat{\phi}_m(k)]^T \in R^m$ is the estimation of PG $\boldsymbol{\phi}_c(k)$ in the CFDL data model, $\hat{\phi}_i(1)$ is the initial value of $\hat{\phi}_i(k)$, $i = 1, \ldots, m$, $\lambda > 0$, $\mu > 0$, $\eta \in (0,2]$, and $\rho \in (0,1]$.

Obviously, scheme (5.14)–(5.16) is a special case of the CFDL–MFAC scheme (5.9)–(5.12) for MIMO discrete-time nonlinear systems.

5.2.2 Stability Analysis

The following lemma will be used for stability analysis.

Lemma 5.1 [182]

Let $A = (\alpha_{ij}) \in C^{n\times n}$. For each $1 \leq i \leq n$, the Gerschgorin disk is defined as $D_i = \{z \,\|\, z - \alpha_{ii}| \leq \sum_{j=1,j\neq i}^{n} |\alpha_{ij}|\}$, $z \in C$, and the Gerschgorin domain is defined as a union of all the Gerschgorin disks $D_A = \bigcup_{i=1}^{n} D_i$. All eigenvalues of matrix A lie in the Gerschgorin domain D_A.

Theorem 5.1

If nonlinear system (5.1), satisfying Assumptions 3.7, 3.8, and 5.1, is controlled by the CFDL–MFAC scheme (5.9)–(5.12), for a regulation problem $\boldsymbol{y}^*(k+1) = \boldsymbol{y}^*$, then there exists a constant $\lambda_{\min} > 0$ such that the following two properties hold for any $\lambda > \lambda_{\min}$:

a. The tracking error sequence is convergent, that is, $\lim_{k\to\infty} \|\boldsymbol{y}(k+1) - \boldsymbol{y}^*\|_v = 0$, where $\|\cdot\|_v$ is the consistent norm.
b. The closed-loop system is BIBO stable, that is, $\{\boldsymbol{y}(k)\}$ and $\{\boldsymbol{u}(k)\}$ are bounded sequences.

Proof

This proof consists of two steps. The first step is to prove the boundedness of PJM estimation. The second step is to prove the convergence of the tracking error and the BIBO stability of the MFAC system.

Step 1 Let $\hat{\Phi}_c(k) = [\hat{\boldsymbol{\phi}}_1^T(k),\ldots,\hat{\boldsymbol{\phi}}_m^T(k)]^T$, $\hat{\boldsymbol{\phi}}_i(k) = [\hat{\phi}_{i1}(k),\ldots,\hat{\phi}_{im}(k)], i = 1,\ldots,m.$ PJM estimation algorithm (5.9) can be rewritten as

$$\hat{\boldsymbol{\phi}}_i(k) = \hat{\boldsymbol{\phi}}_i(k-1) + \frac{\eta(\Delta y_i(k) - \hat{\boldsymbol{\phi}}_i(k-1)\Delta\boldsymbol{u}(k-1))\Delta\boldsymbol{u}^T(k-1)}{\mu + \|\Delta\boldsymbol{u}(k-1)\|^2}, \tag{5.17}$$

where $\Delta y_i(k) = \boldsymbol{\phi}_i(k-1)\Delta\boldsymbol{u}(k-1)$, $i = 1,\ldots,m$.

Let $\tilde{\boldsymbol{\phi}}_i(k) = \hat{\boldsymbol{\phi}}_i(k) - \boldsymbol{\phi}_i(k)$. Subtracting $\boldsymbol{\phi}_i(k)$ from both sides of (5.17) yields

$$\tilde{\boldsymbol{\phi}}_i(k) = \tilde{\boldsymbol{\phi}}_i(k-1) + \boldsymbol{\phi}_i(k-1) - \boldsymbol{\phi}_i(k) - \frac{\eta\tilde{\boldsymbol{\phi}}_i(k-1)\Delta\boldsymbol{u}(k-1)\Delta\boldsymbol{u}^T(k-1)}{\mu + \|\Delta\boldsymbol{u}(k-1)\|^2}. \tag{5.18}$$

From Theorem 3.5, $\|\boldsymbol{\Phi}_c(k)\|$ is bounded by a constant $\bar{b}$, that is, $\|\boldsymbol{\Phi}_c(k)\| \le \bar{b}$. Consequently, $\|\boldsymbol{\phi}_i(k-1) - \boldsymbol{\phi}_i(k)\| \le 2\bar{b}$ holds.

Taking the norm on both sides of (5.18) leads to

$$\begin{aligned}\|\tilde{\boldsymbol{\phi}}_i(k)\| &\le \left\|\tilde{\boldsymbol{\phi}}_i(k-1)\left(\mathbf{I} - \frac{\eta\Delta\boldsymbol{u}(k-1)\Delta\boldsymbol{u}^T(k-1)}{\mu + \|\Delta\boldsymbol{u}(k-1)\|^2}\right)\right\| + \|\boldsymbol{\phi}_i(k-1) - \boldsymbol{\phi}_i(k)\| \\ &\le \left\|\tilde{\boldsymbol{\phi}}_i(k-1)\left(\mathbf{I} - \frac{\eta\Delta\boldsymbol{u}(k-1)\Delta\boldsymbol{u}^T(k-1)}{\mu + \|\Delta\boldsymbol{u}(k-1)\|^2}\right)\right\| + 2\bar{b}.\end{aligned} \tag{5.19}$$

Squaring the first term on the right side of (5.19) gives

$$\left\|\tilde{\boldsymbol{\phi}}_i(k-1)\left(\mathbf{I}-\frac{\eta\Delta\boldsymbol{u}(k-1)\Delta\boldsymbol{u}^T(k-1)}{\mu+\left\|\Delta\boldsymbol{u}(k-1)\right\|^2}\right)\right\|^2$$

$$=\left\|\tilde{\boldsymbol{\phi}}_i(k-1)\right\|^2+\left(-2+\frac{\eta\left\|\Delta\boldsymbol{u}(k-1)\right\|^2}{\mu+\left\|\Delta\boldsymbol{u}(k-1)\right\|^2}\right)\frac{\eta\left\|\tilde{\boldsymbol{\phi}}_i(k-1)\Delta\boldsymbol{u}(k-1)\right\|^2}{\mu+\left\|\Delta\boldsymbol{u}(k-1)\right\|^2}. \quad (5.20)$$

Since $0<\eta\leq 2$ and $\mu>0$, the following inequality holds:

$$-2+\frac{\eta\left\|\Delta\boldsymbol{u}(k-1)\right\|^2}{\mu+\left\|\Delta\boldsymbol{u}(k-1)\right\|^2}<0. \quad (5.21)$$

Combining (5.20) and (5.21) implies that there exists a constant $0<d_1<1$ such that

$$\left\|\tilde{\boldsymbol{\phi}}_i(k-1)\left(\mathbf{I}-\frac{\eta\Delta\boldsymbol{u}(k-1)\Delta\boldsymbol{u}^T(k-1)}{\mu+\left\|\Delta\boldsymbol{u}(k-1)\right\|^2}\right)\right\|\leq d_1\left\|\tilde{\boldsymbol{\phi}}_i(k-1)\right\|. \quad (5.22)$$

Note that we only need the existence of d_1 instead of its exact value. Substituting (5.22) into (5.19) yields

$$\begin{aligned}\left\|\tilde{\boldsymbol{\phi}}_i(k)\right\| &\leq d_1\left\|\tilde{\boldsymbol{\phi}}_i(k-1)\right\|+2b\leq d_1{}^2\left\|\tilde{\boldsymbol{\phi}}_i(k-2)\right\|+2d_1b+2b\\ &\leq\cdots\leq d_1{}^{k-1}\left\|\tilde{\boldsymbol{\phi}}_i(1)\right\|+\frac{2b(1-d_1{}^{k-1})}{1-d_1}.\end{aligned} \quad (5.23)$$

Inequality (5.23) implies that $\tilde{\boldsymbol{\phi}}_i(k)$ is bounded. Since $\boldsymbol{\phi}_i(k)$ is bounded from Theorem 3.5, $\hat{\boldsymbol{\phi}}_i(k)$ and $\hat{\boldsymbol{\Phi}}_c(k)$ are bounded too.

Step 2 Define the tracking error as

$$\boldsymbol{e}(k)=\boldsymbol{y}^*-\boldsymbol{y}(k). \quad (5.24)$$

Substituting CFDL data model (5.2) and controller algorithm (5.12) into (5.24) yields

$$\boldsymbol{e}(k+1)=\boldsymbol{e}(k)-\boldsymbol{\Phi}_c(k)\Delta\boldsymbol{u}(k)=\left[\mathbf{I}-\frac{\rho\boldsymbol{\Phi}_c(k)\hat{\boldsymbol{\Phi}}_c^T(k)}{\lambda+\|\hat{\boldsymbol{\Phi}}_c(k)\|^2}\right]\boldsymbol{e}(k). \quad (5.25)$$

From Lemma 5.1, we have

$$D_j = \left\{ z \left| \left| z - \left| 1 - \frac{\rho \sum_{i=1}^{m} \phi_{ji}(k)\hat{\phi}_{ji}(k)}{\lambda + \| \hat{\mathbf{\Phi}}_c(k) \|^2} \right| \right| \le \sum_{l=1, l\neq j}^{m} \left| \frac{\rho \sum_{i=1}^{m} \phi_{ji}(k)\hat{\phi}_{li}(k)}{\lambda + \| \hat{\mathbf{\Phi}}_c(k) \|^2} \right| \right. \right\}, \tag{5.26}$$

where z is the eigenvalue of the matrix $\mathbf{I} - \rho \mathbf{\Phi}_c(k)\hat{\mathbf{\Phi}}_c^T(k)/(\lambda + \|\hat{\mathbf{\Phi}}_c(k)\|^2)$, D_j, $j = 1,\ldots,m$, is the Gerschgorin disk.

Using triangle inequality, Equation (5.26) can be rewritten as

$$D_j = \left\{ z \left| |z| \le \left| 1 - \frac{\rho \sum_{i=1}^{m} \phi_{ji}(k)\hat{\phi}_{ji}(k)}{\lambda + \| \hat{\mathbf{\Phi}}_c(k) \|^2} \right| + \sum_{l=1, l\neq j}^{m} \left| \frac{\rho \sum_{i=1}^{m} \phi_{ji}(k)\hat{\phi}_{li}(k)}{\lambda + \| \hat{\mathbf{\Phi}}_c(k) \|^2} \right| \right. \right\} \tag{5.27}$$

From resetting algorithms (5.10) and (5.11), we have $b_2 \le |\hat{\phi}_{ii}(k)| \le \alpha b_2$ and $|\hat{\phi}_{ij}(k)| \le b_1$, $i = 1,\ldots,m$, $j = 1,\ldots,m$, $i \neq j$. Assumption 5.1 gives that $b_2 \le |\phi_{ii}(k)| \le \alpha b_2$ and $|\phi_{ij}(k)| \le b_1$, $i = 1, \ldots,m$, $j = 1,\ldots,m$, $i \neq j$. Thus, the following two inequalities hold:

$$1 - \frac{\rho \sum_{i=1}^{m} |\phi_{ji}(k)||\hat{\phi}_{ji}(k)|}{\lambda + \|\hat{\mathbf{\Phi}}_c(k)\|^2} \le 1 - \frac{\rho |\phi_{jj}(k)||\hat{\phi}_{jj}(k)|}{\lambda + \|\hat{\mathbf{\Phi}}_c(k)\|^2} \le 1 - \frac{\rho b_2^2}{\lambda + \|\hat{\mathbf{\Phi}}_c(k)\|^2}, \tag{5.28}$$

$$\begin{aligned}
\sum_{l=1, l\neq j}^{m} \left| \frac{\rho \sum_{i=1}^{m} \phi_{ji}(k)\hat{\phi}_{li}(k)}{\lambda + \|\hat{\mathbf{\Phi}}_c(k)\|^2} \right| &\le \rho \sum_{l=1, l\neq j}^{m} \frac{\sum_{i=1}^{m} |\phi_{ji}(k)|\,|\hat{\phi}_{li}(k)|}{\lambda + \|\hat{\mathbf{\Phi}}_c(k)\|^2} \\
&= \rho \frac{\sum_{l=1, l\neq j}^{m} |\phi_{jj}(k)|\,|\hat{\phi}_{lj}(k)|}{\lambda + \|\hat{\mathbf{\Phi}}_c(k)\|^2} + \rho \sum_{l=1, l\neq j}^{m} \frac{\sum_{i=1, i\neq j}^{m} |\phi_{ji}(k)|\,|\hat{\phi}_{li}(k)|}{\lambda + \|\hat{\mathbf{\Phi}}_c(k)\|^2} \\
&= \rho \frac{\sum_{l=1, l\neq j}^{m} |\phi_{jj}(k)|\,|\hat{\phi}_{lj}(k)|}{\lambda + \|\hat{\mathbf{\Phi}}_c(k)\|^2} + \rho \frac{\sum_{l=1, l\neq j}^{m} |\phi_{jl}(k)|\,|\hat{\phi}_{ll}(k)|}{\lambda + \|\hat{\mathbf{\Phi}}_c(k)\|^2} \\
&\quad + \rho \sum_{l=1, l\neq j}^{m} \frac{\sum_{i=1, i\neq j, l}^{m} |\phi_{ji}(k)|\,|\hat{\phi}_{li}(k)|}{\lambda + \|\hat{\mathbf{\Phi}}_c(k)\|^2} \\
&\le \rho \frac{2\alpha b_1 b_2 (m-1) + b_1^2 (m-1)(m-2)}{\lambda + \|\hat{\mathbf{\Phi}}_c(k)\|^2}.
\end{aligned} \tag{5.29}$$

Also from Assumption 5.1, we have $b_2 > b_1(2\alpha + 1)(m-1)$.
Summing (5.28) and (5.29) yields

$$
\begin{aligned}
&1-\frac{\rho\sum_{i=1}^{m}|\phi_{ji}(k)||\hat{\phi}_{ji}(k)|}{\lambda+\|\hat{\mathbf{\Phi}}_c(k)\|^2}+\sum_{h=1,h\neq j}^{m}\left|\frac{\rho\sum_{i=1}^{m}\phi_{ji}(k)\hat{\phi}_{hi}(k)}{\lambda+\|\hat{\mathbf{\Phi}}_c(k)\|^2}\right| \\
&\le 1-\rho\frac{b_2^2-2\alpha b_1 b_2(m-1)-b_1^2(m-1)(m-2)}{\lambda+\|\hat{\mathbf{\Phi}}_c(k)\|^2} \\
&= 1-\rho\frac{b_2\left(b_2-2\alpha b_1(m-1)\right)-b_1^2(m-1)(m-2)}{\lambda+\|\hat{\mathbf{\Phi}}_c(k)\|^2} \\
&< 1-\rho\frac{b_2 b_1(m-1)-b_1^2(m-1)(m-2)}{\lambda+\|\hat{\mathbf{\Phi}}_c(k)\|^2} \\
&< 1-\rho\frac{b_2 b_1(m-1)-b_1^2(m-1)(m-1)}{\lambda+\|\hat{\mathbf{\Phi}}_c(k)\|^2} \\
&= 1-\rho\frac{b_1(m-1)\left(b_2-b_1(m-1)\right)}{\lambda+\|\hat{\mathbf{\Phi}}_c(k)\|^2} \\
&< 1-\rho\frac{2\alpha b_1^2(m-1)^2}{\lambda+\|\hat{\mathbf{\Phi}}_c(k)\|^2}.
\end{aligned}
\tag{5.30}
$$

By resetting algorithm (5.11) and Assumption 5.1, $\phi_{ji}(k)\hat{\phi}_{ji}(k) > 0$, $i = 1, \ldots, m$, $j = 1, \ldots, m$. Thus, there exists a constant $\lambda_{\min} > 0$ such that the following equation holds for $\lambda > \lambda_{\min}$:

$$
\begin{aligned}
&\frac{\sum_{i=1}^{m}\phi_{ji}(k)\hat{\phi}_{ji}(k)}{\lambda+\|\hat{\mathbf{\Phi}}_c(k)\|^2}=\frac{\sum_{i=1}^{m}|\phi_{ji}(k)||\hat{\phi}_{ji}(k)|}{\lambda+\|\hat{\mathbf{\Phi}}_c(k)\|^2}\le\frac{\alpha^2 b_2^2+b_1^2(m-1)}{\lambda+\|\hat{\mathbf{\Phi}}_c(k)\|^2} \\
&<\frac{\alpha^2 b_2^2+b_1^2(m-1)}{\lambda_{\min}+\|\hat{\mathbf{\Phi}}_c(k)\|^2}<1.
\end{aligned}
\tag{5.31}
$$

Properly selecting $0 < \rho \leq 1$ and $\lambda > \lambda_{\min}$ such that

$$\left|1 - \frac{\rho \sum_{i=1}^{m} \phi_{ji}(k)\hat{\phi}_{ji}(k)}{\lambda + \|\hat{\mathbf{\Phi}}_c(k)\|^2}\right| = 1 - \frac{\rho \sum_{i=1}^{m} \|\hat{\phi}_{ji}(k)\|}{\lambda + \|\hat{\mathbf{\Phi}}_c(k)\|^2}. \tag{5.32}$$

Obviously, the following inequality holds for $\lambda > \lambda_{\min}$:

$$\begin{aligned} 0 < M_1 &\leq \frac{2\alpha b_1^2(m-1)^2}{\lambda + \|\hat{\mathbf{\Phi}}_c(k)\|^2} < \frac{b_2^2}{\lambda + \|\hat{\mathbf{\Phi}}_c(k)\|^2} \\ &\leq \frac{\alpha^2 b_2^2 + b_1^2(m-1)}{\lambda + \|\hat{\mathbf{\Phi}}_c(k)\|^2} < \frac{\alpha^2 b_2^2 + b_1^2(m-1)}{\lambda + \|\hat{\mathbf{\Phi}}_c(k)\|^2} < 1. \end{aligned} \tag{5.33}$$

From (5.30), (5.32), and (5.33), we have

$$\left|1 - \frac{\rho \sum_{i=1}^{m} \phi_{ji}(k)\hat{\phi}_{ji}(k)}{\lambda + \|\hat{\mathbf{\Phi}}_c(k)\|^2}\right| + \sum_{l=1,l\neq j}^{m} \left|\frac{\rho \sum_{i=1}^{m} \phi_{ji}(k)\hat{\phi}_{li}(k)}{\lambda + \|\hat{\mathbf{\Phi}}_c(k)\|^2}\right| < 1 - \rho M_1 < 1. \tag{5.34}$$

According to (5.27) and (5.34), one obtains

$$s\left(I - \frac{\rho \mathbf{\Phi}_c(k)\hat{\mathbf{\Phi}}_c^T(k)}{\lambda + \|\hat{\mathbf{\Phi}}_c(k)\|^2}\right) < 1 - \rho M_1, \tag{5.35}$$

where $s(A)$ is the spectral radius of matrix A, that is, $s(A) = \max_{i\in\{1,2,\ldots,m\}} |z_i|$, and z_i, $i = 1,2, \ldots ,m$ is the eigenvalue of matrix A.

Using the conclusion on spectral radius in Ref. [183], we know that there exists an arbitrarily small positive constant ε_1 such that

$$\left\|I - \frac{\rho \mathbf{\Phi}_c(k)\hat{\mathbf{\Phi}}_c^T(k)}{\lambda + \|\hat{\mathbf{\Phi}}_c(k)\|^2}\right\|_v < s\left(I - \frac{\rho \mathbf{\Phi}_c(k)\hat{\mathbf{\Phi}}_c^T(k)}{\lambda + \|\hat{\mathbf{\Phi}}_c(k)\|^2}\right) + \varepsilon_1 \leq 1 - \rho M_1 + \varepsilon_1 < 1, \tag{5.36}$$

where $\|A\|_v$ is the consistent norm of matrix A.

Let $d_2 = 1 - \rho M_1 + \varepsilon_1$. Taking the norm on both sides of (5.25) yields

$$\left\|e(k+1)\right\|_v \leq \left\| I - \frac{\rho \boldsymbol{\Phi}_c(k)\hat{\boldsymbol{\Phi}}_c^T(k)}{\lambda + \|\hat{\boldsymbol{\Phi}}_c(k)\|^2} \right\|_v \left\|e(k)\right\|_v \leq d_2 \left\|e(k)\right\|_v \leq \cdots \leq d_2^k \left\|e(1)\right\|_v . \quad (5.37)$$

Conclusion (a) of Theorem 5.1 is the direct result of (5.37).
Since $\boldsymbol{y}^*$ is a given bounded and $\boldsymbol{e}(k)$ is bounded, $\boldsymbol{y}(k)$ is also bounded.
Since $\hat{\boldsymbol{\Phi}}_c(k)$ is bounded, there exists a positive constant M_2 such that

$$\left\| \frac{\rho \hat{\boldsymbol{\Phi}}_c^T(k)}{\lambda + \|\hat{\boldsymbol{\Phi}}_c(k)\|^2} \right\|_v \leq M_2 . \quad (5.38)$$

Using (5.12), (5.37), and (5.38) yields

$$\begin{aligned}
\left\|\boldsymbol{u}(k)\right\|_v &\leq \left\|\boldsymbol{u}(k) - \boldsymbol{u}(k-1)\right\|_v + \left\|\boldsymbol{u}(k-1)\right\|_v \\
&\leq \left\|\boldsymbol{u}(k) - \boldsymbol{u}(k-1)\right\|_v + \left\|\boldsymbol{u}(k-1) - \boldsymbol{u}(k-2)\right\|_v + \left\|\boldsymbol{u}(k-2)\right\|_v \\
&\leq \left\|\Delta \boldsymbol{u}(k)\right\|_v + \left\|\Delta \boldsymbol{u}(k-1)\right\|_v + \cdots + \left\|\Delta \boldsymbol{u}(1)\right\|_v + \left\|\boldsymbol{u}(0)\right\|_v \\
&\leq M_2 \left(\left\|e(k)\right\| + \left\|e(k-1)\right\| + \cdots + \left\|e(2)\right\| + \left\|e(1)\right\|\right) + \left\|\boldsymbol{u}(0)\right\|_v \\
&\leq M_2 \left(d_2^{k-1}\left\|e(1)\right\| + d_2^{k-2}\left\|e(1)\right\| + \cdots + d_2\left\|e(1)\right\| + \left\|e(1)\right\|\right) + \left\|\boldsymbol{u}(0)\right\|_v \\
&< M_2 \frac{1}{1-d_2}\left\|e(1)\right\| + \left\|\boldsymbol{u}(0)\right\|_v .
\end{aligned} \quad (5.39)$$

Thus, conclusion (b) of Theorem 5.1 is obtained. ■

Remark 5.4

It is worth pointing out that $b_1 = 0$ holds for the systems without input couplings. In this case, error dynamics (5.25) can be rewritten as

$$\boldsymbol{e}(k+1) = \begin{bmatrix} 1-\dfrac{\rho\phi_{11}(k)\hat{\phi}_{11}(k)}{\lambda+\|\hat{\boldsymbol{\Phi}}_c(k)\|^2} & & & \\ & 1-\dfrac{\rho\phi_{22}(k)\hat{\phi}_{22}(k)}{\lambda+\|\hat{\boldsymbol{\Phi}}_c(k)\|^2} & & \\ & & \ddots & \\ & & & 1-\dfrac{\rho\phi_{mm}(k)\hat{\phi}_{mm}(k)}{\lambda+\|\hat{\boldsymbol{\Phi}}_c(k)\|^2} \end{bmatrix} \boldsymbol{e}(k). \tag{5.40}$$

By virtue of Assumption 5.1 and resetting algorithm (5.10), both PJM and its estimation are bounded. Thus, there exists a constant $\lambda_{\min} > 0$, such that the following inequality holds for $\lambda > \lambda_{\min}$:

$$0 < M_1 \le \frac{\rho\phi_{ii}(k)\hat{\phi}_{ii}(k)}{\lambda+\|\hat{\boldsymbol{\Phi}}_c(k)\|^2} < 1, \quad i = 1,2,\ldots,m.$$

Thus, the convergence of the tracking error is a direct result of (5.40).

Remark 5.5

For discrete-time MISO nonlinear system (5.13), the corresponding assumption and conclusions are listed as follows.

Assumption 5.1′

The sign of all the elements of $\boldsymbol{\phi}_c(k)$ is unchanged.

Corollary 5.1

If nonlinear system (5.13), satisfying Assumptions 3.7′, 3.8′, and 5.1′, is controlled by the CFDL–MFAC scheme (5.14)–(5.16), for a regulation problem $y^*(k+1) = y^* = \text{const.}$, then there exists a constant $\lambda_{\min} > 0$ such that the following two properties hold for any $\lambda > \lambda_{\min}$:

a. The tracking error sequence is convergent, that is, $\lim_{k\to\infty}\left|y(k+1) - y^*\right| = 0$.
b. The closed-loop system is BIBO stable, that is, $\{y(k)\}$ and $\{\boldsymbol{u}(k)\}$ are bounded sequences.

5.2.3 Simulation Results

Two numerical simulations are given to verify the correctness and effectiveness of CFDL–MFAC scheme (5.14)–(5.16) for MISO systems and CFDL–MFAC scheme (5.9)–(5.12) for MIMO systems, respectively. The models in the following two examples are merely used to generate the I/O data of the controlled plant, but not for MFAC design.

Example 5.1

Consider a discrete-time MISO nonlinear system with two inputs and one output

$$y(k+1) = \frac{5y(k) + 2u_1(k) - 3u_2^2(k) + 2u_1^2(k)}{5 + u_1(k) + 5u_2(k)}. \tag{5.41}$$

The desired trajectory is given as

$$y^*(k+1) = (-1)^{\text{round}(k/100)}. \tag{5.42}$$

The system's initial values are $\boldsymbol{u}(1) = \boldsymbol{u}(2) = [1,1]^T$, $y(1) = 1$, $y(2) = 0.5$, and $\varepsilon = 10^{-5}$. The parameters in the CFDL–MFAC scheme are set to $\eta = \rho = 1$, $\lambda = 3$, $\mu = 1$, and $\hat{\boldsymbol{\phi}}_c(1) = [0.5, -0.2]^T$.

The simulation results are shown in Figure 5.1. Figure 5.1a, b, and c shows the tracking performance of the output y, the control inputs u_1 and u_2, and the PG estimation, respectively. The rationality of Assumption 5.1′ can be seen from Figure 5.1c. The simulation results illustrate that the control performance of the CFDL–MFAC scheme is quite good. Furthermore, the computational burden and realization difficulty of the CFDL–MFAC scheme are much lower than that of model-based control methods or neural-network-based control methods.

Example 5.2

Consider a discrete-time MIMO nonlinear system with two inputs and two outputs

$$\begin{cases} x_{11}(k+1) = \dfrac{x_{11}^2(k)}{1 + x_{11}^2(k)} + 0.3x_{12}(k), \\ x_{12}(k+1) = \dfrac{x_{11}^2(k)}{1 + x_{12}^2(k) + x_{21}^2(k) + x_{22}^2(k)} + a(k)u_1(k), \\ x_{21}(k+1) = \dfrac{x_{21}^2(k)}{1 + x_{21}^2(k)} + 0.2x_{22}(k), \\ x_{22}(k+1) = \dfrac{x_{21}^2(k)}{1 + x_{11}^2(k) + x_{12}^2(k) + x_{22}^2(k)} + b(k)u_2(k), \\ y_1(k+1) = x_{11}(k+1), \\ y_2(k+1) = x_{21}(k+1). \end{cases} \tag{5.43}$$

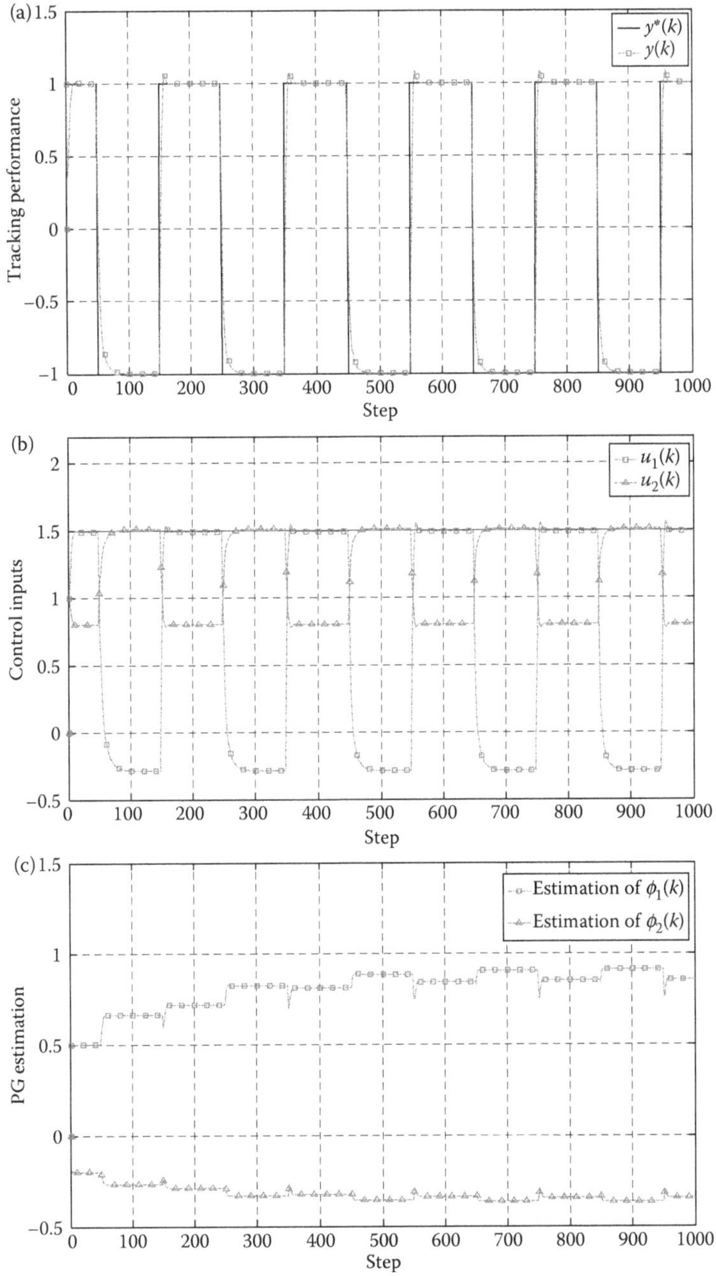

Figure 5.1 Simulation results of applying the CFDL–MFAC scheme to discrete-time MISO nonlinear system (5.41). (a) Tracking performance. (b) Control inputs $u_1(k)$ and $u_2(k)$. (c) PG estimation $\hat{\phi}_c(k)$.

where $a(k) = 1 + 0.1\sin(2\pi k/1500)$ and $b(k) = 1 + 0.1\cos(2\pi k/1500)$ are two time-varying parameters. When $a(k) = 1$ and $b(k) = 1$, system (5.43) is the same as that of Ref. [184]. Obviously, it is a time-varying coupled nonlinear system.

The desired trajectories are given as follows:

$$\begin{cases} y_1^*(k) = 0.5 + 0.25\cos(0.25\pi k/100) + 0.25\sin(0.5\pi k/100), \\ y_2^*(k) = 0.5 + 0.25\sin(0.25\pi k/100) + 0.25\sin(0.5\pi k/100). \end{cases} \tag{5.44}$$

The initial values are $x_{1,1}(j) = x_{2,1}(j) = 0.5$, $x_{1,2}(j) = x_{2,2}(j) = 0$, $j = 1,2$, $\boldsymbol{u}(1) = \boldsymbol{u}(2) = [0, 0]^T$. The controller parameters are $\hat{\boldsymbol{\Phi}}_c(1) = \hat{\boldsymbol{\Phi}}_c(2) = \begin{bmatrix} 0.5 & 0 \\ 0 & 0.5 \end{bmatrix}^T$, $\eta = \rho = 1$, $\mu = 1$, $\lambda = 0.5$.

The simulation results are shown in Figure 5.2. Figure 5.2a, b, c, and d shows the tracking performance of the output y_1 and y_2, the control inputs u_1 and u_2, and the PJM estimation, respectively. Simulation results show that the tracking performance with the CFDL–MFAC scheme is quite good, even better than that of the neural-network-based control method [184], although the unknown time-varying parameters and unknown couplings exist in the unknown discrete-time MIMO nonlinear system. Figure 5.2d shows that the variation of PJM estimation is very slow despite the time-varying and coupled factors of the original system. Thus, Assumption 5.1 is reasonable. In addition, this simulation also demonstrates that the MFAC scheme has a very strong ability to deal with MIMO coupled nonlinear systems.

5.3 PFDL Data Model Based MFAC

5.3.1 Control System Design

From Theorem 3.6, we can see that the MIMO nonlinear system (5.1), satisfying Assumptions 3.9 and 3.10 with $\|\Delta\bar{U}_L(k)\| \neq 0$ for all k, can be transformed into the following PFDL data model:

$$\Delta \boldsymbol{y}(k+1) = \boldsymbol{\Phi}_{p,L}(k)\Delta\bar{U}_L(k), \tag{5.45}$$

where $\boldsymbol{\Phi}_{p,L}(k) = \begin{bmatrix} \boldsymbol{\Phi}_1(k) & \cdots & \boldsymbol{\Phi}_L(k) \end{bmatrix} \in R^{m\times mL}$ is the unknown and bounded PPJM of system (5.1):

$$\boldsymbol{\Phi}_i(k) = \begin{bmatrix} \phi_{11i}(k) & \phi_{12i}(k) & \cdots & \phi_{1mi}(k) \\ \phi_{21i}(k) & \phi_{22i}(k) & \cdots & \phi_{2mi}(k) \\ \vdots & \vdots & \vdots & \vdots \\ \phi_{m1i}(k) & \phi_{m2i}(k) & \cdots & \phi_{mmi}(k) \end{bmatrix} \in R^{m\times m}, \quad i = 1,\dots,L,$$

and $\Delta\bar{U}_L(k) = [\Delta\boldsymbol{u}^T(k),\dots,\Delta\boldsymbol{u}^T(k-L+1)]^T$.

Assumption 5.2

$\mathbf{\Phi}_1(k)$ of the PPJM $\mathbf{\Phi}_{p,L}(k)$ is a diagonally dominant matrix in the following sense: $|\phi_{ij1}(k)| \le b_1$, $b_2 \le |\phi_{ii1}(k)| \le \alpha b_2$, $\alpha \ge 1$, $b_2 > b_1(2\alpha+1)(m-1)$, $i = 1, \ldots, m$, $j = 1, \ldots, m$, $i \ne j$, and the sign of all the elements of $\mathbf{\Phi}_1(k)$ is unchanged.

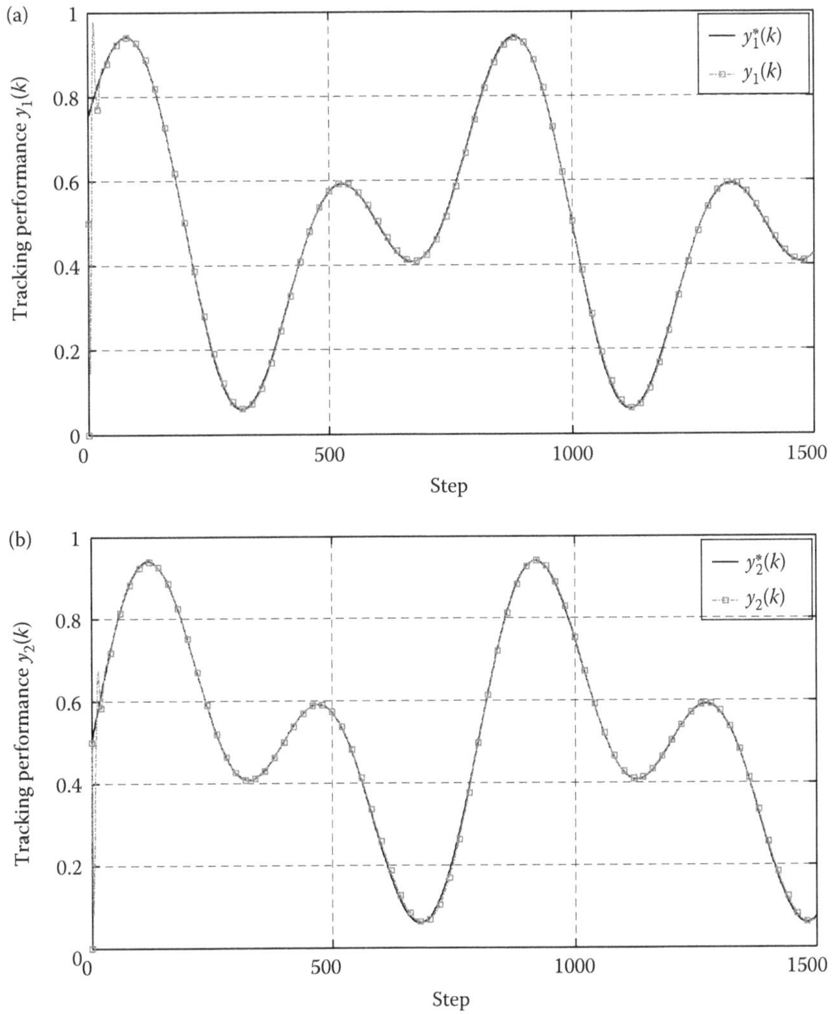

Figure 5.2 Simulation results of applying the CFDL–MFAC scheme to discrete-time MISO nonlinear system (5.43). (a) Tracking performance $y_1(k)$. (b) Tracking performance $y_2(k)$. (c) Control inputs $u_1(k)$ and $u_2(k)$. (d) PJM estimation.

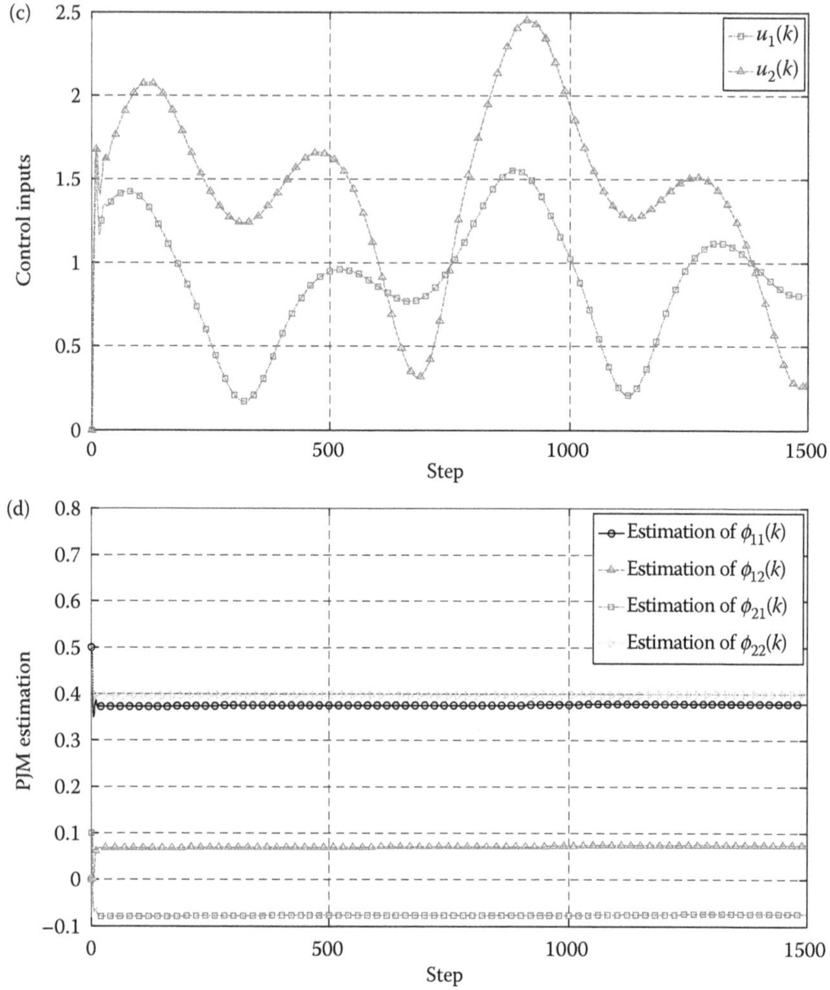

Figure 5.2 (continued) Simulation results of applying the CFDL–MFAC scheme to discrete-time MISO nonlinear system (5.43). (a) Tracking performance $y_1(k)$. (b) Tracking performance $y_2(k)$. (c) Control inputs $u_1(k)$ and $u_2(k)$. (d) PJM estimation.

Remark 5.6

Similar discussions as in Remark 5.1 could be given here.

5.3.1.1 Controller Algorithm

Consider the following cost function of control input:

$$J(\boldsymbol{u}(k)) = \|\boldsymbol{y}^*(k+1) - \boldsymbol{y}(k+1)\|^2 + \lambda \|\boldsymbol{u}(k) - \boldsymbol{u}(k-1)\|^2, \tag{5.46}$$

where $\lambda > 0$ is a weighting factor.

Substituting PFDL data model (5.45) into cost function (5.46), differentiating cost function (5.46) with respect to $\boldsymbol{u}(k)$ and setting it to be zero, we have

$$\begin{aligned}\boldsymbol{u}(k) &= \boldsymbol{u}(k-1) \\ &\quad + \left(\lambda \mathbf{I} + \boldsymbol{\Phi}_1^T(k)\boldsymbol{\Phi}_1(k)\right)^{-1} \boldsymbol{\Phi}_1^T(k)\left(\left(\boldsymbol{y}^*(k+1) - \boldsymbol{y}(k)\right) - \sum_{i=2}^{L} \boldsymbol{\Phi}_i(k)\Delta\boldsymbol{u}(k-i+1)\right).\end{aligned} \tag{5.47}$$

Similar to the discussion in Section 5.2, a simplified controller algorithm without matrix inversion is given as follows:

$$\boldsymbol{u}(k) = \boldsymbol{u}(k-1) + \frac{\boldsymbol{\Phi}_1^T(k)\left(\rho_1\left(\boldsymbol{y}^*(k+1) - \boldsymbol{y}(k)\right) - \sum_{i=2}^{L} \rho_i \boldsymbol{\Phi}_i(k)\Delta\boldsymbol{u}(k-i+1)\right)}{\lambda + \left\|\boldsymbol{\Phi}_1(k)\right\|^2}, \tag{5.48}$$

where $\rho_i \in (0,1]$, $i = 1,2,\ldots,L$, is a step factor.

5.3.1.2 PPJM Estimation Algorithm

Consider a cost function of the PPJM estimation

$$J(\boldsymbol{\Phi}_{p,L}(k)) = \left\|\Delta\boldsymbol{y}(k) - \boldsymbol{\Phi}_{p,L}(k)\Delta\bar{\boldsymbol{U}}_L(k-1)\right\|^2 + \mu\left\|\boldsymbol{\Phi}_{p,L}(k) - \hat{\boldsymbol{\Phi}}_{p,L}(k-1)\right\|^2, \tag{5.49}$$

where $\mu > 0$ is a weighting factor.

Minimizing (5.49) yields the following PPJM estimation algorithm:

$$\begin{aligned}\hat{\boldsymbol{\Phi}}_{p,L}(k) &= \hat{\boldsymbol{\Phi}}_{p,L}(k-1) + (\Delta\boldsymbol{y}(k) - \hat{\boldsymbol{\Phi}}_{p,L}(k-1)\Delta\bar{\boldsymbol{U}}_L(k-1))\Delta\bar{\boldsymbol{U}}_L^T(k-1) \\ &\quad \times \left(\mu I + \Delta\bar{\boldsymbol{U}}_L(k-1)\Delta\bar{\boldsymbol{U}}_L^T(k-1)\right)^{-1}.\end{aligned} \tag{5.50}$$

Similarly, the following simplified PPJM estimation algorithm without matrix inversion is used:

$$\hat{\boldsymbol{\Phi}}_{p,L}(k) = \hat{\boldsymbol{\Phi}}_{p,L}(k-1) + \frac{\eta(\Delta \boldsymbol{y}(k) - \hat{\boldsymbol{\Phi}}_{p,L}(k-1)\Delta \bar{\boldsymbol{U}}_L(k-1))\Delta \bar{\boldsymbol{U}}_L^T(k-1)}{\mu + \left\|\Delta \bar{\boldsymbol{U}}_L(k-1)\right\|^2}, \quad (5.51)$$

where $\eta \in (0,2]$ is an added step factor, $\hat{\boldsymbol{\Phi}}_{p,L}(k) = [\hat{\boldsymbol{\Phi}}_1(k)\ \hat{\boldsymbol{\Phi}}_2(k)\ \cdots\ \hat{\boldsymbol{\Phi}}_L(k)] \in R^{m\times mL}$ is an estimation of $\Phi_{p,L}(k)$, and

$$\hat{\boldsymbol{\Phi}}_i(k) = \begin{bmatrix} \hat{\phi}_{11_i}(k) & \hat{\phi}_{12_i}(k) & \cdots & \hat{\phi}_{1m_i}(k) \\ \hat{\phi}_{21_i}(k) & \hat{\phi}_{22_i}(k) & \cdots & \hat{\phi}_{2m_i}(k) \\ \vdots & \vdots & \vdots & \vdots \\ \hat{\phi}_{m1_i}(k) & \hat{\phi}_{m2_i}(k) & \cdots & \hat{\phi}_{mm_i}(k) \end{bmatrix} \in R^{m\times m}, \quad i = 1,\ldots,L.$$

5.3.1.3 System Control Scheme

Combining PJM estimation algorithm (5.51) and controller algorithm (5.48), the PFDL–MFAC scheme for discrete-time MIMO nonlinear systems is constructed as follows:

$$\hat{\boldsymbol{\Phi}}_{p,L}(k) = \hat{\boldsymbol{\Phi}}_{p,L}(k-1) + \frac{\eta(\Delta \boldsymbol{y}(k) - \hat{\boldsymbol{\Phi}}_{p,L}(k-1)\Delta \bar{\boldsymbol{U}}_L(k-1))\Delta \bar{\boldsymbol{U}}_L^T(k-1)}{\mu + \left\|\Delta \bar{\boldsymbol{U}}_L(k-1)\right\|^2}, \quad (5.52)$$

$$\hat{\phi}_{ii1}(k) = \hat{\phi}_{ii1}(1), \text{ if } |\hat{\phi}_{ii1}(k)| < b_2 \text{ or } |\hat{\phi}_{ii1}(k)| > \alpha b_2 \text{ or } \operatorname{sign}(\hat{\phi}_{ii1}(k)) \neq \operatorname{sign}(\hat{\phi}_{ii1}(1)),$$
$$i = 1,\ldots,m \quad (5.53)$$

$$\hat{\phi}_{ij1}(k) = \hat{\phi}_{ij1}(1), \quad \text{if } |\hat{\phi}_{ij1}(k)| > b_1 \text{ or } \operatorname{sign}(\hat{\phi}_{ij1}(k)) \neq \operatorname{sign}(\hat{\phi}_{ij1}(1)),$$
$$i,j = 1,\ldots,m, i \neq j \quad (5.54)$$

$$\boldsymbol{u}(k) = \boldsymbol{u}(k-1) + \frac{\hat{\boldsymbol{\Phi}}_1^T(k)(\rho_1(\boldsymbol{y}^*(k+1) - \boldsymbol{y}(k)) - \sum_{i=2}^{L}\rho_i \hat{\boldsymbol{\Phi}}_i(k)\Delta \boldsymbol{u}(k-i+1))}{\lambda + \|\hat{\boldsymbol{\Phi}}_1(k)\|^2}, \quad (5.55)$$

where $\hat{\phi}_{ij1}(1)$ is the initial value of $\hat{\phi}_{ij1}(k)$, $i = 1,\ldots,m$, $j = 1,\ldots,m$, $\rho_1,\rho_2,\ldots\rho_L$, $\eta \in (0,2]$ $\lambda > 0$, $\mu > 0$.

Remark 5.7

Using the PFDL data model $y(k+1) = y(k) + \bar{\boldsymbol{\phi}}_{p,L}^T(k)\Delta\bar{U}_L(k)$, the PFDL–MFAC scheme for discrete-time MISO nonlinear system (5.13) is given as follows:

$$\hat{\bar{\boldsymbol{\phi}}}_{p,L}(k) = \hat{\bar{\boldsymbol{\phi}}}_{p,L}(k-1) + \frac{\eta\left(\Delta y(k) - \hat{\bar{\boldsymbol{\phi}}}_{p,L}^T(k-1)\Delta\bar{U}_L(k-1)\right)\Delta\bar{U}_L(k-1)}{\mu + \left\|\Delta\bar{U}_L(k-1)\right\|^2}, \quad (5.56)$$

$$\hat{\phi}_{i1}(k) = \hat{\phi}_{i1}(1), \quad \text{if } |\hat{\phi}_{i1}(k)| < b_2 \text{ or } \operatorname{sign}(\hat{\phi}_{i1}(k)) \neq \operatorname{sign}(\hat{\phi}_{i1}(1)),\ i = 1,\ldots,m \quad (5.57)$$

$$\boldsymbol{u}(k) = \boldsymbol{u}(k-1) + \frac{\hat{\bar{\boldsymbol{\phi}}}_1(k)\left(\rho_1(y^*(k+1) - y(k)) - \sum_{i=2}^{L}\rho_i\hat{\bar{\boldsymbol{\phi}}}_i^T(k)\Delta\boldsymbol{u}(k-i+1)\right)}{\lambda + \|\hat{\bar{\boldsymbol{\phi}}}_1(k)\|^2}, \quad (5.58)$$

where $\hat{\bar{\boldsymbol{\phi}}}_{p,L}(k) = [\hat{\bar{\boldsymbol{\phi}}}_1^T(k) \ \cdots \ \hat{\bar{\boldsymbol{\phi}}}_L^T(k)]^T \in R^{mL}$ is the estimation of PPG $\bar{\boldsymbol{\phi}}_{p,L}(k)$ in PFDL data model (3.43), $\hat{\bar{\boldsymbol{\phi}}}_i(k) = [\hat{\phi}_{1i}(k), \hat{\phi}_{2i}(k), \ldots, \hat{\phi}_{mi}(k)]^T$, $i = 1,\ldots, L$. $\hat{\bar{\boldsymbol{\phi}}}_{p,L}(1)$ is the initial value of $\hat{\bar{\boldsymbol{\phi}}}_{p,L}(k)$, $\rho_1, \rho_2, \ldots, \rho_L$, $\eta \in (0,2]$, $\lambda > 0$, $\mu > 0$.

5.3.2 Stability Analysis

Theorem 5.2

If nonlinear system (5.1), satisfying Assumptions 3.9, 3.10, and 5.2, is controlled by PFDL–MFAC scheme (5.51)–(5.55), for a regulation problem $\boldsymbol{y}^*(k+1) = \boldsymbol{y}^*$, then there exists a constant $\lambda_{\min} > 0$ such that the following two properties hold for $\lambda > \lambda_{\min}$:

a. The tracking error sequence is convergent, that is, $\lim_{k\to\infty}\left\|\boldsymbol{y}(k+1) - \boldsymbol{y}^*\right\|_v = 0$, where $\|\cdot\|_v$ is the consistent norm.
b. The closed-loop system is BIBO stable, that is, $\{\boldsymbol{y}(k)\}$ and $\{\boldsymbol{u}(k)\}$ are bounded sequences.

Proof

The proof of boundedness of $\hat{\boldsymbol{\Phi}}_{p,L}(k)$ is similar to that of Theorem 5.1. On the basis of the boundedness of $\hat{\boldsymbol{\Phi}}_{p,L}(k)$, the convergence of the tracking error and BIBO stability is proved as follows:

Define the tracking error as

$$e(k) = y^* - y(k). \tag{5.59}$$

Let

$$\mathbf{A}_1(k) = \begin{bmatrix} -\dfrac{\rho_2 \hat{\boldsymbol{\Phi}}_1^T(k)\hat{\boldsymbol{\Phi}}_2(k)}{\lambda + \|\hat{\boldsymbol{\Phi}}_1(k)\|^2} & \cdots & -\dfrac{\rho_L \hat{\boldsymbol{\Phi}}_1^T(k)\hat{\boldsymbol{\Phi}}_L(k)}{\lambda + \|\hat{\boldsymbol{\Phi}}_1(k)\|^2} & 0 \\ \mathrm{I} & \cdots & 0 & 0 \\ \vdots & \ddots & \vdots & \vdots \\ 0 & \cdots & \mathrm{I} & 0 \end{bmatrix}_{mL\times mL},$$

$$\mathbf{C}_1(k) = \begin{bmatrix} \dfrac{\hat{\boldsymbol{\Phi}}_1(k)}{\lambda + \|\hat{\boldsymbol{\Phi}}_1(k)\|^2} & 0 \\ 0 & 0 \end{bmatrix}_{mL\times mL}, \text{ and } \boldsymbol{E}(k) = \begin{bmatrix} e(k) \\ 0_{mL-m} \end{bmatrix} \in R^{mL}.$$

Then, controller algorithm (5.55) is rewritten as

$$\Delta\bar{U}_L(k) = \mathbf{A}_1(k)\Delta\bar{U}_L(k-1) + \rho_1 \mathbf{C}_1(k)E(k). \tag{5.60}$$

Substituting PFDL data models (5.45) and (5.60) into (5.59) yields

$$\begin{aligned} e(k+1) &= y^* - y(k) - \big(y(k+1) - y(k)\big) \\ &= e(k) - \boldsymbol{\Phi}_{p,L}(k)\Delta\bar{U}_L(k) \\ &= \left(\mathrm{I} - \frac{\rho_1 \boldsymbol{\Phi}_1(k)\hat{\boldsymbol{\Phi}}_1^T(k)}{\lambda + \|\hat{\boldsymbol{\Phi}}_1(k)\|^2} \right) e(k) - \boldsymbol{\Phi}_{p,L}(k)\mathbf{A}_1(k)\Delta\bar{U}_L(k-1). \end{aligned} \tag{5.61}$$

Let

$$\mathbf{A}_2(k) = \begin{bmatrix} \mathrm{I} - \dfrac{\rho_1 \boldsymbol{\Phi}_1(k)\hat{\bar{\phi}}_1(k)}{\lambda + \|\hat{\boldsymbol{\Phi}}_1(k)\|^2} & 0 \\ 0 & 0 \end{bmatrix}_{mL\times mL}, \text{ and } \mathbf{C}_2(k) = \begin{bmatrix} \boldsymbol{\Phi}_{p,L}(k) \\ 0 \end{bmatrix}_{mL\times mL}.$$

Equation (5.61) can be rewritten as

$$\begin{aligned} E(k+1) &= E(k) - \mathbf{C}_2(k)\Delta\bar{U}_L(k) \\ &= \mathbf{A}_2(k)E(k) - \mathbf{C}_2(k)\mathbf{A}_1(k)\Delta\bar{U}_L(k-1). \end{aligned} \tag{5.62}$$

Since $\mathbf{\Phi}_{p,L}(k)$ and $\hat{\mathbf{\Phi}}_{p,L}(k)$ are bounded, there exist four positive constants M_1, M_3, M_4, and $\lambda_{\min} > 0$, such that the following six inequalities hold for any $\lambda > \lambda_{\min}$:

$$0 < M_1 \le \frac{2\alpha b_1^2(m-1)^2}{\lambda + \|\hat{\mathbf{\Phi}}_1(k)\|^2} < \frac{\alpha^2 b_2^2 + b_1^2(m-1)^2}{\lambda_{\min} + \|\hat{\mathbf{\Phi}}_1(k)\|^2} < 1, \tag{5.63}$$

$$s(\mathbf{A}_1(k)) \le 1, \tag{5.64}$$

$$\|\mathbf{C}_1(k)\|_v \le M_3 < 1, \tag{5.65}$$

$$\|\mathbf{C}_2(k)\|_v \le M_4, \tag{5.66}$$

$$M_1 + M_3 M_4 < 1, \tag{5.67}$$

$$s\left(\mathbf{A}_2(k)\right) < 1 - \rho_1 M_1 < 1, \tag{5.68}$$

where $s(\mathbf{A})$ is the spectral radius of matrix $\mathbf{A}$.

Note that (5.63), (5.64), and (5.68) can be derived by a similar procedure as in (5.26)–(5.36).

The characteristic equation of $\mathbf{A}_1(k)$ is

$$z^m \det\left(z^{L-1} I + \sum_{i=2}^{L} z^{L-i} \frac{\rho_i \hat{\mathbf{\Phi}}_1^T(k)\hat{\mathbf{\Phi}}_i(k)}{\lambda + \|\hat{\mathbf{\Phi}}_1(k)\|^2} \right) = 0. \tag{5.69}$$

Note that $\det(z^{L-1}I + \sum_{i=2}^{L} z^{L-i}(\rho_i \hat{\mathbf{\Phi}}_1^T(k)\hat{\mathbf{\Phi}}_i(k)/(\lambda + \|\hat{\mathbf{\Phi}}_1(k)\|^2)))$ is an $(mL-m)$-order monic polynomial with respect to z, which can be rewritten as $z^{(L-1)m} + ((\chi(z))/(\lambda + \|\hat{\mathbf{\Phi}}_1(k)\|^2)) \max_{i=2,\ldots,L} \rho_i$, where $\chi(z)$ is an $(mL-m-1)$-order monic polynomial with respect to z.

Since $\mathbf{\Phi}_{p,L}(k)$ and $\hat{\mathbf{\Phi}}_{p,L}(k)$ are bounded, and $s\left(\mathbf{A}_1(k)\right) \le 1$, there exists a positive constant M_2, such that

$$\frac{|\chi(z_i)|}{\lambda + \|\hat{\mathbf{\Phi}}_1(k)\|^2} \le M_2, \quad \forall z_1,\ldots,z_{mL}, \tag{5.70}$$

where z_i is the eigenvalue of the matrix of $\mathbf{A}_1(k)$, $i = 1,2,\ldots,mL$.

Select $\max_{i=2,\ldots,L} \rho_i$ such that the following inequality holds for any z_i, $i = 1,2,\ldots,mL$,

$$|z_i|^{(L-1)m} \leq \max_{i=2,\ldots,L} \rho_i M_2 < 1. \tag{5.71}$$

From (5.71), we have

$$s(\mathbf{A}_1(k)) \leq \left(\max_{i=2,\ldots,L} \rho_i M_2 \right)^{\frac{1}{(L-1)m}} < 1. \tag{5.72}$$

The conclusion on spectral radius means that there exists an arbitrary small positive constant ε such that

$$\|\mathbf{A}\|_v < s(\mathbf{A}) + \varepsilon. \tag{5.73}$$

From (5.68), (5.72), and (5.73), we have

$$\|\mathbf{A}_1(k)\|_v < s(\mathbf{A}_1(k)) + \varepsilon \leq \left(\max_{i=2,\ldots,L} \rho_i M_2 \right)^{\frac{1}{(L-1)m}} + \varepsilon < 1, \tag{5.74}$$

and

$$\|\mathbf{A}_2(k)\|_v < s(\mathbf{A}_2(k)) + \varepsilon \leq 1 - \rho_1 M_1 + \varepsilon < 1. \tag{5.75}$$

Let

$$d_1 = \left(\max_{i=2,\ldots,L} \{\rho_i\} M_2 \right)^{\frac{1}{(L-1)m}} + \varepsilon, \tag{5.76}$$

$$d_2 = 1 - \rho_1 M_1 + \varepsilon. \tag{5.77}$$

Taking the norm on both sides of (5.60) and using the initial condition $\|\Delta \bar{U}_L(0)\|_v = 0$ yield

$$\begin{aligned} \|\Delta \bar{U}_L(k)\|_v &= \|\mathbf{A}_1(k)\|_v \|\Delta \bar{U}_L(k-1)\|_v + \rho_1 \|\mathbf{C}_1(k)\|_v \|E(k)\|_v \\ &< d_1 \|\Delta \bar{U}_L(k-1)\|_v + \rho_1 M_3 \|E(k)\|_v \\ &< \cdots < \rho_1 M_3 \sum_{i=1}^{k} d_1^{k-i} \|E(i)\|_v. \end{aligned} \tag{5.78}$$

Taking the norm on both sides of (5.62) gives

$$\begin{aligned}\|E(k+1)\|_v &= \|\mathbf{A}_2(k)\|_v\|E(k)\|_v + \|\mathbf{C}_2(k)\|_v\|\mathbf{A}_1(k)\|_v\|\Delta\bar{U}_L(k-1)\|_v\\ &< d_2\|E(k)\|_v + d_1M_4\|\Delta\bar{U}_L(k-1)\|_v\\ &< \cdots < d_2^{\,k}\|E(1)\|_v + M_4\sum_{j=1}^{k-1} d_2^{\,k-1-j}d_1\|\Delta\bar{U}_L(j)\|_v. \end{aligned} \tag{5.79}$$

Let $d_3 = \rho_1 M_3 M_4$. Substituting (5.78) into (5.79) gives

$$\begin{aligned}\|E(k+1)\|_v &< d_2^{\,k}\|E(1)\|_v + M_4\sum_{j=1}^{k-1} d_2^{\,k-1-j}d_1\|\Delta\bar{U}_L(j)\|_v\\ &< d_2^{\,k}\|E(1)\|_v + M_4\sum_{j=1}^{k-1} d_2^{\,k-1-j}d_1\left(\rho_1M_3\sum_{i=1}^{j} d_1^{\,j-i}\|E(i)\|_v\right)\\ &= d_2^{\,k}\|E(1)\|_v + d_1d_3\sum_{j=1}^{k-1} d_2^{\,k-1-j}\left(\sum_{i=1}^{j} d_1^{\,j-i}\|E(i)\|_v\right)\\ &= g(k+1), \end{aligned} \tag{5.80}$$

where $g(k+1) = d_2^{\,k}\|E(1)\|_v + d_1d_3\sum_{j=1}^{k-1} d_2^{\,k-1-j}(\sum_{i=1}^{j} d_1^{\,j-i}\|E(i)\|_v)$.

Obviously, $\|E(k+1)\|_v$ will converge to zero if $g(k+1)$ converges to zero.

Computing $g(k+2)$ yields

$$\begin{aligned}g(k+2) &= d_2^{\,k+1}\|E(1)\|_v + d_1d_3\sum_{j=1}^{k} d_2^{\,k-j}\left(\sum_{i=1}^{j} d_1^{\,j-i}\|E(i)\|_v\right)\\ &= d_2g(k+1) + d_3d_1^{\,k}\|E(1)\|_v + d_3d_1^{\,k-1}\|E(2)\|_v + \cdots + d_3d_1\|E(k)\|_v\\ &< d_2g(k+1) + d_3d_1^{\,k}\|E(1)\|_v + d_3d_1^{\,k-1}\|E(2)\|_v + \cdots + d_3d_1g(k)\\ &= d_2g(k+1) + h(k), \end{aligned} \tag{5.81}$$

where $h(k) = d_3d_1^{\,k}\|E(1)\|_v + d_3d_1^{\,k-1}\|E(2)\|_v + \cdots + d_3d_1g(k)$.

Using (5.67), we have

$$d_2 = 1-\rho_1M_1+\varepsilon > \rho_1(M_1+M_3M_4)-\rho_1M_1+\varepsilon = \rho_1M_3M_4+\varepsilon > d_3.$$

Rewriting $h(k)$ gives

$$\begin{aligned}
h(k) &< d_3 d_1^{k} \|E(1)\|_v + d_3 d_1^{k-1} \|E(2)\|_v + \cdots + d_3 d_1^{2} \|E(k-1)\|_v + d_2 d_1 g(k) \\
&< d_3 d_1^{k} \|E(1)\|_v + d_3 d_1^{k-1} \|E(2)\|_v + \cdots + d_3 d_1^{2} \|E(k-1)\|_v \\
&\quad + d_2 d_1 \left(d_2^{k-1} \|E(1)\|_v + d_1 d_3 \sum_{j=1}^{k-2} d_2^{k-2-j} \left(\sum_{i=1}^{j} d_1^{j-i} \|E(i)\|_v \right)\right) \\
&< d_1 \left(d_2^{k} \|E(1)\|_v + d_1 d_3 \sum_{j=1}^{k-1} d_2^{k-1-j} \left(\sum_{i=1}^{j} d_1^{j-i} \|E(i)\|_v \right)\right) \\
&= d_1 g(k+1).
\end{aligned} \tag{5.82}$$

Substituting (5.82) into (5.81) yields

$$g(k+2) = d_2 g(k+1) + h(k) < (d_2 + d_1) g(k+1). \tag{5.83}$$

Properly select $0 < \rho_1 \le 1, \ldots, 0 < \rho_L \le 1$ such that

$$0 < \left\{ \max_{i=2,\ldots,L} \{\rho_i\} M_2 \right\}^{\frac{1}{(L-1)m}} < \rho_1 M_1 < 1$$

and

$$0 < 1 - \rho_1 M_1 + \left\{ \max_{i=2,\ldots,L} \{\rho_i\} M_2 \right\}^{\frac{1}{(L-1)m}} < 1 \tag{5.84}$$

hold.

Since ε is an arbitrarily small positive constant, the following equation holds:

$$d_2 + d_1 = 1 - \rho_1 M_1 + \varepsilon + \left\{ \max_{i=2,\ldots,L} \{\rho_i\} M_2 \right\}^{\frac{1}{(L-1)m}} + \varepsilon < 1. \tag{5.85}$$

Substituting (5.85) into (5.83) yields

$$\lim_{k\to\infty} g(k+2) < \lim_{k\to\infty} (d_2 + d_1) g(k+1) < \cdots < \lim_{k\to\infty} (d_2 + d_1)^{k-1} g(2) = 0, \tag{5.86}$$

where $g(2) = d_2 \|E(1)\|$.

Then, conclusion (a) of Theorem 5.2 is the direct result of (5.80) and (5.86).

Since $\boldsymbol{y}^*$ is a given constant vector and $\boldsymbol{e}(k)$ is a bounded vector, $\boldsymbol{y}(k)$ is also bounded.

From (5.78), (5.80), and (5.86), we have

$$\begin{aligned}
\left\|\bar{U}_L(k)\right\|_v &\le \sum_{i=1}^{k}\left\|\Delta\bar{U}_L(i)\right\|_v \le \rho_1 M_3 \sum_{i=1}^{k}\sum_{j=1}^{i} d_1^{i-j}\left\|E(j)\right\|_v \\
&< \frac{\rho_1 M_3}{1-d_1}\left(\left\|E(1)\right\|_v + \cdots + \left\|E(k)\right\|_v\right) \\
&< \frac{\rho_1 M_3}{1-d_1}\left(\left\|E(1)\right\|_v + g(2) + \cdots + g(k)\right) \\
&< \frac{\rho_1 M_3}{1-d_1}\left(\left\|E(1)\right\|_v + \frac{g(2)}{1-d_1-d_2}\right).
\end{aligned} \tag{5.87}$$

Thus, conclusion (b) of Theorem 5.2 is obtained. ■

Remark 5.8

For discrete-time MISO nonlinear system (5.13), the corresponding assumption and corollary are listed as follows.

Assumption 5.2′

The sign of all the elements of $\bar{\boldsymbol{\phi}}_1^T(k)$ of PPG $\bar{\boldsymbol{\phi}}_{p,L}(k)$ is unchanged.

Corollary 5.2

If nonlinear system (5.13), satisfying Assumptions 3.9′, 3.10′, and 5.2′, is controlled by the PFDL–MFAC scheme (5.56)–(5.58), for the regulation problem of $y^*(k+1) = y^* = \text{const.}$, then there exists a constant $\lambda_{\min} > 0$ such that the following two properties hold for $\lambda > \lambda_{\min}$:

a. The tracking error sequence is convergent, that is, $\lim_{k\to\infty}\left|y(k+1) - y^*\right| = 0$.
b. The closed-loop system is BIBO stable, that is, $\{y(k)\}$ and $\{\boldsymbol{u}(k)\}$ are bounded sequences.

5.3.3 Simulation Results

Two numerical simulations are given to verify the correctness and effectiveness of the proposed control scheme (5.56)–(5.58) for MISO systems and control scheme (5.51)–(5.55) for the MIMO system, respectively. The models in the following two examples are merely used to generate the I/O data of the controlled plant, but not for MFAC design.

Example 5.3

Consider a discrete-time MISO nonlinear system

$$y(k+1) = y(k)u_1^2(k) + a(k)u_2(k) - b(k)u_1(k-1)y^2(k-1) + u_2^2(k-1), \quad (5.88)$$

where $a(k) = \text{round}(k/50)$ is an unknown fast time-varying parameter, and $b(k) = \sin(k/100)$ is an unknown slowly time-varying parameter.

The desired trajectory is given as follows:

$$y^*(k+1) = 0.5(-1)^{\text{round}(k/100)}. \quad (5.89)$$

The initial values are $\boldsymbol{u}(1) = [0.1, 0.2]^T$, $y(1) = 0.1$, $y(2) = 1$, $\varepsilon = 10^{-5}$, and $M = 50$. The controller parameters are $L = 3$, $\eta = \rho_1 = \rho_2 = \rho_3 = 1$, $\lambda = 22$, $\mu = 1$, and $\hat{\boldsymbol{\phi}}_{p,L}(1) = [-0.5, 1, -0.5, 0, 0, 0]^T$.

Figure 5.3a, b, and c shows the tracking performances of the output y, the control inputs u_1 and u_2, and the PPG estimation, respectively. The simulation results clearly show that the control performance of the PFDL–MFAC scheme is quite good, and the spikes in Figure 5.3a are caused by fast time-varying parameters.

Example 5.4

Consider a discrete-time MIMO coupled nonlinear nonminimum phase system with 2 inputs and 2 outputs

$$\begin{cases} y_1(k+1) = \dfrac{2.5y_1(k)y_1(k-1) + 0.09u_1(k)u_1(k-1)}{1 + y_1^2(k) + y_1^2(k-1)} \\ \qquad + 1.2u_1(k) + 1.6u_1(k-2) + 0.09u_1(k)u_2(k-1) + 0.5u_2(k) \\ \qquad + 0.7\sin\big(0.5\big(y_1(k) + y_1(k-1)\big)\big)\cos\big(0.5\big(y_1(k) + y_1(k-1)\big)\big), \\ y_2(k+1) = \dfrac{5y_2(k)y_2(k-1)}{1 + y_1^2(k) + y_1^2(k-1) + y_1^2(k-2)} + u_2(k) + 1.1u_2(k-1) \\ \qquad + 1.4u_2(k) + 0.5u_1(k). \end{cases} \quad (5.90)$$

The desired trajectories are given as follows:

$$\begin{cases} y_1^*(k+1) = 5\sin(\pi/50) + 2\cos(\pi/20), \\ y_2^*(k+1) = 2\sin(\pi/50) + 5\cos(\pi/20). \end{cases} \quad (5.91)$$

The simulation results are given to compare the control performance of CFDL–MFAC scheme (5.9)–(5.12) with that of PFDL–MFAC scheme (5.51)–(5.55). The initial values are $y_1(1) = y_1(3) = 0$, $y_1(2) = 1$, $y_2(1) = y_2(3) = 0$, $y_2(2) = 1$, $u_1(1) = u_1(2) = 1$, $u_2(1) = 1$, and $u_2(2) = 0$. The step factors and weighting factors of the CFDL–MFAC and PFDL–MFAC schemes are $\eta = \rho = \rho_1 = \rho_2 = \rho_3 = 0.5$, $\mu = 1$,

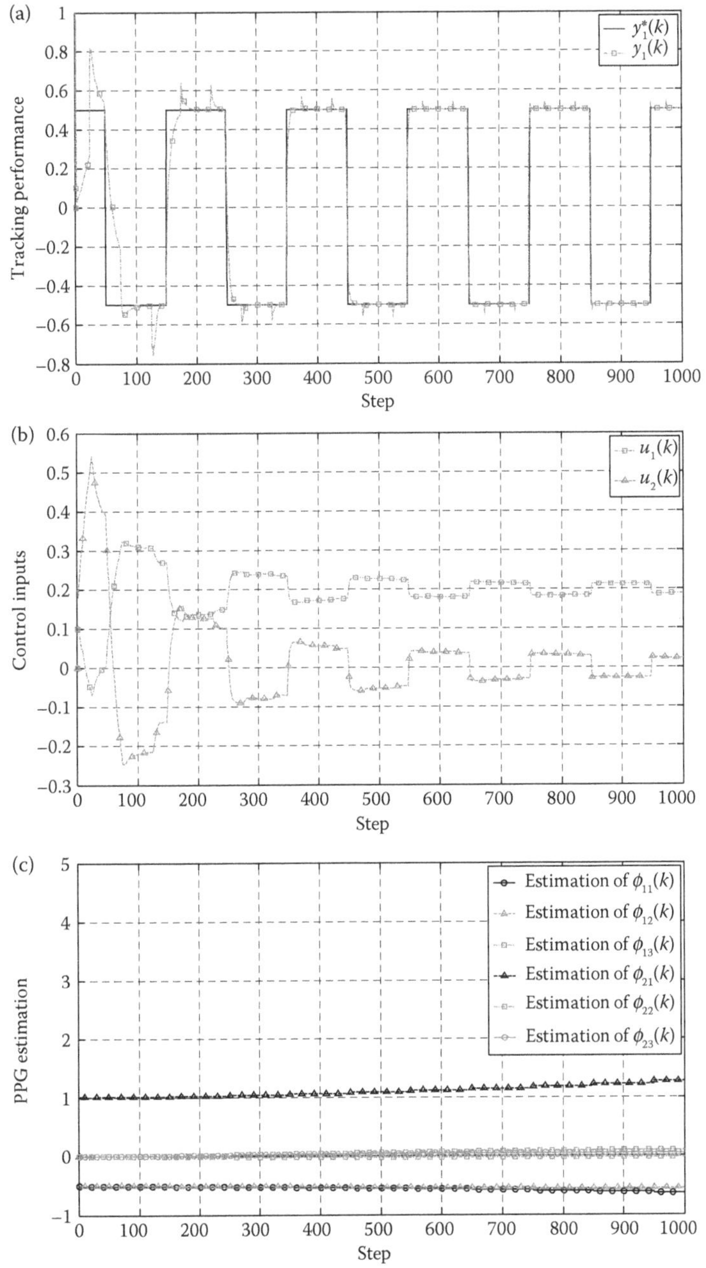

Figure 5.3 Simulation results of applying the PFDL–MFAC scheme to discrete-time MISO nonlinear system (5.88). (a) Tracking performance. (b) Control inputs $u_1(k)$ and $u_2(k)$. (c) PPG estimation $\hat{\phi}_{p,L}(k)$.

$\lambda = 0.01$, and the initial values of PJM and PPJM are $\hat{\mathbf{\Phi}}_c(1) = \hat{\mathbf{\Phi}}_c(2) = \begin{bmatrix} 0.5 & 0 \\ 0 & 0.5 \end{bmatrix}$, and $\hat{\mathbf{\Phi}}_{p,L}(1) = \hat{\mathbf{\Phi}}_{p,L}(2) = \begin{bmatrix} 0.5 & 0 & 0 & 0 & 0 & 0 \\ 0 & 0 & 0 & 0.5 & 0 & 0 \end{bmatrix}^T$, respectively.

Figure 5.4a, b, c, and d shows the tracking performances of output $y_1(k)$ and $y_2(k)$, the control inputs of $u_1(k)$ and $u_2(k)$ by applying the two control schemes, respectively.

From the simulation results, we can observe that the two control schemes can work well for such a discrete-time MIMO nonlinear system with unknown couplings. However, the control effect of the PFDL–MFAC scheme is better than that of the CFDL–MFAC scheme.

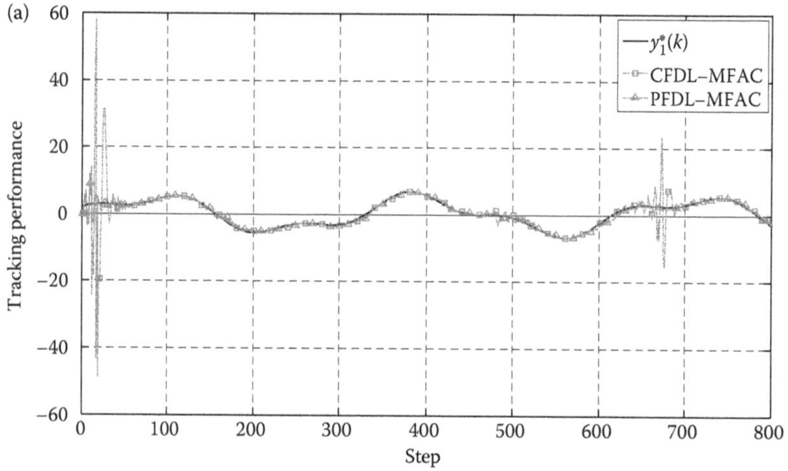

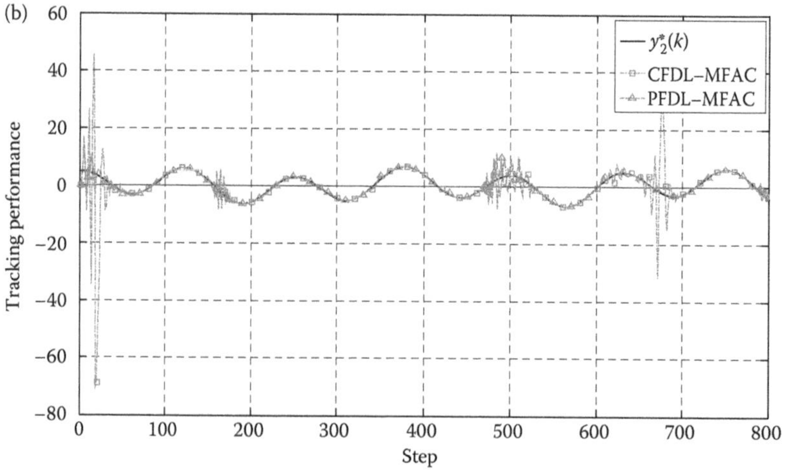

Figure 5.4 Simulation comparison between the CFDL–MFAC scheme and the PFDL–MFAC scheme. (a) Tracking performance $y_1(k)$. (b) Tracking performance $y_2(k)$. (c) Control input $u_1(k)$. (d) Control input $u_2(k)$.

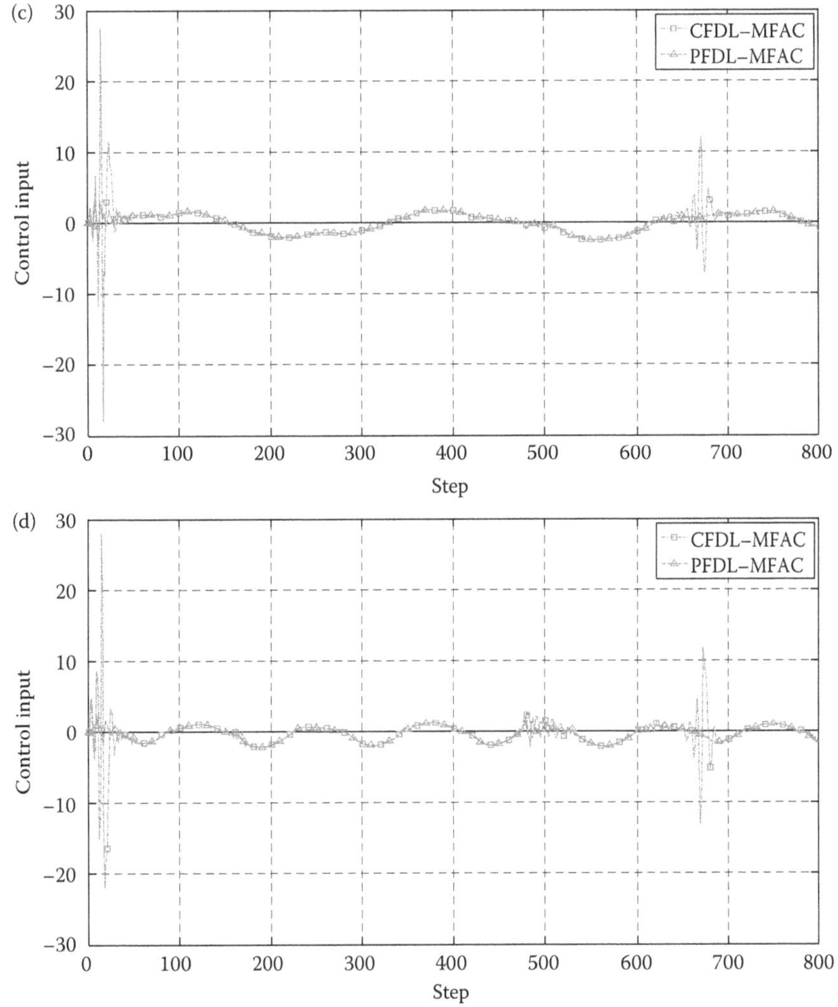

Figure 5.4 (continued) Simulation comparison between the CFDL–MFAC scheme and the PFDL–MFAC scheme. (a) Tracking performance $y_1(k)$. (b) Tracking performance $y_2(k)$. (c) Control input $u_1(k)$. (d) Control input $u_2(k)$.

5.4 FFDL Data Model Based MFAC

5.4.1 Control System Design

From Theorem 3.7, we know that system (5.1), satisfying Assumptions 3.11 and 3.12 with $\|\Delta \bar{H}_{L_y,L_u}(k)\| \neq 0$ for all k, can be transformed into the following FFDL data model:

$$\Delta \boldsymbol{y}(k+1) = \boldsymbol{\Phi}_{f,L_y,L_u}(k)\Delta \bar{\boldsymbol{H}}_{L_y,L_u}(k), \tag{5.92}$$

where $\boldsymbol{\Phi}_{f,L_y,L_u}(k) = \left[\boldsymbol{\Phi}_1(k) \quad \cdots \quad \boldsymbol{\Phi}_{L_y+L_u}(k)\right] \in R^{m\times(mL_y+mL_u)}$ is the unknown and bounded PPJM of system (5.1),

$$\boldsymbol{\Phi}_i(k) = \begin{bmatrix} \phi_{11i}(k) & \phi_{12i}(k) & \cdots & \phi_{1mi}(k) \\ \phi_{21i}(k) & \phi_{22i}(k) & \cdots & \phi_{2mi}(k) \\ \vdots & \vdots & \vdots & \vdots \\ \phi_{m1i}(k) & \phi_{m2i}(k) & \cdots & \phi_{mmi}(k) \end{bmatrix} \in R^{m\times m}, \quad i = 1\ldots, L_y + L_u,$$

and $\Delta \bar{\boldsymbol{H}}_{L_y,L_u}(k) = [\Delta \boldsymbol{y}^T(k),\ldots,\Delta \boldsymbol{y}^T(k - L_y + 1), \Delta \boldsymbol{u}^T(k),\ldots,\Delta \boldsymbol{u}^T(k - L_u + 1)]^T$.

5.4.1.1 Controller Algorithm

Consider the following cost function of control input:

$$J(\boldsymbol{u}(k)) = \|\boldsymbol{y}^*(k+1) - \boldsymbol{y}(k+1)\|^2 + \lambda\, \|\boldsymbol{u}(k) - \boldsymbol{u}(k-1)\|^2, \tag{5.93}$$

where $\lambda > 0$ is a weighting factor.

Substituting FFDL data model (5.92) into cost function (5.93), differentiating cost function (5.93) with respect to $\boldsymbol{u}(k)$ and setting it to zero, yields

$$\begin{aligned} \boldsymbol{u}(k) &= \boldsymbol{u}(k-1) + \left(\lambda \mathbf{I} + \boldsymbol{\Phi}_{L_y+1}^T(k)\boldsymbol{\Phi}_{L_y+1}(k)\right)^{-1} \boldsymbol{\Phi}_{L_y+1}^T(k) \\ &\quad \times \left(\left(\boldsymbol{y}^*(k+1) - \boldsymbol{y}(k)\right) - \sum_{i=1}^{L_y} \boldsymbol{\Phi}_i(k)\Delta \boldsymbol{y}(k-i+1) - \sum_{i=L_y+2}^{L_y+L_u} \boldsymbol{\Phi}_i(k)\Delta \boldsymbol{u}(k+L_y-i+1) \right). \end{aligned} \tag{5.94}$$

The following simplified controller algorithm without matrix inversion is used:

$$\begin{aligned} \boldsymbol{u}(k) &= \boldsymbol{u}(k-1) + \frac{\boldsymbol{\Phi}_{L_y+1}^T(k)\left(\rho_{L_y+1}\left(\boldsymbol{y}^*(k+1) - \boldsymbol{y}(k)\right)\right)}{\lambda + \|\hat{\boldsymbol{\Phi}}_{L_y+1}(k)\|^2} \\ &\quad - \frac{\boldsymbol{\Phi}_{L_y+1}^T(k)\left(\sum_{i=1}^{L_y} \rho_i \boldsymbol{\Phi}_i(k)\Delta \boldsymbol{y}(k-i+1) + \sum_{i=L_y+2}^{L_y+L_u} \rho_i \boldsymbol{\Phi}_i(k)\Delta \boldsymbol{u}(k+L_y-i+1)\right)}{\lambda + \left\|\boldsymbol{\Phi}_{L_y+1}(k)\right\|^2}, \end{aligned} \tag{5.95}$$

where $\rho_i \in (0,1], i = 1,2,\ldots, L_y + L_u$ is a step factor.

5.4.1.2 PPJM Estimation Algorithm

Consider a cost function of the PPJM estimation:

$$J(\boldsymbol{\Phi}_{f,L_y,L_u}(k)) = \left\|\Delta \boldsymbol{y}(k) - \boldsymbol{\Phi}_{f,L_y,L_u}(k)\Delta \bar{\boldsymbol{H}}_{L_y,L_u}(k-1)\right\|^2 + \mu\left\|\boldsymbol{\Phi}_{f,L_y,L_u}(k) - \hat{\boldsymbol{\Phi}}_{f,L_y,L_u}(k-1)\right\|^2, \tag{5.96}$$

where $\mu > 0$ is a weighting factor.

Minimizing cost function (5.96) yields the following PPJM estimation algorithm:

$$\begin{aligned}\hat{\boldsymbol{\Phi}}_{f,L_y,L_u}(k) &= \hat{\boldsymbol{\Phi}}_{f,L_y,L_u}(k-1) \\ &\quad + \eta\left(\Delta \boldsymbol{y}(k) - \hat{\boldsymbol{\Phi}}_{f,L_y,L_u}(k-1)\Delta \bar{\boldsymbol{H}}_{L_y,L_u}(k-1)\right)\Delta \bar{\boldsymbol{H}}^T_{L_y,L_u}(k-1) \\ &\quad \times\left(\mu + \Delta \bar{\boldsymbol{H}}_{L_y,L_u}(k-1)\Delta \bar{\boldsymbol{H}}^T_{L_y,L_u}(k-1)\right)^{-1},\end{aligned} \tag{5.97}$$

The following simplified PPJM estimation algorithm without matrix inversion is used:

$$\hat{\boldsymbol{\Phi}}_{f,L_y,L_u}(k) = \hat{\boldsymbol{\Phi}}_{f,L_y,L_u}(k-1) + \frac{\eta\left(\Delta \boldsymbol{y}(k) - \hat{\boldsymbol{\Phi}}_{f,L_y,L_u}(\boldsymbol{k}-1)\Delta \bar{\boldsymbol{H}}_{L_y,L_u}(k-1)\right)\Delta \bar{\boldsymbol{H}}^T_{L_y,L_u}(k-1)}{\mu + \left\|\Delta \bar{\boldsymbol{H}}_{L_y,L_u}(k-1)\right\|^2}, \tag{5.98}$$

where $\eta \in (0,2]$ is an added step factor, and $\hat{\boldsymbol{\Phi}}_{f,L_y,L_u}(k) = [\hat{\boldsymbol{\Phi}}_1(k) \ \ \hat{\boldsymbol{\Phi}}_2(k) \cdots \hat{\boldsymbol{\Phi}}_{L_y+L_u}(k)] \in R^{m\times m(L_y+L_u)}$ is an estimation of the unknown PPJM $\boldsymbol{\Phi}_{f,L_y+L_u}(k)$,

$$\hat{\boldsymbol{\Phi}}_i(k) = \begin{bmatrix} \hat{\phi}_{11i}(k) & \hat{\phi}_{12i}(k) & \cdots & \hat{\phi}_{1mi}(k) \\ \hat{\phi}_{21i}(k) & \hat{\phi}_{22i}(k) & \cdots & \hat{\phi}_{2mi}(k) \\ \vdots & \vdots & \vdots & \vdots \\ \hat{\phi}_{m1i}(k) & \hat{\phi}_{m2i}(k) & \cdots & \hat{\phi}_{mmi}(k) \end{bmatrix} \in R^{m\times m}, \quad i = 1,\ldots,L_y + L_u.$$

5.4.1.3 System Control Scheme

Combining PPJM estimation algorithm (5.98) and controller algorithm (5.95), the FFDL–MFAC scheme for discrete-time MIMO nonlinear systems is constructed as follows:

$$\hat{\boldsymbol{\Phi}}_{f,L_y,L_u}(k)=\hat{\boldsymbol{\Phi}}_{f,L_y,L_u}(k-1)+\frac{\eta\left(\Delta\boldsymbol{y}(k)-\hat{\boldsymbol{\Phi}}_{f,L_y,L_u}(k-1)\Delta\bar{\boldsymbol{H}}_{L_y,L_u}(k-1)\right)\Delta\bar{\boldsymbol{H}}^T_{L_y,L_u}(k-1)}{\mu+\left\|\Delta\bar{\boldsymbol{H}}_{L_y,L_u}(k-1)\right\|^2}, \tag{5.99}$$

$$\hat{\phi}_{ii(L_y+1)}(k)=\hat{\phi}_{ii(L_y+1)}(1),\quad \text{if } |\hat{\phi}_{ii(L_y+1)}(k)|<b_2 \text{ or } |\hat{\phi}_{ii(L_y+1)}(k)|>\alpha b_2 \text{ or}$$
$$\text{sign}(\hat{\phi}_{ii(L_y+1)}(k))\neq\text{sign}(\hat{\phi}_{ii(L_y+1)}(1)),\ i=1,\ldots,m, \tag{5.100}$$

$$\hat{\phi}_{ij(L_y+1)}(k)=\hat{\phi}_{ij(L_y+1)}(1),\quad \text{if } |\hat{\phi}_{ij(L_y+1)}(k)|>b_1 \text{ or}$$
$$\text{sign }(\hat{\phi}_{ij(L_y+1)}(k))\neq\text{sign}(\hat{\phi}_{ij(L_y+1)}(1)),\ i,j=1,\ldots,m, i\neq j, \tag{5.101}$$

$$\boldsymbol{u}(k)=\boldsymbol{u}(k-1)+\frac{\hat{\boldsymbol{\Phi}}^T_{L_y+1}(k)\left(\rho_{L_y+1}\left(\boldsymbol{y}^*(k+1)-\boldsymbol{y}(k)\right)\right)}{\lambda+\|\hat{\boldsymbol{\Phi}}_{L_y+1}(k)\|^2}-\frac{\hat{\boldsymbol{\Phi}}^T_{L_y+1}(k)\left(\sum_{i=1}^{L_y}\rho_i\hat{\boldsymbol{\Phi}}_i(k)\Delta\boldsymbol{y}(k-i+1)+\sum_{i=L_y+2}^{L_y+L_u}\rho_i\hat{\boldsymbol{\Phi}}_i(k)\Delta\boldsymbol{u}(k+L_y-i+1)\right)}{\lambda+\|\hat{\boldsymbol{\Phi}}_{L_y+1}(k)\|^2}, \tag{5.102}$$

where $\hat{\phi}_{ijl}(1)$ is the initial value of $\hat{\phi}_{ijl}(k)$, $i=1,\ldots,m$, $j=1,\ldots,m$, $\rho_1,\rho_2,\ldots,\rho_{L_y+L_u}\in(0,1]$, $\eta\in(0,2]$, $\lambda>0$, $\mu>0$.

Remark 5.9

For the discrete-time MISO nonlinear systems (5.13), by using the FFDL data model $y(k+1)=y(k)+\boldsymbol{\phi}^T_{f,L_y,L_u}(k)\Delta\breve{\boldsymbol{H}}_{L_y,L_u}(k)$, the corresponding FFDL–MFAC scheme is given as

$$\hat{\bar{\boldsymbol{\phi}}}_{f,L_y,L_u}(k)=\hat{\bar{\boldsymbol{\phi}}}_{f,L_y,L_u}(k-1)+\frac{\eta\left(\Delta y(k)-\hat{\bar{\boldsymbol{\phi}}}^T_{f,L_y,L_u}(k-1)\Delta\breve{\boldsymbol{H}}_{L_y,L_u}(k-1)\right)\Delta\breve{\boldsymbol{H}}_{L_y,L_u}(k-1)}{\mu+\left\|\Delta\breve{\boldsymbol{H}}_{L_y,L_u}(k-1)\right\|^2}, \tag{5.103}$$

$$\hat{\phi}_{i(L_y+1)}(k)=\hat{\phi}_{i(L_y+1)}(1),\quad \text{if } |\hat{\phi}_{i(L_y+1)}(k)|<b_2 \quad \text{or} \quad \text{sign}(\hat{\phi}_{i(L_y+1)}(k))\neq \text{sign}(\hat{\phi}_{i(L_y+1)}(1)),\ i=1,\ldots,m \tag{5.104}$$

$$u(k)=u(k-1)+\frac{\rho_{L_y+1}\hat{\bar{\boldsymbol{\phi}}}_{L_y+1}(k)\big(y^*(k+1)-y(k)\big)}{\lambda+\left\|\hat{\bar{\boldsymbol{\phi}}}_{L_y+1}(k)\right\|^2}-\frac{\hat{\bar{\boldsymbol{\phi}}}_{L_y+1}(k)\sum_{i=1}^{L_y}\rho_i\hat{\bar{\phi}}_i(k)\Delta y(k-i+1)}{\lambda+\left\|\hat{\bar{\boldsymbol{\phi}}}_{L_y+1}(k)\right\|^2}-\frac{\hat{\bar{\boldsymbol{\phi}}}_{L_y+1}(k)\sum_{i=L_y+2}^{L_y+L_u}\rho_i\hat{\bar{\boldsymbol{\phi}}}_i(k)\Delta u(k+L_y-i+1)}{\lambda+\left\|\hat{\bar{\boldsymbol{\phi}}}_{L_y+1}(k)\right\|^2}, \tag{5.105}$$

where $\hat{\bar{\boldsymbol{\phi}}}_{f,L_y,L_u}(k)=[\hat{\phi}_1\ \cdots\ \hat{\phi}_{L_y}\ \hat{\bar{\boldsymbol{\phi}}}^T_{L_y+1}(k)\ \cdots\ \hat{\bar{\boldsymbol{\phi}}}^T_{L_y+L_u}(k)]^T\in R^{L_y+mL_u}$ is the estimation of $\bar{\boldsymbol{\phi}}_{f,L_y,L_u}(k)$ in the FFDL data model (3.50), $\hat{\bar{\boldsymbol{\phi}}}_i(k)=[\hat{\phi}_{1i}(k),\hat{\phi}_{2i}(k),\ldots,\hat{\phi}_{mi}(k)]^T$, $i=L_y+1,\ldots,L_y+L_u$, $\hat{\bar{\boldsymbol{\phi}}}_{f,L_y,L_u}(1)$ is the initial value of $\hat{\bar{\boldsymbol{\phi}}}_{f,L_y,L_u}(k)$, $\rho_1,\rho_2,\ldots,\rho_{L_y+L_u}\in(0,1]$, $\eta\in(0,2]$, $\lambda>0$, $\mu>0$.

It is worth pointing out that the stability analysis of the FFDL–MFAC scheme is essentially equivalent to studying the stability problem of adaptive control for MIMO time-varying linear systems. Thus, it is still an open problem.

5.4.2 Simulation Results

Two numerical simulations are given to verify the correctness and effectiveness of FFDL–MFAC scheme (5.103)–(5.105) for MISO systems and FFDL–MFAC

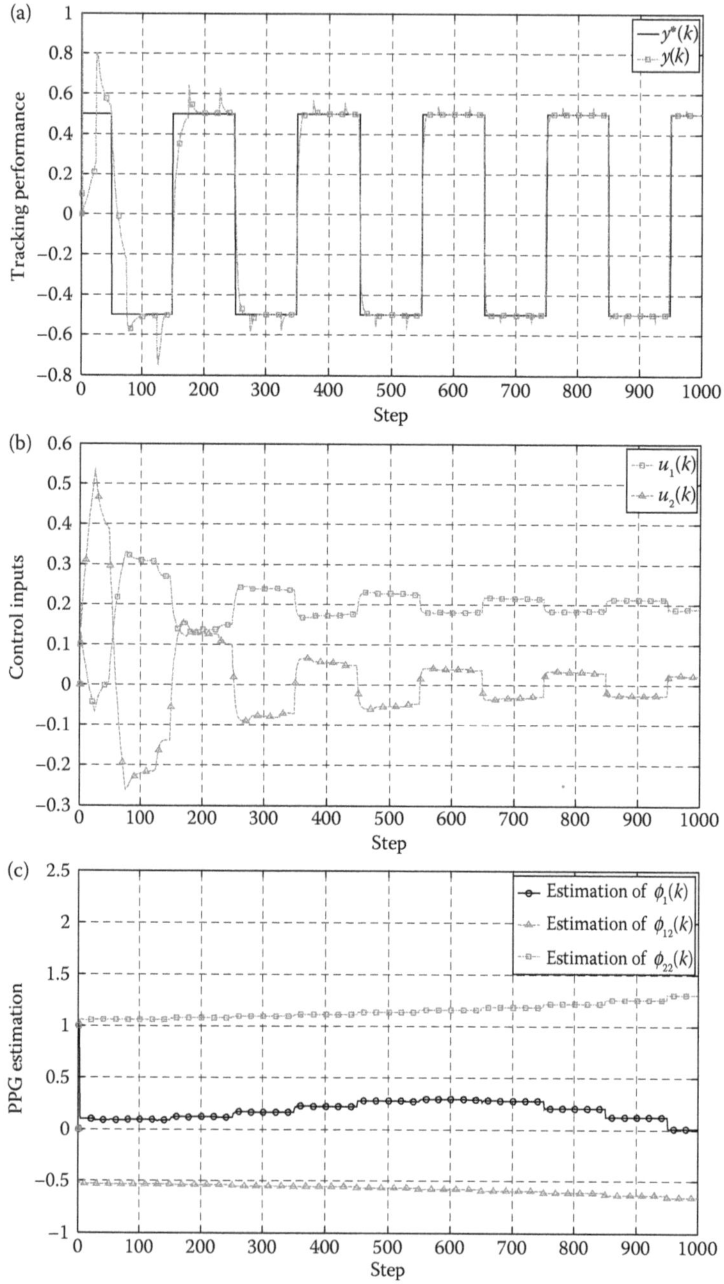

Figure 5.5 Simulation results of applying the FFDL–MFAC scheme to discrete-time MISO nonlinear system (5.88). (a) Tracking performance. (b) Control inputs $u_1(k)$ and $u_2(k)$. (c) PPG estimation $\hat{\boldsymbol{\phi}}_{f,L_y,L_u}(k)$.

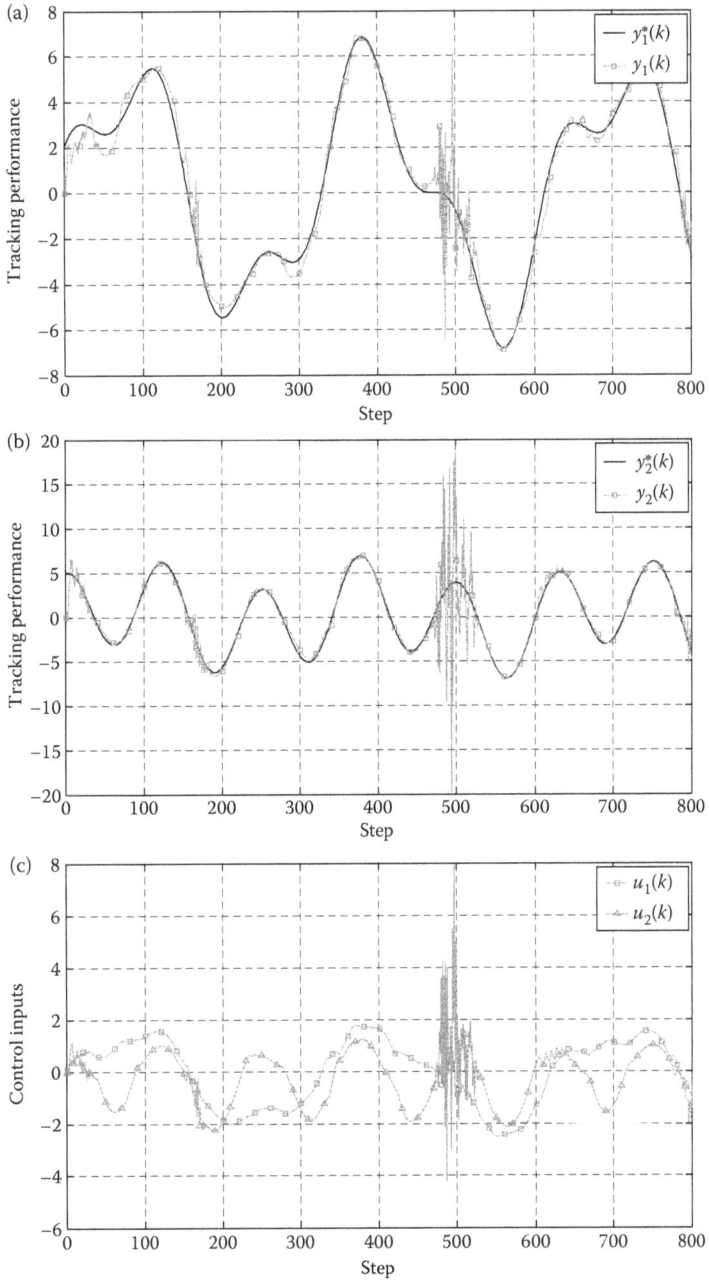

Figure 5.6 Simulation results of applying the FFDL–MFAC scheme to discrete-time MIMO nonlinear system (5.90). (a) Tracking performance $y_1(k)$. (b) Tracking performance $y_2(k)$. (c) Control inputs $u_1(k)$ and $u_2(k)$.

scheme (5.99)–(5.102) for MIMO discrete-time nonlinear systems, respectively. The models in the following two examples are merely used to generate the I/O data of the controlled plant, but not for MFAC design.

Example 5.5

The nonlinear system, the desired trajectory, and the initial values are the same as Example 5.3.

The parameters in FFDL–MFAC scheme (5.103)–(5.105) are $L_y = 1$, $L_u = 1$, $\eta = \rho_1 = \rho_2 = 1$, $\lambda = 22$, $\mu = 1$, and $\hat{\boldsymbol{\phi}}_{f,L_y,L_u}(1) = [1,-0.5,1]^T$.

Figure 5.5a, b, and c shows the tracking performance, the control inputs u_1 and u_2, and the PPG estimation, respectively. The simulation results show that the control performance of the FFDL–MFAC scheme is quite good.

Example 5.6

The nonlinear system, the desired trajectory, and the initial values are the same as Example 5.4.

The parameters in FFDL–MFAC scheme (5.99)–(5.102) are set as $L_y = 1$, $L_u = 3$, $\eta = 0.1$, $\rho_1 = \rho_2 = \rho_3 = \rho_4 = 0.5$, $\mu = 1$, $\lambda = 1$, and $\hat{\boldsymbol{\Phi}}_{f,L_y,L_u}(1) = \hat{\boldsymbol{\Phi}}_{f,L_y,L_u}(2) = \begin{bmatrix} 0 & 0 & 0.1 & 0 & 0 & 0 & 0 & 0 \\ 0 & 0 & 0 & 0.1 & 0 & 0 & 0 & 0 \end{bmatrix}^T$.

The simulation results are shown in Figure 5.6. Figure 5.6a, b, and c is the tracking performances of the outputs $y_1(k)$ and $y_2(k)$, and the control inputs $u_1(k)$ and $u_2(k)$, respectively. From the simulation results, one can observe that the control performance of the FFDL–MFAC scheme of unknown discrete-time MIMO coupled nonlinear systems is quite satisfactory.

5.5 Conclusions

In this chapter, the MFAC design and analysis approaches for discrete-time SISO nonlinear systems are directly extended to the MIMO cases. Three MFAC schemes, including CFDL–MFAC, PFDL–MFAC, and FFDL–MFAC, are presented for a class of discrete-time MIMO nonlinear nonaffine systems based on the novel dynamic linearization approach presented in Chapter 3. The corresponding results for the MISO case are also provided.

The MFAC approaches only use the online I/O data of the controlled plant to design the controller directly, and thus it is independent of the mathematical model of the controlled plant. The MFAC controllers have a few parameters to update online, which can be designed more easily than that of traditional adaptive control systems. BIBO stability and tracking error convergence analyses of closed-loop systems are also given under some suitable assumptions. The effectiveness and applicability of the MFAC schemes are verified via numerical simulations.

Chapter 6

Model-Free Adaptive Predictive Control

6.1 Introduction

Model predictive control (MPC) was first developed from industrial practices in the late 1970s. Up to now, MPC has been the most widely applied control method besides PID, and it is also the hot research topic in the control theory community for a long time. The fundamental principles of MPC for both linear systems and nonlinear systems are the same, that is, the prediction based on a model, the receding horizon optimization, and the feedback correction. In more detail, at each control interval, the predictive control first explicitly uses a model to predict the plant output within a prespecified prediction horizon, and then calculates an optimal control sequence by minimizing a given objective function, and only applies the first control signal in the optimal control input sequence calculated at each step to the plant according to the receding horizon principle, and finally, updates the output error prediction by using the new available measurements in order to improve the tracking performance. The representative MPC algorithms are model predictive heuristic control (MPHC) [185], dynamic matrix control (DMC) [186], and generalized predictive control (GPC) [187]. MPC methods have been successfully applied in many practical industrial processes, especially in the oil industry, the chemical industry [147,188], and so on. Although the MPC has many advantages, such as high control performance and robustness to uncertainties and/or disturbances, the plant model or its structure is still required to be known for controller design, and the accuracy of the model will directly affect the control performance. Furthermore, most of the existing MPC methods were proposed to deal with linear systems [128,129,185,189–194], and only a few works have been carried out for nonlinear systems [147,195–197]. Therefore, there is still a lot of work that needs to be done for the MPC of nonlinear systems.

It is of great significance to study model-free adaptive predictive control (MFAPC) for unknown nonlinear systems. MFAPC has the advantages of MPC and MFAC; that is, only the I/O data of the controlled plant are used to predict the system output sequence within the prediction horizon and to calculate the control input sequence within the control input horizon, without the controlled plant model involved.

On the basis of the three dynamic linearization approaches proposed in Chapter 3, the corresponding MFAPC schemes are studied for a class of unknown discrete-time nonaffine nonlinear systems in this chapter. The presented schemes merely depend on the I/O data of the controlled system and have a low computational burden [23,24,29,198]. It is worth pointing out that different predictive control methods can be developed by using different predictive control methods and the equivalent dynamical linearization data models, such as model-free adaptive predictive functional control [30,31], model-free adaptive predictive PI control [199,200], and so on. Owing to page limitation, only MFAPC schemes of discrete-time SISO nonlinear systems are presented in this chapter, and the results can easily be extended to MISO and MIMO nonlinear systems.

This chapter is organized as follows. In Section 6.2, the CFDL data model based MFAPC (CFDL–MFAPC), including controller design, convergence analysis, and simulation study, is shown in detail. The PFDL data model based MFAPC (PFDL–MFAPC) scheme, the FFDL data model based MFAPC (FFDL–MFAPC) scheme, and their simulation results are presented in Sections 6.3 and 6.4, respectively. Finally, the conclusions are presented in Section 6.5.

6.2 CFDL Data Model Based MFAPC

6.2.1 Control System Design

Consider a class of SISO discrete-time nonlinear systems described by

$$y(k+1) = f\left(y(k), y(k-1), \ldots, y(k-n_y), u(k), u(k-1), \ldots, u(k-n_u)\right), \quad (6.1)$$

where $y(k) \in R$ and $u(k) \in R$ are the output and input at sample time k, respectively, n_y and n_u are two unknown integers, and $f(\cdots) : R^{n_u+n_y+2} \mapsto R$ denotes an unknown scalar nonlinear function.

It is difficult to predict the future output sequence of system (6.1) if nonlinear function $f(\ldots)$ is unknown. However, by virtue of Theorem 3.1, discrete-time nonlinear system (6.1) can be transformed into the following CFDL data model:

$$\Delta y(k+1) = \phi_c(k)\Delta u(k),$$

where $\phi_c(k) \in R$ is the PPD.

On the basis of the above incremental data model which is equivalent to the unknown discrete time nonlinear system (6.1), the following one-step-ahead prediction equation can be obtained straightforwardly:

$$y(k+1) = y(k) + \phi_c(k)\Delta u(k). \tag{6.2}$$

According to (6.2), N-step-ahead prediction equations are given as follows:

$$\begin{cases} y(k+1) = y(k) + \phi_c(k)\Delta u(k), \\ y(k+2) = y(k+1) + \phi_c(k+1)\Delta u(k+1) \\ \qquad = y(k) + \phi_c(k)\Delta u(k) + \phi_c(k+1)\Delta u(k+1), \\ \qquad \vdots \\ y(k+N) = y(k+N-1) + \phi_c(k+N-1)\Delta u(k+N-1) \\ \qquad = y(k+N-2) + \phi_c(k+N-2)\Delta u(k+N-2) \\ \qquad\quad + \phi_c(k+N-1)\Delta u(k+N-1) \\ \qquad \vdots \\ \qquad = y(k) + \phi_c(k)\Delta u(k) + \cdots \\ \qquad\quad + \phi_c(k+N-1)\Delta u(k+N-1). \end{cases} \tag{6.3}$$

Let

$$\begin{cases} \boldsymbol{Y}_N(k+1) = \left[y(k+1),\ldots,y(k+N)\right]^T, \\ \Delta \boldsymbol{U}_N(k) = \left[\Delta u(k),\ldots,\Delta u(k+N-1)\right]^T, \\ \boldsymbol{E}(k) = \left[1,1,\ldots,1\right]^T, \\ \mathbf{A}(k) = \begin{bmatrix} \phi_c(k) & 0 & 0 & 0 & 0 & 0 \\ \phi_c(k) & \phi_c(k+1) & 0 & 0 & & \\ \vdots & \vdots & \ddots & \vdots & & \vdots \\ \phi_c(k) & \cdots & & \phi_c(k+N_u-1) & & \\ \vdots & & & \vdots & \ddots & 0 \\ \phi_c(k) & \phi_c(k+1) & \cdots & \phi_c(k+N_u-1) & \cdots & \phi_c(k+N-1) \end{bmatrix}_{N\times N}, \end{cases}$$

where $\boldsymbol{Y}_N(k+1)$ denotes the N-step-ahead prediction vector of the system output, $\Delta \boldsymbol{U}_N(k)$ is the control input increment vector, and N_u is the control input horizon.

Then, Equation (6.3) can be rewritten in a compact form:

$$\boldsymbol{Y}_N(k+1) = \boldsymbol{E}(k)y(k) + \mathbf{A}(k)\Delta\boldsymbol{U}_N(k). \tag{6.4}$$

If $\Delta u(k+j-1)=0, j>N_u$, then prediction equation (6.4) becomes

$$\boldsymbol{Y}_N(k+1) = \boldsymbol{E}(k)y(k) + \mathbf{A}_1(k)\Delta\boldsymbol{U}_{N_u}(k), \tag{6.5}$$

where

$$\mathbf{A}_1(k) = \begin{bmatrix} \phi_c(k) & 0 & 0 & 0 \\ \phi_c(k) & \phi_c(k+1) & 0 & 0 \\ \vdots & \vdots & \ddots & \vdots \\ \phi_c(k) & \phi_c(k+1) & \cdots & \phi_c(k+N_u-1) \\ \vdots & \vdots & \cdots & \vdots \\ \phi_c(k) & \phi_c(k+1) & \cdots & \phi_c(k+N_u-1) \end{bmatrix}_{N\times N_u},$$

and $\Delta\boldsymbol{U}_{N_u}(k) = [\Delta u(k),\ldots,\Delta u(k+N_u-1)]^T$.

6.2.1.1 Controller Algorithm

Consider the following cost function of the control input:

$$J = \sum_{i=1}^{N}\left(y(k+i) - y^*(k+i)\right)^2 + \lambda\sum_{j=0}^{N_u-1}\Delta u^2(k+j), \tag{6.6}$$

where $\lambda > 0$ is a weighting factor, and $y^*(k+i)$ is the desired output at time $k+i$, $i=1,\ldots,N$.

Let $\boldsymbol{Y}_N^*(k+1) = [y^*(k+1),\ldots,y^*(k+N)]^T$; cost function (6.6) becomes

$$J = [\boldsymbol{Y}_N^*(k+1) - \boldsymbol{Y}_N(k+1)]^T[\boldsymbol{Y}_N^*(k+1) - \boldsymbol{Y}_N(k+1)] + \lambda\Delta\boldsymbol{U}_{N_u}^T(k)\Delta\boldsymbol{U}_{N_u}(k). \tag{6.7}$$

Substituting (6.5) into (6.7) and using the optimality condition $\partial J/(\partial\boldsymbol{U}_{N_u}(k)) = 0$ yield the control law

$$\Delta\boldsymbol{U}_{N_u}(k) = \left[\mathbf{A}_1^T(k)\mathbf{A}_1(k) + \lambda\mathbf{I}\right]^{-1}\mathbf{A}_1^T(k)\left[\boldsymbol{Y}_N^*(k+1) - \boldsymbol{E}(k)y(k)\right]. \tag{6.8}$$

Thus, the control input at current time k is obtained according to the receding horizon principle as follows:

$$u(k) = u(k-1) + \boldsymbol{g}^T \Delta \boldsymbol{U}_{N_u}(k), \tag{6.9}$$

where $\boldsymbol{g} = [1,0,\ldots,0]^T$.

When $N_u = 1$, Equation 6.9 becomes

$$u(k) = u(k-1) + \frac{1}{\phi_c^2(k) + \lambda/N} \frac{1}{N}\left[\phi_c(k)\sum_{i=1}^{N}\left(y^*(k+i) - y(k)\right)\right]. \tag{6.10}$$

Remark 6.1

In Remark 4.2, we have pointed out that λ is an important parameter. The proper selection of λ can guarantee stability and improve the tracking performance. Compared with the control law in CFDL–MFAC scheme (4.4), the weighting factor λ in control law (6.10) is divided by N, which indicates the control law (6.10) is insensitive to the selection of λ, so it can be seen as a "roughly tuning" manner or can be regarded as an "averaged form" of control law (4.4); therefore, a more smooth transient process can be achieved.

6.2.1.2 PPD Estimation Algorithm and Prediction Algorithm

Since $\mathbf{A}_1(k)$ in (6.9) contains unknown PPD parameters $\phi_c(k)$, $\phi_c(k+1), \ldots,$ $\phi_c(k+N_u-1)$, some time-varying parameter estimation or prediction algorithm should be developed when it is used in applications. Theoretically speaking, any estimation algorithm for time-varying parameters can be applied to PPD parameter estimation, but we still use the modified projection algorithm to estimate $\phi_c(k)$ here in order to facilitate the theoretical analysis for the control system, that is,

$$\hat{\phi}_c(k) = \hat{\phi}_c(k-1) + \frac{\eta \Delta u(k-1)}{\mu + \Delta u(k-1)^2}\left[\Delta y(k) - \hat{\phi}_c(k-1)\Delta u(k-1)\right], \tag{6.11}$$

where $\mu > 0$ is a weighting factor, and $0 < \eta \leq 1$ is a step size factor.

Since $\phi_c(k+1),\ldots,\phi_c(k+N_u-1)$ cannot be directly calculated from the I/O data till sample time k, $\phi_c(k+1),\ldots,\phi_c(k+N_u-1)$ need to be predicted according to the past estimated sequence $\hat{\phi}_c(1),\ldots,\hat{\phi}_c(k)$.

There exist many methods for parameter prediction, such as the Aström prediction method [131], the self-tuning method [201], and the multilevel hierarchical forecasting method [78,79]. According to the simulation results in Refs. [78,79], the multilevel hierarchical forecasting method possesses the best predictive error, and thus the multilevel hierarchical forecasting method is applied here to predict unknown parameters $\phi_c(k+1),\ldots,\phi_c(k+N_u-1)$. For simplicity, only the simple two-level prediction method is cited as follows.

Assume that the estimated values $\hat{\phi}_c(1),\ldots,\hat{\phi}_c(k)$ have been calculated by (6.11) at time k. Using these estimated values, an autoregressive (AR) model for prediction is constructed as follows:

$$\hat{\phi}_c(k+1) = \theta_1(k)\hat{\phi}_c(k) + \theta_2(k)\hat{\phi}_c(k-1) + \cdots + \theta_{n_p}(k)\hat{\phi}_c(k-n_p+1), \quad (6.12)$$

where θ_i, $i = 1,\ldots,n_p$ is the coefficient and n_p is the fixed model order, which is usually set to $2 \sim 7$ as recommended by Refs. [78,79].

Using (6.12), the prediction equation becomes

$$\hat{\phi}_c(k+j) = \theta_1(k)\hat{\phi}_c(k+j-1) + \theta_2(k)\hat{\phi}_c(k+j-2) + \cdots + \theta_{n_p}(k)\hat{\phi}_c(k+j-n_p), \quad (6.13)$$

where $j = 1,\ldots,N_u - 1$.

Defining $\boldsymbol{\theta}(k) = [\theta_1(k),\ldots,\theta_{n_p}(k)]^T$, it is determined by the following equation:

$$\boldsymbol{\theta}(k) = \boldsymbol{\theta}(k-1) + \frac{\hat{\boldsymbol{\varphi}}(k-1)}{\delta + \|\hat{\boldsymbol{\varphi}}(k-1)\|^2}\left[\hat{\phi}_c(k) - \hat{\boldsymbol{\varphi}}^T(k-1)\boldsymbol{\theta}(k-1)\right], \quad (6.14)$$

where $\hat{\boldsymbol{\varphi}}(k-1) = [\hat{\phi}_c(k-1),\ldots,\hat{\phi}_c(k-n_p)]^T$, and $\delta \in (0,1]$ is a positive constant.

6.2.1.3 Control Scheme

By integrating controller algorithm (6.9), parameter estimation algorithm (6.11), and prediction algorithm (6.13)–(6.14), the CFDL–MFAPC scheme is designed as follows:

$$\hat{\phi}_c(k) = \hat{\phi}_c(k-1) + \frac{\eta\Delta u(k-1)}{\mu + \Delta u(k-1)^2}\left[\Delta y(k) - \hat{\phi}_c(k-1)\Delta u(k-1)\right], \quad (6.15)$$

$$\hat{\phi}_c(k) = \hat{\phi}_c(1), \quad \text{if } |\hat{\phi}_c(k)| \le \varepsilon \text{ or } |\Delta u(k-1)| \le \varepsilon \text{ or } \operatorname{sign}(\hat{\phi}_c(k)) \ne \operatorname{sign}(\hat{\phi}_c(1)), \quad (6.16)$$

$$\boldsymbol{\theta}(k) = \boldsymbol{\theta}(k-1) + \frac{\hat{\boldsymbol{\varphi}}(k-1)}{\delta + \|\hat{\boldsymbol{\varphi}}(k-1)\|^2}[\hat{\phi}_c(k) - \hat{\boldsymbol{\varphi}}^T(k-1)\boldsymbol{\theta}(k-1)], \quad (6.17)$$

$$\boldsymbol{\theta}(k) = \boldsymbol{\theta}(1), \quad \text{if } \|\boldsymbol{\theta}(k)\| \ge M, \quad (6.18)$$

$$\hat{\phi}_c(k+j) = \theta_1(k)\hat{\phi}_c(k+j-1) + \theta_2(k)\hat{\phi}_c(k+j-2) + \cdots + \theta_{n_p}(k)\hat{\phi}_c(k+j-n_p)$$
$$j = 1,2,\ldots,N_u - 1 \tag{6.19}$$

$$\hat{\phi}_c(k+j) = \hat{\phi}_c(1), \quad \text{if } |\hat{\phi}_c(k+j)| < \varepsilon \text{ or } \text{sign}(\hat{\phi}_c(k+j)) \neq \text{sign}(\hat{\phi}_c(1)),$$
$$j = 1,2,\ldots,N_u - 1 \tag{6.20}$$

$$\Delta \boldsymbol{U}_{N_u}(k) = \left[\hat{\mathbf{A}}_1^T(k)\hat{\mathbf{A}}_1(k) + \lambda \mathbf{I}\right]^{-1} \hat{\mathbf{A}}_1^T(k)\left[\boldsymbol{Y}_N^*(k+1) - \mathbf{E}(k)y(k)\right], \tag{6.21}$$

$$u(k) = u(k-1) + \boldsymbol{g}^T \Delta \boldsymbol{U}_{N_u}(k), \tag{6.22}$$

where ε and M are positive constants, $\hat{\mathbf{A}}_1(k)$ and $\hat{\phi}_c(k+j)$ are the estimated values of $\mathbf{A}_1(k)$ and $\phi_c(k+j)$, $j = 1, \ldots, (N_u - 1)$, respectively; $\lambda > 0$, $\mu > 0$, $\eta \in (0,1]$, $\delta \in (0,1]$.

Remark 6.2

To make parameter estimation algorithm (6.15) have a strong ability to track the time-varying parameter, a reset mechanism is taken as (6.16). The boundedness of $\hat{\mathbf{A}}_1(k)$ is guaranteed by (6.18). The sign of the predicted parameters is ensured by (6.20).

Remark 6.3

The CFDL–MFAPC scheme has N_u parameters to be tuned online, and it can be implemented merely using the I/O data of the controlled system without the plant model and order involved. It is totally different from the traditional predictive control method.

In addition, the control horizon should satisfy $N_u \leq N$. N_u should be chosen to be 1 for simple systems. N_u should be chosen as a larger value to improve transient and tracking performance for complex systems. The larger the control horizon, the higher the online computational burden.

The initial value of PPD estimation is generally set to $\hat{\phi}_c(1) > 0$, since $\phi_c(k) > 0$ in many practical industrial systems, such as the temperature control system and the pressure control system (see Remark 4.6 for details).

The prediction horizon N should be chosen sufficiently large such that the dynamics of the plant can be covered. For a time-delay system, N should be selected larger than the dead time at least. In real applications, it is usually set to $4 \sim 10$.

Generally, the larger the λ, the slower the response, and the smaller the overshoot, and vice versa. Therefore, it is an important parameter which will affect the dynamics of the closed-loop system.

The selection of prediction order n_p refers to Refs. [78,79], in which $2 \sim 7$ is recommended.

6.2.2 *Stability Analysis*

In this section, we will discuss the stability and convergence of CFDL–MFAPC scheme (6.15)–(6.22).

Theorem 6.1

If discrete-time nonlinear system (6.1), satisfying Assumptions 3.1, 3.2, and 4.1, is controlled by CFDL–MFAPC scheme (6.15)–(6.22) for a regulation problem, that is, $y^*(k+1) = y^* = \text{const.}$, then there exists a constant $\lambda_{\min} > 0$ such that the following properties hold for any $\lambda > \lambda_{\min}$:

a. The tracking error of the system converges, that is, $\lim_{k\to\infty} \left|y^* - y(k+1)\right| = 0$.
b. $\{y(k)\}$ and $\{u(k)\}$ are bounded sequences.

Proof

First, we prove the boundedness of the PPD estimation $\hat{\phi}_c(k)$ and the PPD predictions $\hat{\phi}_c(k+1),\ldots,\hat{\phi}_c(k+N_u-1)$.

If $|\hat{\phi}_c(k)| \le \varepsilon$, $\Delta u(k-1) \le \varepsilon$, or $\text{sign}(\hat{\phi}_c(k)) \ne \text{sign}(\hat{\phi}_c(1))$, then the boundedness of $\hat{\phi}_c(k)$ is obvious.

In the other case, define the parameter estimation error as $\tilde{\phi}_c(k) = \hat{\phi}_c(k) - \phi_c(k)$. Subtracting $\phi_c(k)$ from both sides of parameter estimation algorithm (6.15) and using (6.2) yield

$$
\begin{aligned}
\tilde{\phi}_c(k) &= \tilde{\phi}_c(k-1) - \Delta\phi_c(k) + \frac{\eta\Delta u(k-1)}{\mu + \Delta u(k-1)^2}\left[\Delta y(k) - \hat{\phi}_c(k-1)\Delta u(k-1)\right] \\
&= \left(1 - \frac{\eta\Delta u(k-1)^2}{\mu + \Delta u(k-1)^2}\right)\tilde{\phi}_c(k-1) - \Delta\phi_c(k). \qquad (6.23)
\end{aligned}
$$

The conclusion $\left|\phi_c(k)\right| \le \bar{b}$ in Theorem 3.1 leads to $\left|\phi_c(k-1) - \phi_c(k)\right| \le 2\bar{b}$.

Taking the absolute value on both sides of (6.23) yields

$$\begin{aligned}\left|\tilde{\phi}_c(k)\right| &\le \left|1-\frac{\eta\Delta u(k-1)^2}{\mu+\Delta u(k-1)^2}\right|\left|\tilde{\phi}_c(k-1)\right|+\left|\Delta\phi_c(k)\right| \\ &\le \left|1-\frac{\eta\Delta u(k-1)^2}{\mu+\Delta u(k-1)^2}\right|\left|\tilde{\phi}_c(k-1)\right|+2\bar{b}. \end{aligned} \quad (6.24)$$

Since $\mu > 0$ and $\eta \in (0,1]$, there exists a positive constant d_1 such that the following inequality holds:

$$0 < 1-\frac{\eta\Delta u(k-1)^2}{\mu+\Delta u(k-1)^2} \le d_1 < 1. \quad (6.25)$$

Noting (6.24) and (6.25), we have

$$\begin{aligned}\left|\tilde{\phi}_c(k)\right| &\le d_1\left|\tilde{\phi}_c(k-1)\right|+2\bar{b} \le d_1^2\left|\tilde{\phi}_c(k-2)\right|+2d_1\bar{b}+2\bar{b} \\ &\le \cdots \le d_1^{k-1}\left|\tilde{\phi}_c(1)\right|+\frac{2\bar{b}}{1-d_1}. \end{aligned} \quad (6.26)$$

This means that $\tilde{\phi}_c(k)$ is bounded. Since $\phi_c(k)$ is bounded, the boundedness of $\hat{\phi}_c(k)$ can be guaranteed. The boundedness of the prediction values $\hat{\phi}_c(k+j)$ $(j = 1,\ldots,(N_u-1))$ is the direct result of algorithms (6.17)–(6.20).

Second, we will prove the convergence of the tracking error and BIBO stability.

Define the tracking error as $e(k+1) = y^* - y(k+1)$. Substituting (6.2) into the tracking error equation and using (6.21) and (6.22), we have

$$\begin{aligned} e(k+1) &= y^* - y(k+1) = y^* - y(k) - \phi_c(k)\Delta u(k) \\ &= \left(1-\phi_c(k)\left[\boldsymbol{g}^T\left(\hat{\mathbf{A}}_1^T(k)\hat{\mathbf{A}}_1(k)+\lambda\mathbf{I}\right)^{-1}\hat{\mathbf{A}}_1^T(k)\boldsymbol{E}(k)\right]\right)\left(y^*-y(k)\right). \end{aligned} \quad (6.27)$$

Taking the absolute value on both sides of (6.27) yields

$$\left|e(k+1)\right| \le \left|1-\phi_c(k)\left[\boldsymbol{g}^T\left(\hat{\mathbf{A}}_1^T(k)\hat{\mathbf{A}}_1(k)+\lambda\mathbf{I}\right)^{-1}\hat{\mathbf{A}}_1^T(k)\boldsymbol{E}(k)\right]\right|\left|e(k)\right|. \quad (6.28)$$

Let $\boldsymbol{P} = (\hat{\mathbf{A}}_1^T(k)\hat{\mathbf{A}}_1(k)+\lambda\mathbf{I})$. Since $\hat{\mathbf{A}}_1^T(k)\hat{\mathbf{A}}_1(k)$ is a semipositive definite matrix, $\boldsymbol{P}$ and $\boldsymbol{P}^{-1}$ are positive definite matrixes for any $\lambda > 0$.

Since $\boldsymbol{P}^{-1} = \dfrac{\boldsymbol{P}^*}{\det(\boldsymbol{P})}$, where $\boldsymbol{P}^* = \begin{bmatrix} P_{11} & \cdots & P_{Nu1} \\ \vdots & \ddots & \vdots \\ P_{1Nu} & \cdots & P_{NuNu} \end{bmatrix}$ is the adjoint matrix of $\boldsymbol{P}$ and P_{ij} is the algebraic cofactor of $\boldsymbol{P}$, the following equation holds:

$$\begin{aligned}
&\boldsymbol{g}^T(\hat{\mathbf{A}}_1^T(k)\hat{\mathbf{A}}_1(k)+\lambda\mathbf{I})^{-1}\hat{\mathbf{A}}_1^T(k)\boldsymbol{E}(k)\\
&=\boldsymbol{g}^T\boldsymbol{P}^{-1}\hat{\mathbf{A}}_1^T(k)\boldsymbol{E}(k)\\
&=\boldsymbol{g}^T\frac{\boldsymbol{P}^*}{\det(\boldsymbol{P})}\hat{\mathbf{A}}_1^T(k)\boldsymbol{E}(k)\\
&=\frac{N\hat{\phi}_c(k)P_{11}}{\det(\boldsymbol{P})}+\frac{(N-1)\hat{\phi}_c(k+1)P_{21}}{\det(\boldsymbol{P})}+\cdots+\frac{(N-N_u+1)\hat{\phi}_c(k+N_u-1)P_{Nu1}}{\det(\boldsymbol{P})}.
\end{aligned} \tag{6.29}$$

Equation (6.29) is bounded as $\hat{\phi}_c(k)$ is bounded for any time k, and its upper bound is a constant independent of k.

Since $\boldsymbol{P}$ is a positive definite matrix, $\det(\boldsymbol{P}) > 0$ is a monic polynomial in λ of degree N_u, $P_{11} > 0$ is a monic polynomial in λ of degree $(N_u - 1)$, and P_{i1} $(i = 2,3,\ldots,N_u)$ is a monic polynomial in λ of degree $(N_u - 2)$. Thus, there exists a constant $\lambda_{\min} > 0$, such that (6.29) has the same positive sign as $P_{11}/(\det(\boldsymbol{P}))$ for any $\lambda \geq \lambda_{\min}$. In the sequel, there exists a positive constant d_2 such that

$$0 < 1 - \phi_c(k)[\boldsymbol{g}^T(\hat{\mathbf{A}}_1^T(k)\hat{\mathbf{A}}_1(k)+\lambda\mathbf{I})^{-1}\hat{\mathbf{A}}_1^T(k)\boldsymbol{E}(k)] \leq d_2 < 1. \tag{6.30}$$

Combining (6.28) and (6.30) gives

$$|e(k+1)| \leq d_2|e(k)| \leq \cdots \leq d_2^k|e(1)|. \tag{6.31}$$

Therefore, $\lim_{k\to\infty}|e(k+1)| = 0$.

Since $y^*(k)$ is a constant, sequence $\{y(k)\}$ is bounded.

In the following, we prove the boundedness of the input sequence. From (6.22) and (6.21), we have

$$\begin{aligned}
|\Delta u(k)| &\leq \left|\boldsymbol{g}^T\left(\hat{\mathbf{A}}_1^T(k)\hat{\mathbf{A}}_1(k)+\lambda\mathbf{I}\right)^{-1}\hat{\mathbf{A}}_1^T(k)\boldsymbol{E}(k)\right||e(k)\\
&\leq \chi|e(k)|.
\end{aligned} \tag{6.32}$$

where χ is a bounded constant.

Using (6.32) recursively, it gives

$$\begin{aligned}|u(k)| &\le |\Delta u(k)| + |\Delta u(k-1)| + \cdots + |\Delta u(2)| + |u(1)| \\ &\le \chi\left(|e(k)| + |e(k-1)| + \cdots + |e(2)|\right) + |u(1)| \\ &\le \chi\left(d_2^{k-1}|e(1)| + \cdots + d_2|e(1)|\right) + |u(1)| \\ &\le \chi\frac{d_2|e(1)|}{1-d_2} + |u(1)|. \end{aligned} \tag{6.33}$$

This equation implies the boundedness of sequence $\{u(k)\}$. ■

6.2.3 Simulation Results

In this section, two numerical simulations are given to verify the correctness and effectiveness of CFDL–MFAPC scheme (6.15)–(6.22) and show its superiority to PID and CFDL–MFAC (4.7)–(4.9). The following two examples are controlled by the same CFDL–MFAPC scheme with the same initial conditions. The parameters of the CFDL–MFAPC scheme are $\varepsilon = 10^{-5}$, $\delta = 1$, $\eta = 1$, $M = 10$, $\hat{\phi}_c(1) = 1$, $\hat{\phi}_c(2) = 0.5$, $\hat{\phi}_c(3) = 1$, $\hat{\phi}_c(4) = 0.3$, $\hat{\phi}_c(5) = 2$, and $\theta(4) = [0.3,0.2,0.4]^T$. The parameters of the CFDL–MFAC scheme are $\varepsilon = 10^{-5}$, $\hat{\phi}_c(1) = 0.5$, $\eta = 1$, and $\rho = 0.25$.

Example 6.1

SISO discrete-time nonlinear system

$$y(k+1) = \begin{cases} \dfrac{2.5y(k)y(k-1)}{1+y(k)^2+y(k-1)^2} + 0.7\sin\left(0.5\left(y(k)+y(k-1)\right)\right) \\ \quad\times\cos\left(0.5\left(y(k)+y(k-1)\right)\right) + 1.2u(k) + 1.4u(k-1), \\ \qquad\qquad\qquad\qquad\qquad\qquad 1 \le k \le 500 \\ -0.1y(k) - 0.2y(k-1) - 0.3y(k-3) + 0.1u(k) \\ \quad + 0.02u(k-1) + 0.03u(k-2), \qquad 500 < k \le 1000. \end{cases} \tag{6.34}$$

Nonlinear system (6.34) consists of a nonminimum-phase nonlinear subsystem and a nonminimum-phase linear subsystem. Obviously, the structure and orders of the controlled system are time-varying.

To demonstrate the effectiveness and superiority of the CFDL–MFAPC scheme, the system using PID control and CFDL–MFAC are also simulated. The classical PID controller is given as follows:

$$u(k) = K_P\left[e(k) + \frac{1}{T_I}\sum_{j=0}^{k} e(j) + T_D\left(e(k)-e(k-1)\right)\right],$$

where $e(k) = y^*(k) - y(k)$ is the output error and K_P, T_I, T_D are the PID controller parameters.

The desired trajectory is given as

$$y^*(k+1) = 5 \times (-1)^{\text{round}(k/200)}.$$

Considering oscillation and the transient process of the second subsystem, the parameters of the three control methods are carefully tuned so that the closed-loop system has a satisfying control performance. Figure 6.1 shows the simulation

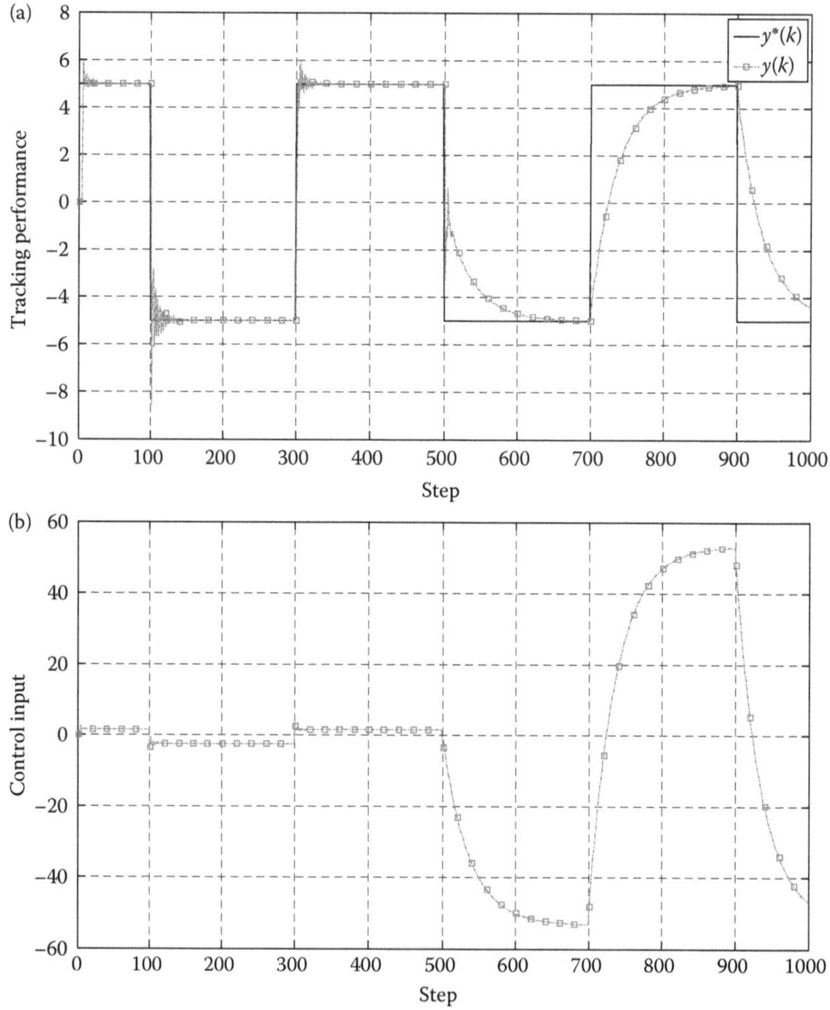

Figure 6.1 Simulation results of PID in Example 6.1. (a) Tracking performance. (b) Control input.

results using the PID controller with $K_P = 0.1$, $T_I = 0.33$, and $T_D = 1$, which is finely tuned. Figure 6.2 gives the simulation results using the CFDL–MFAC controller with $\lambda = 0.01$ and $\mu = 2$. Figure 6.3 shows the simulation results of the CFDL–MFAPC scheme with $N = 5$, $N_u = 1$, $\lambda = 0.01$, and $\mu = 2$.

The effectiveness and superiorities of the proposed CFDL–MFAPC scheme have been illustrated by the above simulation results. From Figure 6.3, the effect of the prediction in the proposed control scheme has been observed, which leads to improvement of the tracking performance. It is well known that

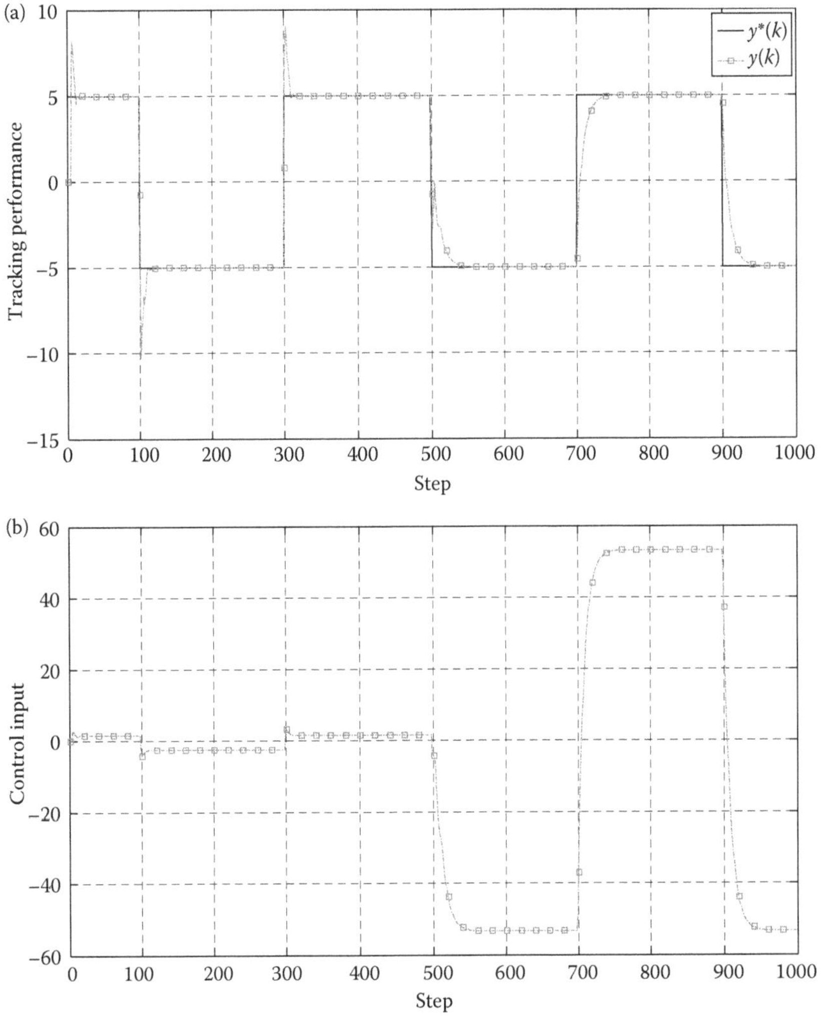

Figure 6.2 Simulation results of CFDL–MFAC scheme in Example 6.1. (a) Tracking performance. (b) Control input.

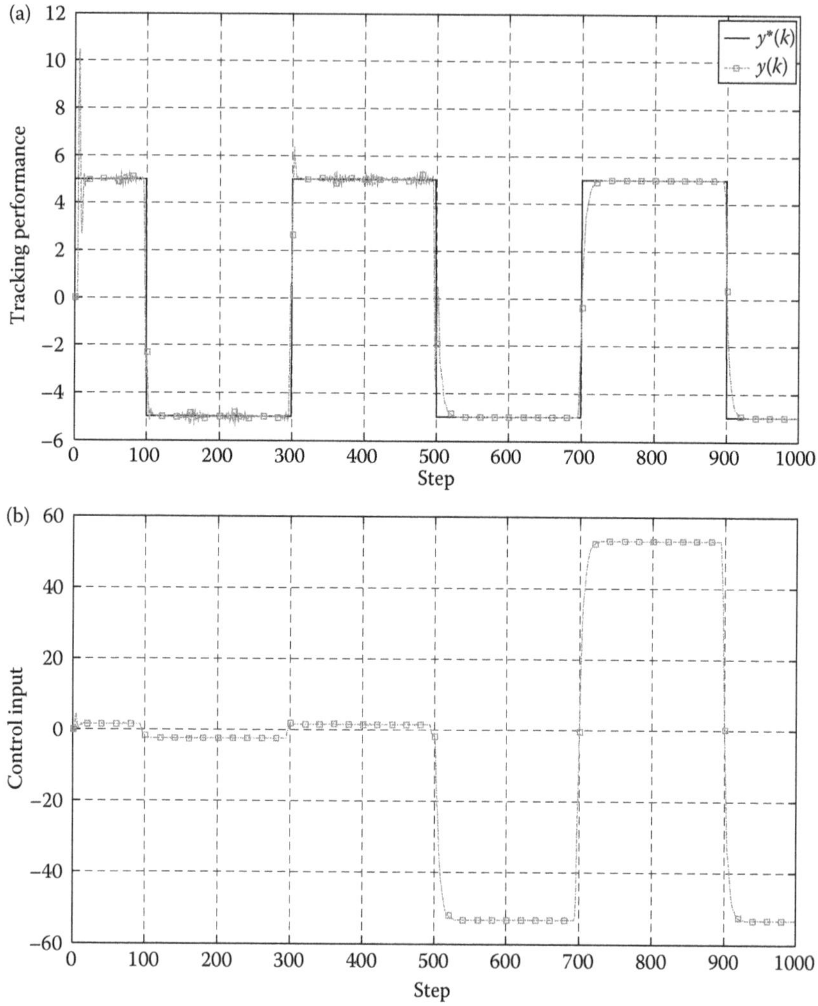

Figure 6.3 Simulation results of CFDL–MFAPC scheme in Example 6.1. (a) Tracking performance. (b) Control input.

tuning the parameters in the PID controller is quite time consuming, and the control performance is very sensitive to the variation of the PID parameters. A small change in the parameters may lead to a dramatic change in tracking performance, and even lead to instability of the controlled system. Compared with the PID control method, tuning the CFDL–MFAC scheme and the CFDL–MFAPC scheme is relatively easy, and the control performance is robust to λ tuning.

Example 6.2

SISO discrete-time nonlinear system

$$y(k+1)=\begin{cases}\dfrac{y(k)y(k-1)y(k-2)u(k-1)\big(y(k)-1\big)+u(k)}{1+y(k)^2+y(k-1)^2+y(k-2)^2}, \\ \qquad\qquad\qquad\qquad\qquad\qquad 1\le k\le 500 \\ -0.1y(k)-0.2y(k-1)-0.3y(k-3)+0.1u(k) \\ \quad +0.02u(k-1)+0.03u(k-2), \qquad 500<k\le 1000, \end{cases} \tag{6.35}$$

where the two subsystems are taken from Refs. [165,168], respectively. In Refs. [165,168], these two subsystems are controlled by different neural network control methods. Obviously, system (6.35) is a nonminimum-phase nonlinear system with a time-varying structure and a time-varying order.

Applying the CFDL–MFAPC scheme to system (6.35) with the control parameters $N=5$, $N_u=1$, $\lambda=0.5$, and $\mu=1$ gives the simulation results shown in Figure 6.4. Compared with the neural network control method, the CFDL–MFAPC scheme yields a better tracking performance, a smoother transient process, and non-overshoot.

6.3 PFDL Data Model Based MFAPC

6.3.1 Control System Design

On the basis of the conclusion in Section 3.2.2, discrete-time SISO nonlinear system (6.1) can be transformed into the following equivalent PFDL data model under certain assumptions:

$$\Delta y(k+1)=\boldsymbol{\phi}_{p,L}^T(k)\Delta \boldsymbol{U}_L(k).$$

According to the above equivalent incremental data model, the one-step-ahead prediction equation is given as follows:

$$y(k+1)=y(k)+\boldsymbol{\phi}_{p,L}^T(k)\Delta \boldsymbol{U}_L(k), \tag{6.36}$$

where

$$\boldsymbol{\phi}_{p,L}(k)=[\phi_1(k),\ldots,\phi_L(k)]^T \text{ and } \Delta\boldsymbol{U}_L(k)=\big[\Delta u(k),\ldots,\Delta u(k-L+1)\big]^T.$$

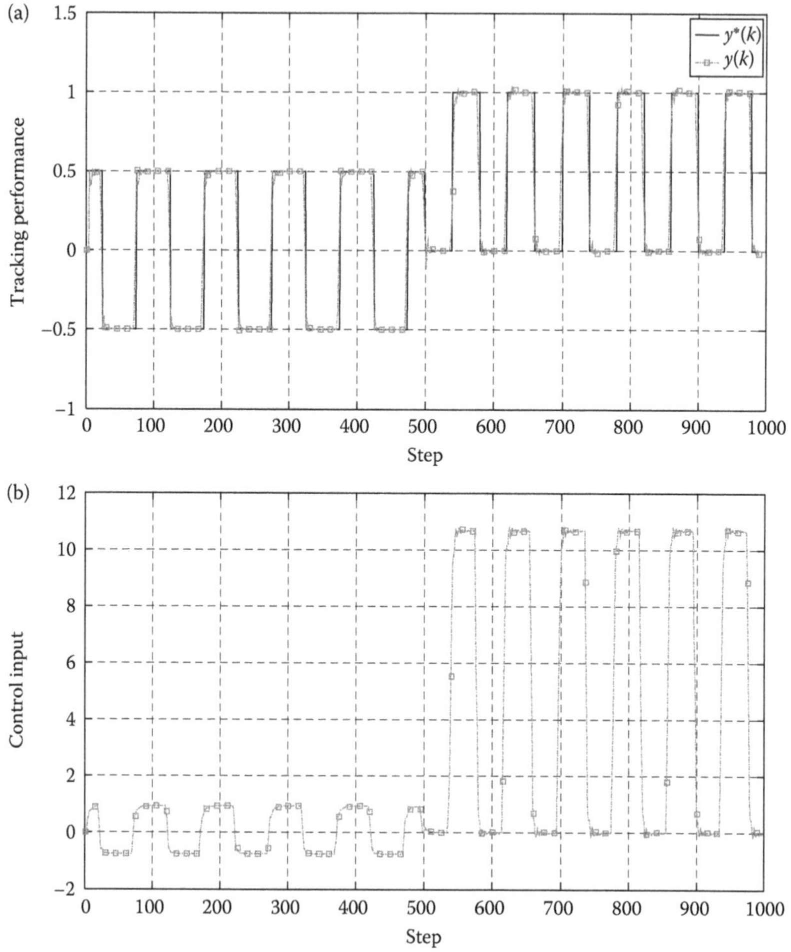

Figure 6.4 Simulation results of CFDL–MFAPC scheme in Example 6.2. (a) Tracking performance. (b) Control input.

Let $\mathbf{A} = \begin{bmatrix} 0 & & & \\ 1 & 0 & & \\ & \ddots & \ddots & \\ & & 1 & 0 \end{bmatrix}_{L\times L}$ and $\mathbf{B} = \begin{bmatrix} 1 \\ 0 \\ \vdots \\ 0 \end{bmatrix}_{L\times 1}$, then (6.36) can be rewritten as

$$y(k+1) = y(k) + \boldsymbol{\phi}_{p,L}^T(k)\Delta \boldsymbol{U}_L(k) = y(k) + \boldsymbol{\phi}_{p,L}^T(k)\mathbf{A}\Delta \boldsymbol{U}_L(k-1) + \boldsymbol{\phi}_{p,L}^T(k)\mathbf{B}\Delta u(k). \tag{6.37}$$

Similarly, N-step-ahead prediction equations are

$$
\begin{aligned}
y(k+2) &= y(k+1) + \boldsymbol{\phi}_{p,L}^{T}(k+1)\Delta \boldsymbol{U}_L(k+1) \\
&= y(k) + \boldsymbol{\phi}_{p,L}^{T}(k)\mathbf{A}\Delta \boldsymbol{U}_L(k-1) + \boldsymbol{\phi}_{p,L}^{T}(k)\mathbf{B}\Delta u(k) \\
&\quad + \boldsymbol{\phi}_{p,L}^{T}(k+1)\mathbf{A}^2\Delta \boldsymbol{U}_L(k-1) + \boldsymbol{\phi}_{p,L}^{T}(k+1)\mathbf{AB}\Delta u(k) \\
&\quad + \boldsymbol{\phi}_{p,L}^{T}(k+1)\mathbf{B}\Delta u(k+1) \\
&\ \ \vdots \\
y(k+N_u) &= y(k) + \sum_{i=0}^{N_u-1} \boldsymbol{\phi}_{p,L}^{T}(k+i)\mathbf{A}^{i+1}\Delta \boldsymbol{U}_L(k-1) + \sum_{i=0}^{N_u-1} \boldsymbol{\phi}_{p,L}^{T}(k+i)\mathbf{A}^{i}\mathbf{B}\Delta u(k) \\
&\quad + \sum_{i=1}^{N_u-1} \boldsymbol{\phi}_{p,L}^{T}(k+i)\mathbf{A}^{i-1}\mathbf{B}\Delta u(k+1) + \sum_{i=2}^{N_u-1} \boldsymbol{\phi}_{p,L}^{T}(k+i)\mathbf{A}^{i-2}\mathbf{B}\Delta u(k+2) \\
&\quad + \cdots + \boldsymbol{\phi}_{p,L}^{T}(k+N_u-1)\mathbf{B}\Delta u(k+N_u-1) \\
&\ \ \vdots \\
y(k+N) &= y(k) + \sum_{i=0}^{N-1} \boldsymbol{\phi}_{p,L}^{T}(k+i)\mathbf{A}^{i+1}\Delta \boldsymbol{U}_L(k-1) + \sum_{i=0}^{N-1} \boldsymbol{\phi}_{p,L}^{T}(k+i)\mathbf{A}^{i}\mathbf{B}\Delta u(k) \\
&\quad + \sum_{i=1}^{N-1} \boldsymbol{\phi}_{p,L}^{T}(k+i)\mathbf{A}^{i-1}\mathbf{B}\Delta u(k+1) + \sum_{i=2}^{N-1} \boldsymbol{\phi}_{p,L}^{T}(k+i)\mathbf{A}^{i-2}\mathbf{B}\Delta u(k+2) \\
&\quad + \cdots + \sum_{i=N_u-1}^{N-1} \boldsymbol{\phi}_{p,L}^{T}(k+i)\mathbf{A}^{i-N_u+1}\mathbf{B}\Delta u(k+N_u-1).
\end{aligned}
\tag{6.38}
$$

Define $\tilde{\boldsymbol{Y}}_N(k+1) = \left[y(k+1),\ldots,y(k+N)\right]^T$, $\boldsymbol{E} = \left[1,1,\ldots,1\right]^T$,

$$
\tilde{\psi}(k) = \begin{bmatrix}
\boldsymbol{\phi}_{p,L}^{T}(k)\mathbf{B} & & & \\
\sum_{i=0}^{1} \boldsymbol{\phi}_{p}^{T}(k+i)\mathbf{A}^{i}\mathbf{B} & \boldsymbol{\phi}_{p,L}^{T}(k+1)\mathbf{B} & & \\
\vdots & \vdots & \vdots & \vdots \\
\sum_{i=0}^{N_u-1} \boldsymbol{\phi}_{p}^{T}(k+i)\mathbf{A}^{i}\mathbf{B} & \sum_{i=1}^{N_u-1} \boldsymbol{\phi}_{p,L}^{T}(k+i)\mathbf{A}^{i-1}\mathbf{B} & \cdots & \boldsymbol{\phi}_{p,L}^{T}(k+N_u-1)\mathbf{B} \\
\vdots & \vdots & \vdots & \vdots \\
\sum_{i=0}^{N-1} \boldsymbol{\phi}_{p,L}^{T}(k+i)\mathbf{A}^{i}\mathbf{B} & \sum_{i=1}^{N-1} \boldsymbol{\phi}_{p,L}^{T}(k+i)\mathbf{A}^{i-1}\mathbf{B} & \cdots & \sum_{i=N_u-1}^{N-1} \boldsymbol{\phi}_{p,L}^{T}(k+N_u) \times \mathbf{A}^{i-N_u+1}\mathbf{B}
\end{bmatrix}_{N\times Nu},
$$

$$\bar{\boldsymbol{\Psi}}(k)=\begin{bmatrix} \boldsymbol{\phi}_p^T(k)\mathbf{A} \\ \sum_{i=0}^{1}\boldsymbol{\phi}_p^T(k+i)\mathbf{A}^i \\ \vdots \\ \sum_{i=0}^{N_u-1}\boldsymbol{\phi}_p^T(k+i)\mathbf{A}^{i+1} \\ \vdots \\ \sum_{i=0}^{N-1}\boldsymbol{\phi}_p^T(k+i)\mathbf{A}^{i+1} \end{bmatrix}_{N\times L}, \quad \text{and} \quad \Delta\tilde{\boldsymbol{U}}_{N_u}(k)=\left[\Delta u(k),\ldots,\Delta u(k+N_u-1)\right]^T.$$

Then the prediction equations can be rewritten in a compact form:

$$\tilde{\boldsymbol{Y}}_N(k+1) = \boldsymbol{E}y(k) + \tilde{\boldsymbol{\Psi}}(k)\Delta\tilde{\boldsymbol{U}}_{N_u}(k) + \bar{\boldsymbol{\Psi}}(k)\Delta\boldsymbol{U}_L(k-1). \tag{6.39}$$

6.3.1.1 Controller Algorithm

Let $\tilde{\boldsymbol{Y}}_N^*(k+1) = \left[y^*(k+1),\ldots,y^*(k+N)\right]^T$. Using cost function (6.7), substituting (6.39) into cost function (6.7), differentiating cost function (6.7) with respect to $\tilde{\boldsymbol{U}}_L(k)$ and setting it to zero, yield

$$\Delta\tilde{\boldsymbol{U}}_{N_u}(k) = \left(\tilde{\boldsymbol{\Psi}}^T(k)\tilde{\boldsymbol{\Psi}}(k) + \lambda\mathbf{I}\right)^{-1}\tilde{\boldsymbol{\Psi}}^T(k)\left(\tilde{\boldsymbol{Y}}_N^*(k+1) - \boldsymbol{E}y(k) - \bar{\boldsymbol{\Psi}}(k)\Delta\boldsymbol{U}_L(k-1)\right). \tag{6.40}$$

Then, the control input at current time k is obtained as follows:

$$u(k) = u(k-1) + \boldsymbol{g}^T\Delta\tilde{\boldsymbol{U}}_{N_u}(k), \tag{6.41}$$

where $\boldsymbol{g} = [1,0,\ldots,0]^T$.

Since $\boldsymbol{\phi}_{p,L}(k+i)$ in $\tilde{\boldsymbol{\Psi}}(k)$ and $\bar{\boldsymbol{\Psi}}(k)$ are not available for $i = 0,\ldots,N-1$, its estimation and prediction algorithms will be designed here.

6.3.1.2 PG Estimation Algorithm and Prediction Algorithm

From PFDL data model (6.36), we have

$$\Delta y(k+1) = \Delta\boldsymbol{U}_L^T(k)\boldsymbol{\phi}_{p,L}(k). \tag{6.42}$$

In general, any estimation algorithm for time-varying parameters can be used to estimate $\boldsymbol{\phi}_{p,L}(k)$. Here, we take the least-squares algorithm with a time-varying forgetting factor as the estimation algorithm.

$$\hat{\boldsymbol{\phi}}_{p,L}(k) = \hat{\boldsymbol{\phi}}_{p,L}(k-1) + \frac{\mathbf{P}_1(k-2)\Delta \boldsymbol{U}_L(k-1)}{\alpha(k-1) + \Delta \boldsymbol{U}_L^T(k-1)\mathbf{P}_1(k-2)\Delta \boldsymbol{U}_L(k-1)}$$
$$\times [\Delta y(k) - \Delta \boldsymbol{U}_L^T(k-1)\hat{\boldsymbol{\phi}}_{p,L}(k-1)],$$
$$\mathbf{P}_1(k-1) = \frac{1}{\alpha(k-1)}\left[\mathbf{P}_1(k-2) - \frac{\mathbf{P}_1(k-2)\Delta \boldsymbol{U}_L(k-1)\Delta \boldsymbol{U}_L^T(k-1)\mathbf{P}_1(k-2)}{\alpha(k-1) + \Delta \boldsymbol{U}_L^T(k-1)\mathbf{P}_1(k-2)\Delta \boldsymbol{U}_L(k-1)}\right],$$
$$\alpha(k) = \alpha_0\alpha(k-1) + (1-\alpha_0), \tag{6.43}$$

where $\hat{\boldsymbol{\phi}}_{p,L}(k)$ is the estimation of $\boldsymbol{\phi}_{p,L}(k)$, and $\mathbf{P}_1(-1) > 0$, $\alpha(0) = 0.95$, $\alpha_0 = 0.99$.

Remark 6.4

Parameter estimation algorithm (6.43) can generate accurate values of the parameter only if the plant input signal satisfies the persistently exciting condition. In adaptive control, the exciting signals only come from its setting points. If no proper excitation signal acts on the system for a longer time interval, the estimator may forget its tracking task. In the sequel, the "wind-up" and bursting phenomena of adaptive control may be caused with the unmodeled dynamics excited. To overcome these issues, a reset mechanism is the common method, that is, to set $\mathbf{P}(k-1) = \mathbf{P}(-1)$ when trace $(\mathbf{P}(k-1))$ is greater than or equal to a constant M, which has a similar function to the constant-trace algorithm [98]. A similar strategy may be adopted in the coming Equation (6.47) if it is necessary.

By (6.43), only $\hat{\boldsymbol{\phi}}_{p,L}(1), \hat{\boldsymbol{\phi}}_{p,L}(2), \ldots, \hat{\boldsymbol{\phi}}_{p,L}(k)$ are available. Therefore, we should find a prediction algorithm to predict the unknown $\boldsymbol{\phi}_{p,L}(k+1), \ldots, \boldsymbol{\phi}_{p,L}(k+N_u-1)$ including in $\tilde{\boldsymbol{\Psi}}(k)$ and $\bar{\boldsymbol{\Psi}}(k)$ of (6.40) to implement controller algorithm (6.41). The same two-level hierarchical forecasting method as that in Section 6.2.1 is applied here.

Using $\hat{\boldsymbol{\phi}}_{p,L}(1), \ldots, \hat{\boldsymbol{\phi}}_{p,L}(k)$, an AR model is given:

$$\hat{\boldsymbol{\phi}}_{p,L}(k) = \Gamma_1^T(k)\hat{\boldsymbol{\phi}}_{p,L}(k-1) + \Gamma_2^T(k)\hat{\boldsymbol{\phi}}_{p,L}(k-2) + \cdots + \Gamma_{n_p}^T(k)\hat{\boldsymbol{\phi}}_{p,L}(k-n_p), \tag{6.44}$$

where $\boldsymbol{\Gamma}_i^T(k)$, $i = 1, \ldots n_p$ is the time-varying coefficient matrix, and n_p is a fixed integer order.

Let $\boldsymbol{\Lambda}^T(k) = [\boldsymbol{\Gamma}_1^T(k), \ldots, \boldsymbol{\Gamma}_{n_p}^T(k)]$ and $\hat{\boldsymbol{\varsigma}}(k-1) = [\hat{\boldsymbol{\phi}}_{p,L}^T(k-1), \ldots, \hat{\boldsymbol{\phi}}_{p,L}^T(k-n_p)]^T$, then (6.44) can be rewritten in the following compact form:

$$\hat{\boldsymbol{\phi}}_{p,L}^T(k) = \hat{\boldsymbol{\varsigma}}^T(k-1)\boldsymbol{\Lambda}(k), \tag{6.45}$$

where

$$\Lambda(k) = \Lambda(k-1) + \frac{\mathbf{P}_2(k-2)\hat{\boldsymbol{\varsigma}}(k-1)}{\beta(k-1) + \hat{\boldsymbol{\varsigma}}^T(k-1)\mathbf{P}_2(k-2)\hat{\boldsymbol{\varsigma}}(k-1)}$$

$$\times\left[\hat{\boldsymbol{\phi}}_{p,L}^T(k) - \hat{\boldsymbol{\varsigma}}^T(k-1)\boldsymbol{\Lambda}(k\text{-}1)\right],$$

$$\mathbf{P}_2(k-1) = \left[\frac{1}{\beta(k-1)}\mathbf{P}_2(k-2) - \frac{\mathbf{P}_2(k-2)\hat{\boldsymbol{\varsigma}}(k-1)\hat{\boldsymbol{\varsigma}}^T(k-1)\mathbf{P}_2(k-2)}{\beta(k-1) + \hat{\boldsymbol{\varsigma}}^T(k-1)\mathbf{P}_2(k-2)\hat{\boldsymbol{\varsigma}}(k-1)}\right],$$

$$\beta(k) = \beta_0\beta(k-1) + (1-\beta_0), \tag{6.46}$$

and $\mathbf{P}_2(-1) > 0$, $\beta(0) = 0.95$, $\beta_0 = 0.99$.

According to (6.44), the prediction algorithm is given as

$$\hat{\boldsymbol{\phi}}_{p,L}(k+i) = \boldsymbol{\Gamma}_1^T(k)\hat{\boldsymbol{\phi}}_{p,L}(k+i-1) + \boldsymbol{\Gamma}_2^T(k)\hat{\boldsymbol{\phi}}_{p,L}(k+i-2) + \cdots$$

$$+\boldsymbol{\Gamma}_{n_p}^T(k)\hat{\boldsymbol{\phi}}_{p,L}(k+i-n_p), \quad i = 1,\ldots,N-1. \tag{6.47}$$

6.3.1.3 Control Scheme

By integrating controller algorithm (6.40)–(6.41) with parameter estimation algorithm (6.43), and parameter prediction algorithm (6.45)–(6.47), the PFDL–MFAPC scheme is designed as follows:

$$\hat{\boldsymbol{\phi}}_{p,L}(k) = \hat{\boldsymbol{\phi}}_{p,L}(k-1) + \frac{\mathbf{P}_1(k-2)\Delta\boldsymbol{U}_L(k-1)}{\alpha(k-1) + \Delta\boldsymbol{U}_L^T(k-1)\mathbf{P}_1(k-2)\Delta\boldsymbol{U}_L(k-1)}$$

$$\times\left[\Delta y(k) - \Delta\boldsymbol{U}_L^T(k-1)\hat{\boldsymbol{\phi}}_{p,L}(k-1)\right],$$

$$\mathbf{P}_1(k-1) = \frac{1}{\alpha(k-1)}\left[\mathbf{P}_1(k-2) - \frac{\mathbf{P}_1(k-2)\Delta\boldsymbol{U}_L(k-1)\Delta\boldsymbol{U}_L^T(k-1)\mathbf{P}_1(k-2)}{\alpha(k-1) + \Delta\boldsymbol{U}_L^T(k-1)\mathbf{P}_1(k-2)\Delta\boldsymbol{U}_L(k-1)}\right],$$

$$\alpha(k) = \alpha_0\alpha(k-1) + (1-\alpha_0),$$

(6.48)

$$\Lambda(k) = \Lambda(k-1) + \frac{\mathbf{P}_2(k-2)\hat{\boldsymbol{\varsigma}}(k-1)}{\beta(k-1) + \hat{\boldsymbol{\varsigma}}^T(k-1)\mathbf{P}_2(k-2)\hat{\boldsymbol{\varsigma}}(k-1)}$$
$$\times\left[\hat{\boldsymbol{\phi}}^T_{p,L}(k) - \hat{\boldsymbol{\varsigma}}^T(k-1)\boldsymbol{\Lambda}(k-1)\right],$$
$$\mathbf{P}_2(k-1) = \frac{1}{\beta(k-1)}\left[\mathbf{P}_2(k-2) - \frac{\mathbf{P}_2(k-2)\hat{\boldsymbol{\varsigma}}(k-1)\hat{\boldsymbol{\varsigma}}^T(k-1)\mathbf{P}_2(k-2)}{\beta(k-1) + \hat{\boldsymbol{\varsigma}}^T(k-1)\mathbf{P}_2(k-2)\hat{\boldsymbol{\varsigma}}(k-1)}\right],$$
$$\beta(k) = \beta_0\beta(k-1) + (1-\beta_0), \tag{6.49}$$

$$\hat{\boldsymbol{\phi}}_{p,L}(k+i) = \boldsymbol{\Gamma}_1(k)\hat{\boldsymbol{\phi}}_{p,L}(k+i-1) + \boldsymbol{\Gamma}_2(k)\hat{\boldsymbol{\phi}}_{p,L}(k+i-2) + \cdots$$
$$+\boldsymbol{\Gamma}_{np}(k)\hat{\boldsymbol{\phi}}_{p,L}(k+i-n_p). \quad i = 1,\ldots,N-1 \tag{6.50}$$

$$\Delta\tilde{U}_{N_u}(k) = (\hat{\tilde{\Psi}}^T(k)\hat{\tilde{\Psi}}(k) + \lambda I)^{-1}\hat{\tilde{\Psi}}^T(k)(\tilde{Y}_N^*(k+1) - \mathrm{E}y(k) - \hat{\bar{\Psi}}(k)\Delta U_L(k-1)), \tag{6.51}$$

$$u(k) = u(k-1) + \boldsymbol{g}^T\Delta\tilde{\boldsymbol{U}}_{N_u}(k), \tag{6.52}$$

where $\mathbf{P}_1(-1) > 0$, $\alpha(0) = 0.95$, $\alpha_0 = 0.99$, $\mathbf{P}_2(-1) > 0$, $\beta(0) = 0.95$, $\beta_0 = 0.99$, and $\lambda > 0$.

Remark 6.5

Discussions similar to Remark 6.3 could be listed here.

6.3.2 Simulation Results

Simulations are presented to illustrate the effectiveness and the superiorities of PFDL–MFAPC scheme (6.48)–(6.52) to PID, PFDL–MFAC scheme (4.26)–(4.28), and neural-network-based methods for dealing with the adaptive control problem of the controlled system with time-varying structure, time-varying order, and time-varying phase.

In the following two examples, the models are merely used to generate the I/O data rather than to design controllers. The same PFDL–MFAPC scheme and the same initial conditions and parameter settings are used in the following two examples. The parameters in the PFDL–MFAPC scheme are $L = 5$, $N = 5$, $N_u = 10$, $\lambda = 1$, $\alpha(0) = 0.95$, and $\alpha_0 = 0.99$, the initial values of PG are $\hat{\boldsymbol{\phi}}_{p,L}(1) = [0.5,0.1,0.2,0,0.3]^T$,

$\hat{\boldsymbol{\phi}}_{p,L}(2) = [0.3,0.2,0,0.4,0.1]^T$, $\hat{\boldsymbol{\phi}}_{p,L}(3) = [0.6,0,0.1,0.3,0.2]^T$, and $\hat{\boldsymbol{\phi}}_{p,L}(4) = [0.2,0.1,0.2,0.9,0]^T$. The initial values of all the elements in $\boldsymbol{\Lambda}(k)$ are set to a random number in (0,1). The initial variances are $\mathbf{P}_1(-1) = 10\mathbf{I}$ and $\mathbf{P}_2(-1) = 100\mathbf{I}$. The parameters of the PFDL–MFAC scheme are $L = 5$, $\hat{\boldsymbol{\phi}}_{p,L}(1) = [0.5,0,0,0,0]^T$, $\eta = 0.5$, and $\rho = 1$.

Example 6.3

Discrete-time nonlinear system

$$
y(k+1) = \begin{cases}
\dfrac{2.5y(k)y(k-1)}{1+y(k)^2+y(k-1)^2} + 0.7\sin\big(0.5\big(y(k)+y(k-1)\big)\big) \\
\quad \times \cos\big(0.5\big(y(k)+y(k-1)\big)\big) + 1.2u(k) + 1.4u(k-1), \\
\qquad\qquad\qquad\qquad\qquad\qquad\qquad 1 \le k \le 250 \\
\dfrac{2.5y(k)y(k-1)}{1+y(k)^2+y(k-1)^2} + 0.7\sin\big(0.5\big(y(k)+y(k-1)\big)\big) \\
\quad \times \cos\big(0.5\big(y(k)+y(k-1)\big)\big) + 1.2u(k-2) + 1.4u(k-3), \\
\qquad\qquad\qquad\qquad\qquad\qquad\qquad 250 < k \le 500 \\
\dfrac{5y(k)y(k-1)}{1+y(k)^2+y(k-1)^2+y(k-2)^2} + u(k) + 1.1u(k-1), \\
\qquad\qquad\qquad\qquad\qquad\qquad\qquad 500 < k \le 750 \\
-0.1y(k) - 0.2y(k-1) - 0.3y(k-2) + 0.1u(k-2) \\
\quad + 0.02u(k-3) + 0.03u(k-4), \qquad 750 < k \le 1000.
\end{cases} \tag{6.53}
$$

The system consists of four subsystems in series, which is a nonlinear system with time-varying structure, time-varying order, time-varying phase, and time-varying delay. Obviously, the dynamics of the system changes dramatically. What is more, it cannot be controlled effectively by applying any existing traditional model based adaptive control methods, including the neural network control method [162].

The desired trajectory is given as follows:

$$y^*(k+1) = 5 \times (-1)^{\text{round}(k/80)}.$$

Simulation results with PID controller are shown in Figure 6.5. In this simulation, finely considering the trade-off between the oscillation of the second subsystem and the transient behavior of the first subsystem, the PID parameters are tuned carefully to be $K_P = 0.1, T_I = 1$, and $T_D = 1$.

Figures 6.6 and 6.7 show the simulation results of applying the PFDL–MFAC and PFDL–MFAPC schemes with $\lambda = 5$; $\mu = 10$, respectively.

From the simulation results, we can observe that the proposed PFDL–MFAPC scheme has the best control performance, compared with the PID and PFDL–MFAC schemes, especially transition process switching from the third subsystem to the fourth subsystems. The effect of prediction is shown in Figure 6.7 evidently. In addition, it is difficult to tune the PID controller parameters, since the control performance is sensitive to these parameters. In contrast, tuning λ in the PFDL–MFAC scheme and the PFDL–MFAPC scheme is relatively easier, and the control performance is robust to it.

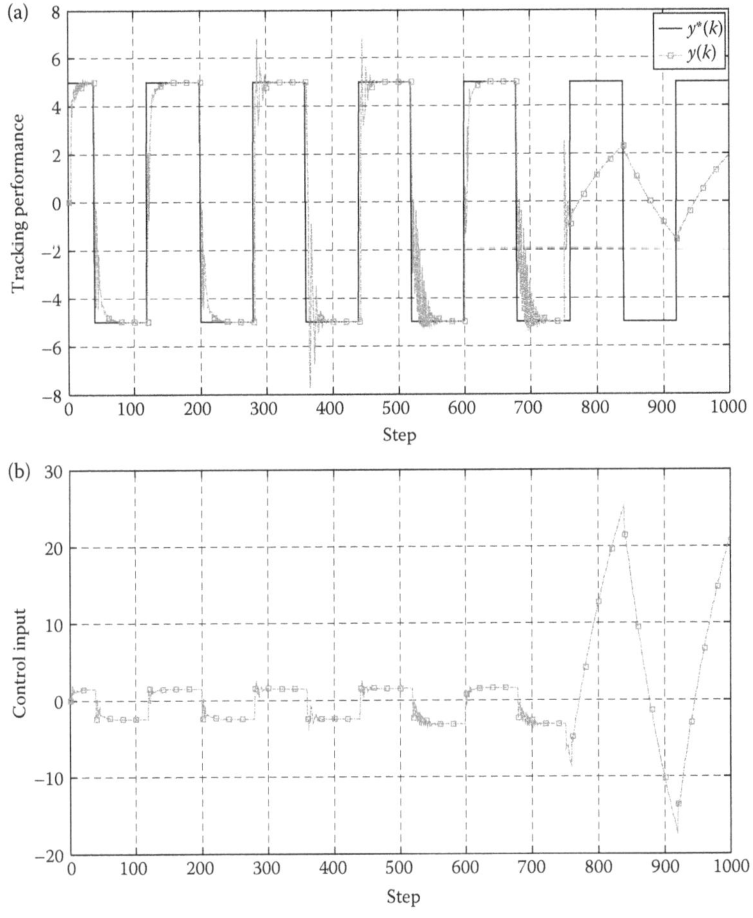

Figure 6.5 Simulation results of PID in Example 6.3. (a) Tracking performance. (b) Control input.

Example 6.4

Discrete-time nonlinear system

$$y(k+1)=\begin{cases}\dfrac{y(k)y(k-1)y(k-2)u(k-1)\big(y(k)-1\big)+u(k)}{1+y(k)^2+y(k-1)^2+y(k-2)^2}, & 1\le k\le 400\\[2mm] \dfrac{y(k)}{1+y(k)^2}+u(k)^3, & 400<k\le 600\\[2mm] 0.55y(k)-0.46y(k-1)-0.07y(k-2)+0.1u(k) & \\ \quad +0.02u(k-1)+0.03u(k-2), & 600<k\le 800\\ -0.1y(k)-0.2y(k-1)-0.3y(k-3)+0.1u(k-2) & \\ \quad +0.02u(k-3)+0.03u(k-4), & 800<k\le 1200\end{cases} \tag{6.54}$$

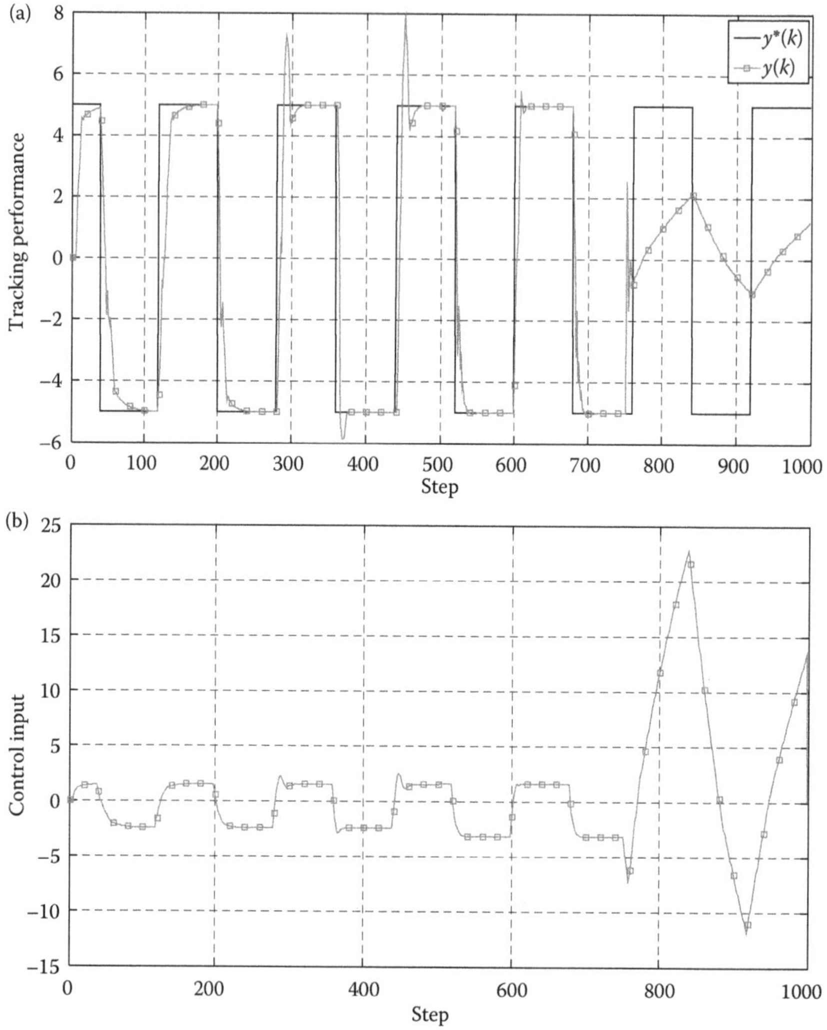

Figure 6.6 Simulation results of PFDL–MFAC scheme in Example 6.3. (a) Tracking performance. (b) Control input.

System (6.54) consists of four subsystems in series. The first two subsystems are taken from Refs. [162,165], respectively. The other two subsystems are taken from Ref. [168]. They are controlled by the neural network control method in the above-mentioned literature individually. The first two subsystems are nonlinear systems with respect to the control input, and the last two subsystems are linear systems, and furthermore the third one is open-loop unstable. Obviously, the system's dynamics, such as structure, order, input delay, and open-loop stability, changes dramatically with time interval.

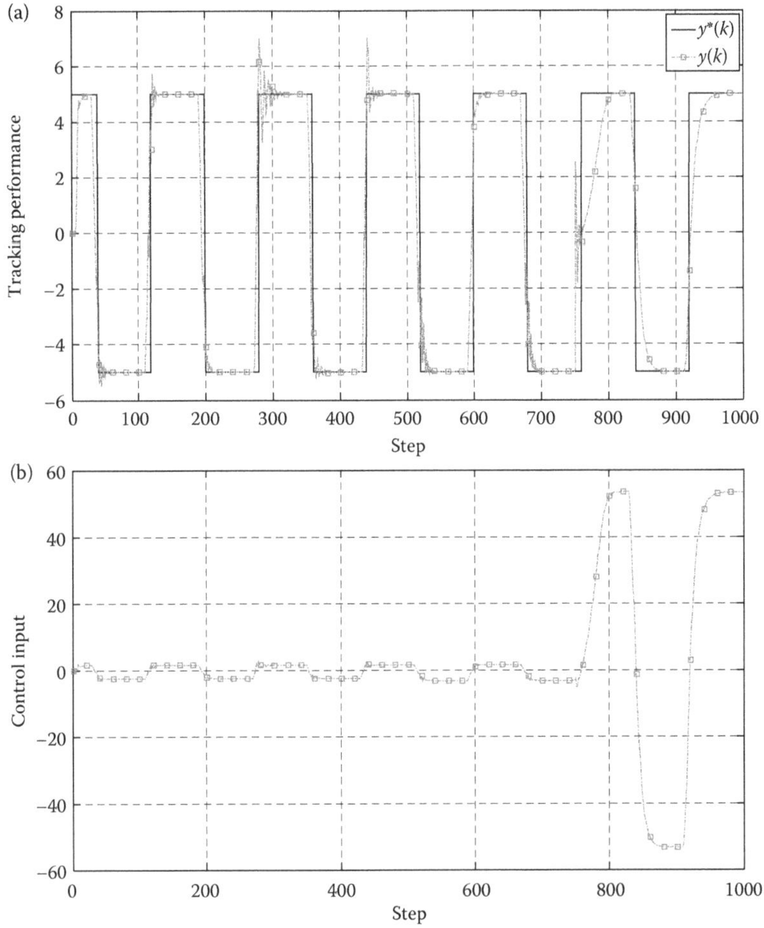

Figure 6.7 Simulation results of PFDL–MFAPC scheme in Example 6.3. (a) Tracking performance. (b) Control input.

The desired trajectory is given as follows:

$$y^*(k+1) = \begin{cases} 0.5 \times (-1)^{\text{round}(k/50)}, & 1 \le k \le 600 \\ 2\sin\left(\dfrac{k}{50}\right) + \cos\left(\dfrac{k}{20}\right), & 600 < k \le 1000 \\ 0.5 \times (-1)^{\text{round}(k/50)}, & 1000 < k \le 1200 \end{cases} \tag{6.55}$$

The control performance of the PFDL–MFAPC scheme is shown in Figure 6.8. It is observed that the overshoot is small, the transient process is relatively smooth, and the control performance in this simulation is slightly superior to that of the neural network control separately. Note that the spikes around the 400th, 600th, and 800th steps are caused by the changing of the model structure.

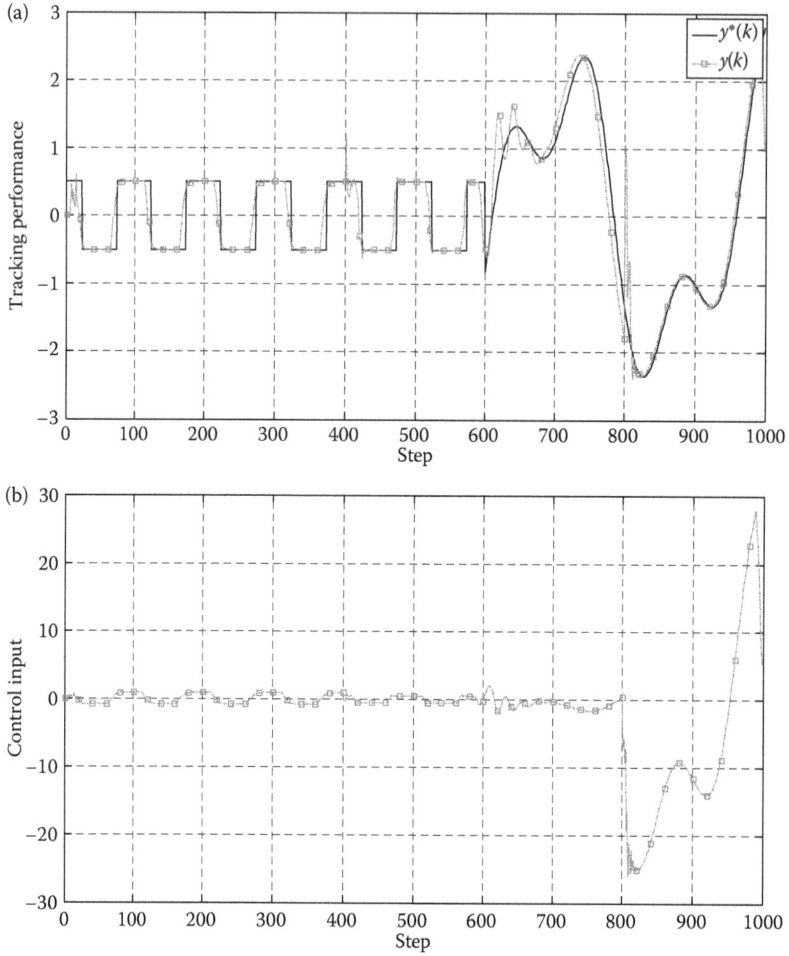

Figure 6.8 Simulation results of PFDL–MFAPC scheme in Example 6.4. (a) Tracking performance. (b) Control input.

6.4 FFDL Data Model Based MFAPC

6.4.1 Control System Design

On the basis of the conclusion in Section 3.2.3, discrete-time nonlinear SISO system (6.1) can be transformed into the following equivalent FFDL data model under certain assumptions:

$$\Delta y(k+1) = \boldsymbol{\phi}^T_{f,L_y,L_u}(k)\Delta \boldsymbol{H}_{L_y,L_u}(k).$$

According to the above equivalent incremental data model, the one-step-ahead prediction equation is given as

$$y(k+1) = y(k) + \boldsymbol{\phi}^T_{f,L_y,L_u}(k)\Delta \boldsymbol{H}_{L_y,L_u}(k), \tag{6.56}$$

where $\Delta \boldsymbol{H}_{L_y,L_u}(k) = \left[\Delta y(k),\ldots,\Delta y(k-L_y+1),\Delta u(k),\ldots,\Delta u(k-L_u+1)\right]^T$, PG $\boldsymbol{\phi}_{f,L_y,L_u}(k) = \left[\phi_1(k) \quad \cdots \quad \phi_{L_y}(k) \quad \phi_{L_y+1}(k) \quad \cdots \quad \phi_{L_y+L_u}(k)\right]^T$. L_y and L_u $(0 \le L_y \le n_y$, $1 \le L_u \le n_u)$ are *pseudo orders.*

Since system (6.56) is time-varying, it is hard to satisfy the assumption that PG is invariant in the prediction horizon. Consequently, the common *Diophantine* equation technique in the GPC cannot be applied in this case. Similar to the prediction method used in the preceding sections, let

$$\mathbf{A} = \begin{bmatrix} 0 & & & \\ 1 & 0 & & \\ & \ddots & \ddots & \\ & & 1 & 0 \end{bmatrix}_{L_u \times L_u}, \mathbf{B} = \begin{bmatrix} 1 \\ 0 \\ \vdots \\ 0 \end{bmatrix}_{L_u \times 1}, \mathbf{C} = \begin{bmatrix} 0 & & & \\ 1 & 0 & & \\ & \ddots & \ddots & \\ & & 1 & 0 \end{bmatrix}_{L_y \times L_y}, \mathbf{D} = \begin{bmatrix} 1 \\ 0 \\ \vdots \\ 0 \end{bmatrix}_{L_y \times 1},$$

$$\Delta Y_{L_y}(k) = \left[\Delta y(k),\ldots,\Delta y(k-L_y+1)\right]^T \in R^{Ly},$$
$$\Delta U_{L_u}(k) = \left[\Delta u(k),\ldots,\Delta u(k-L_u+1)\right]^T \in R^{Lu},$$

$$\boldsymbol{\phi}_{fy}(k) = \left[\phi_1(k) \quad \cdots \quad \phi_{L_y}(k)\right]^T,$$
$$\boldsymbol{\phi}_{fu}(k) = \left[\phi_{L_y+1}(k) \quad \cdots \quad \phi_{L_y+L_u}(k)\right]^T,$$

then (6.56) can be rewritten as

$$\Delta y(k+1) = \boldsymbol{\phi}^T_{fy}(k)\Delta Y_{L_y}(k) + \boldsymbol{\phi}^T_{fu}(k)\mathbf{A}\Delta U_{L_u}(k-1) + \boldsymbol{\phi}^T_{fu}(k)\mathbf{B}\Delta u(k). \tag{6.57}$$

Thus, N-step-ahead prediction equations can be obtained:

$$\begin{aligned}\Delta y(k+2) &= \boldsymbol{\phi}^T_{fy}(k+1)\Delta Y_{L_y}(k+1) + \boldsymbol{\phi}^T_{fu}(k+1)\mathbf{A}\Delta U_{L_u}(k) + \boldsymbol{\phi}^T_{fu}(k+1)\mathbf{B}\Delta u(k+1) \\ &= \boldsymbol{\phi}^T_{fy}(k+1)\mathbf{C}\Delta Y_{L_y}(k) + \boldsymbol{\phi}^T_{fy}(k+1)\mathbf{D}\Delta y(k+1) + \boldsymbol{\phi}^T_{fu}(k+1)\mathbf{A}^2\Delta U_{L_u}(k-1) \\ &\quad + \boldsymbol{\phi}^T_{fu}(k+1)\mathbf{AB}\Delta u(k) + \boldsymbol{\phi}^T_{fu}(k+1)\mathbf{B}\Delta u(k+1)\end{aligned}$$

$$
\begin{aligned}
\Delta y(k+3) &= \boldsymbol{\phi}_{fy}^T(k+2)\Delta \boldsymbol{Y}_{L_y}(k+2) + \boldsymbol{\phi}_{fu}^T(k+2)\mathbf{A}\Delta \boldsymbol{U}_{L_u}(k+1) \\
&\quad + \boldsymbol{\phi}_{fu}^T(k+2)\mathbf{B}\Delta u(k+2) \\
&= \boldsymbol{\phi}_{fy}^T(k+2)\mathbf{C}^2\Delta \boldsymbol{Y}_{L_y}(k) + \boldsymbol{\phi}_{fy}^T(k+2)\mathbf{CD}\Delta y(k+1) \\
&\quad + \boldsymbol{\phi}_{fy}^T(k+2)\mathbf{D}\Delta y(k+2) \\
&\quad + \boldsymbol{\phi}_{fu}^T(k+2)\mathbf{A}^3\Delta \boldsymbol{U}_{L_u}(k-1) + \boldsymbol{\phi}_{fu}^T(k+2)\mathbf{A}^2\mathbf{B}\Delta u(k) \\
&\quad + \boldsymbol{\phi}_{fu}^T(k+2)\mathbf{AB}\Delta u(k+1) \\
&\quad + \boldsymbol{\phi}_{fu}^T(k+2)\mathbf{B}\Delta u(k+2) \\
&\ \ \vdots \\
\Delta y(k+N) &= \boldsymbol{\phi}_{fy}^T(k+N-1)\mathbf{C}^{N-1}\Delta \boldsymbol{Y}_{L_y}(k) + \boldsymbol{\phi}_{fu}^T(k+N-1)\mathbf{A}^N\Delta \boldsymbol{U}_{L_u}(k-1) \\
&\quad + \boldsymbol{\phi}_{fy}^T(k+N-1)\mathbf{C}^{N-2}\mathbf{D}\Delta y(k+1) + \cdots \\
&\quad + \boldsymbol{\phi}_{fy}^T(k+N-1)\mathbf{D}\Delta y(k+N-1) \\
&\quad + \boldsymbol{\phi}_{fu}^T(k+N-1)\mathbf{A}^{N-1}\mathbf{B}\Delta u(k) + \phi_{fu}^T(k+N-1)\mathbf{A}^{N-2}\mathbf{B}\Delta u(k+1) \\
&\quad + \boldsymbol{\phi}_{fu}^T(k+N-1)\mathbf{A}^{N-N_u}\mathbf{B}\Delta u(k+N_u-1).
\end{aligned}
\tag{6.58}
$$

6.4.1.1 Controller Algorithm

Let $\Delta\tilde{\boldsymbol{Y}}_N(k+1) = \tilde{\boldsymbol{Y}}_N(k+1) - \tilde{\boldsymbol{Y}}_N(k)$, $\tilde{\boldsymbol{Y}}_N(k+1) = \left[y(k+1),\ldots,y(k+N)\right]^T$, $\boldsymbol{E} = [1,1,\ldots,1]^T$, $\Delta\tilde{\boldsymbol{U}}_{N_u}(k) = \left[\Delta u(k),\ldots,\Delta u(k+N_u-1)\right]^T$,

$$
\Psi_1(k) = \begin{bmatrix} \boldsymbol{\phi}_{fy}^T(k) \\ \vdots \\ \boldsymbol{\phi}_{fy}^T(k+N-1)C^{N-1} \end{bmatrix}_{N\times L_y}, \quad \Psi_2(k) = \begin{bmatrix} \boldsymbol{\phi}_{fu}^T(k)A \\ \vdots \\ \boldsymbol{\phi}_{fu}^T(k+N-1)A^N \end{bmatrix}_{N\times L_u},
$$

$$
\Psi_3(k) =
\begin{bmatrix}
0 & & & \cdots & 0 \\
\boldsymbol{\phi}_{fy}^T(k+1)\mathbf{D} & 0 & & \cdots & 0 \\
\boldsymbol{\phi}_{fy}^T(k+2)\mathbf{CD} & \boldsymbol{\phi}_{fy}^T(k+2)\mathbf{D} & 0 & \cdots & 0 \\
\vdots & \vdots & \ddots & \ddots & \vdots \\
\boldsymbol{\phi}_{fy}^T(k+N-1)\mathbf{C}^{N-2}\mathbf{D} & \boldsymbol{\phi}_{fy}^T(k+N-1)\mathbf{C}^{N-3}\mathbf{D} & & \boldsymbol{\phi}_{fy}^T(k+N-1)\mathbf{D} & 0
\end{bmatrix}_{N\times N},
$$

$$
\Psi_4(k) = \begin{bmatrix}
\boldsymbol{\phi}_{fu}^T(k)\mathbf{B} & & \\
\boldsymbol{\phi}_{fu}^T(k+1)\mathbf{AB} & \boldsymbol{\phi}_{fu}^T(k+1)\mathbf{B} & \\
\vdots & & \ddots \\
\boldsymbol{\phi}_{fu}^T(k+N_u-1)A^{N_u-1}\mathbf{B} & & \boldsymbol{\phi}_{fu}^T(k+N_u-1)\mathbf{B} \\
\vdots & & \vdots \\
\boldsymbol{\phi}_{fu}^T(k+N-1)\mathbf{A}^{N-1}\mathbf{B} & & \boldsymbol{\phi}_{fu}^T(k+1)\mathbf{A}^{N-N_u}\mathbf{B}
\end{bmatrix}_{N\times N_u},
$$

then the prediction equations can be rewritten in a compact form:

$$\Delta\tilde{\boldsymbol{Y}}_N(k+1) = \boldsymbol{\Psi}_1(k)\Delta\boldsymbol{Y}_{L_y}(k) + \boldsymbol{\Psi}_2(k)\Delta\boldsymbol{U}_{L_u}(k-1) + \boldsymbol{\Psi}_3(k)\Delta\tilde{\boldsymbol{Y}}_N(k+1) + \boldsymbol{\Psi}_4(k)\Delta\tilde{\boldsymbol{U}}_{N_u}(k), \tag{6.59}$$

or

$$\tilde{\boldsymbol{Y}}_N(k+1) = \tilde{\boldsymbol{Y}}_N(k) + \left(I - \boldsymbol{\Psi}_3(k)\right)^{-1} \times \left(\boldsymbol{\Psi}_1(k)\Delta\boldsymbol{Y}_{L_y}(k) + \boldsymbol{\Psi}_2(k)\Delta\boldsymbol{U}_{L_u}(k-1) + \boldsymbol{\Psi}_4(k)\Delta\tilde{\boldsymbol{U}}_{N_u}(k)\right). \tag{6.60}$$

Cost function (6.7) is still applied here for predictive control design. Substituting (6.60) into cost function (6.7), differentiating cost function (6.7) with respect to $\tilde{\boldsymbol{U}}_{N_u}(k)$ and setting it to be zero, yields

$$\begin{aligned}\Delta\tilde{\boldsymbol{U}}_{N_u}(k) = &\left[\left(\left(I - \boldsymbol{\Psi}_3(k)\right)^{-1}\boldsymbol{\Psi}_4(k)\right)^T\left(\left(I - \boldsymbol{\Psi}_3(k)\right)^{-1}\boldsymbol{\Psi}_4(k)\right) + \lambda I\right]^{-1} \\ &\times \left(\left(I - \boldsymbol{\Psi}_3(k)\right)^{-1}\boldsymbol{\Psi}_4(k)\right)^T \\ &\times \left\{\tilde{\boldsymbol{Y}}_N^*(k+1) - \tilde{\boldsymbol{Y}}_N(k) - \left(I - \boldsymbol{\Psi}_3(k)\right)^{-1} \right. \\ &\left. \times \left(\boldsymbol{\Psi}_1(k)\Delta\boldsymbol{Y}_{L_y}(k) + \boldsymbol{\Psi}_2(k)\Delta\boldsymbol{U}_{L_u}(k-1)\right)\right\}.\end{aligned} \tag{6.61}$$

Thus, the control input at the time instant k is given as follows:

$$u(k) = u(k-1) + \boldsymbol{g}^T\Delta\tilde{\boldsymbol{U}}_{N_u}(k), \tag{6.62}$$

where $\boldsymbol{g} = [1,0,\ldots,0]^T$.

6.4.1.2 PG Estimation Algorithm

PG $\boldsymbol{\phi}_{fy}(k)$ and $\boldsymbol{\phi}_{fu}(k)$ in $\boldsymbol{\Psi}_1(k), \boldsymbol{\Psi}_2(k), \boldsymbol{\Psi}_3(k)$, and $\boldsymbol{\Psi}_4(k)$, that is $\boldsymbol{\phi}_{f,L_y,L_u}(k)$, can be estimated by the following projection algorithm:

$$\hat{\boldsymbol{\phi}}_{f,L_y,L_u}(k) = \hat{\boldsymbol{\phi}}_{f,L_y,L_u}(k-1) + \frac{\eta\Delta\boldsymbol{H}_{L_y,L_u}(k-1)}{\mu + \left\|\Delta\boldsymbol{H}_{L_y,L_u}(k-1)\right\|^2} \times \left(\Delta y(k) - \hat{\boldsymbol{\phi}}^T_{f,L_y,L_u}(k-1)\Delta\boldsymbol{H}_{L_y,L_u}(k-1)\right), \tag{6.63}$$

where $\hat{\boldsymbol{\phi}}_{f,L_y,L_u}(k)$ is the estimation of $\boldsymbol{\phi}_{f,L_y,L_u}(k)$, $\mu > 0$, and $\eta \in (0,1]$.

Besides, PG $\phi_{fy}(k+i)$ and $\phi_{fu}(k+i)$ in $\Psi_1(k), \Psi_2(k), \Psi_3(k)$, and $\Psi_4(k)$, that is, $\boldsymbol{\phi}_{f,L_y,L_u}(k+i), i = 1, \ldots, N-1$, are still unknown, which are predicted by the multi-level hierarchical forecasting method, a similar method to the previous two sections.

Define

$$\boldsymbol{\Lambda}(k) = \left[\boldsymbol{\Gamma}_1(k), \ldots, \boldsymbol{\Gamma}_{n_p}(k)\right]^T \quad \text{and} \quad \hat{\boldsymbol{\varsigma}}(k) = \left[\hat{\boldsymbol{\phi}}^T_{f,L_y,L_u}(k-1), \ldots, \hat{\boldsymbol{\phi}}^T_{f,L_y,L_u}(k-n_p)\right]^T.$$

An AR model of the estimation sequence $\hat{\boldsymbol{\phi}}_{f,L_y,L_u}(k)$ can be constructed:

$$\hat{\boldsymbol{\phi}}_{f,L_y,L_u}(k) = \boldsymbol{\Lambda}(k)\,\hat{\boldsymbol{\varsigma}}(k). \tag{6.64}$$

On the basis of (6.64), we design the prediction algorithm for $\hat{\boldsymbol{\phi}}_{f,L_y,L_u}(k+i)$ $(i = 1, \ldots, N-1)$.

$$\begin{aligned}\hat{\boldsymbol{\phi}}_{f,L_y,L_u}(k+i) = {} & \boldsymbol{\Gamma}_1(k)\hat{\boldsymbol{\phi}}_{f,L_y,L_u}(k+i-1) + \boldsymbol{\Gamma}_2(k)\hat{\boldsymbol{\phi}}_{f,L_y,L_u}(k+i-2) + \cdots \\ & + \boldsymbol{\Gamma}_{np}(k)\hat{\boldsymbol{\phi}}_{f,L_y,L_u}(k+i-n_p),\end{aligned} \tag{6.65}$$

where the unknown matrix $\boldsymbol{\Lambda}(k) = \left[\boldsymbol{\Gamma}_1(k), \ldots, \boldsymbol{\Gamma}_{n_p}(k)\right]^T$ can be estimated by the following least-squares algorithm with the forgetting factor:

$$\begin{aligned}\boldsymbol{\Lambda}(k) &= \boldsymbol{\Lambda}(k-1) + \frac{\mathbf{P}(k-2)\hat{\boldsymbol{\varsigma}}(k-1)}{\alpha(k-1) + \hat{\boldsymbol{\varsigma}}^T(k-1)\mathbf{P}(k-2)\hat{\boldsymbol{\varsigma}}(k-1)} \\ &\quad \times \left[\hat{\boldsymbol{\phi}}^T_{f,L_y,L_u}(k) - \hat{\boldsymbol{\varsigma}}^T(k-1)\boldsymbol{\Lambda}(k-1)\right], \\ \mathbf{P}(k-1) &= \frac{1}{\alpha(k-1)}\left[\mathbf{P}(k-2) - \frac{\mathbf{P}(k-2)\hat{\boldsymbol{\varsigma}}(k-1)\hat{\boldsymbol{\varsigma}}^T(k-1)\mathbf{P}(k-2)}{\alpha(k-1) + \hat{\boldsymbol{\varsigma}}^T(k-1)\mathbf{P}(k-2)\hat{\boldsymbol{\varsigma}}(k-1)}\right], \\ \alpha(k) &= \alpha_0\alpha(k-1) + (1-\alpha_0),\end{aligned} \tag{6.66}$$

where $\mathbf{P}(-1) > 0$, $\alpha(0) = 0.95$, $\alpha_0 = 0.99$.

6.4.1.3 Control Scheme

By integrating controller algorithm (6.61)–(6.62) with parameter estimation algorithm (6.63), and parameter predictive algorithm (6.65)–(6.66), the FFDL–MFAPC scheme can be designed:

$$\hat{\boldsymbol{\phi}}_{f,L_y,L_u}(k) = \hat{\boldsymbol{\phi}}_{f,L_y,L_u}(k-1) + \frac{\eta \Delta \boldsymbol{H}_{L_y,L_u}(k-1)}{\mu + \left\| \Delta \boldsymbol{H}_{L_y,L_u}(k-1) \right\|^2} \times \left(\Delta y(k) - \hat{\boldsymbol{\phi}}_{f,L_y,L_u}(k-1) \Delta \boldsymbol{H}_{L_y,L_u}(k-1) \right), \tag{6.67}$$

$$\hat{\boldsymbol{\phi}}_{f,L_y,L_u}(k) = \hat{\boldsymbol{\phi}}_{f,L_y,L_u}(1), \quad \text{if } \left| \hat{\phi}_{L_y+1}(k) \right| \le \varepsilon \ \text{ or } \ \text{sign}(\hat{\phi}_{L_y+1}(k)) \ne \text{sign}(\hat{\phi}_{L_y+1}(1)), \tag{6.68}$$

$$\begin{aligned} \boldsymbol{\Lambda}(k) &= \boldsymbol{\Lambda}(k-1) + \frac{\mathbf{P}(k-2)\hat{\boldsymbol{\varsigma}}(k-1)}{\alpha(k-1) + \hat{\boldsymbol{\varsigma}}^T(k-1)\mathbf{P}(k-2)\hat{\boldsymbol{\varsigma}}(k-1)} \\ &\quad \times \left[\hat{\boldsymbol{\phi}}^T_{f,L_y,L_u}(k) - \hat{\boldsymbol{\varsigma}}^T(k-1)\boldsymbol{\Lambda}(k-1) \right], \\ \mathbf{P}(k-1) &= \frac{1}{\alpha(k-1)} \left[\mathbf{P}(k-2) - \frac{\mathbf{P}(k-2)\hat{\boldsymbol{\varsigma}}(k-1)\hat{\boldsymbol{\varsigma}}^T(k-1)\mathbf{P}(k-2)}{\alpha(k-1) + \hat{\boldsymbol{\varsigma}}^T(k-1)\mathbf{P}(k-2)\hat{\boldsymbol{\varsigma}}(k-1)} \right], \\ \alpha(k) &= \alpha_0 \alpha(k-1) + (1-\alpha_0), \end{aligned} \tag{6.69}$$

$$\hat{\boldsymbol{\phi}}_{f,L_y,L_u}(k+i) = \boldsymbol{\Gamma}_1(k)\hat{\boldsymbol{\phi}}_{f,L_y,L_u}(k+i-1) + \boldsymbol{\Gamma}_2(k)\hat{\boldsymbol{\phi}}_{f,L_y,L_u}(k+i-2) + \cdots + \boldsymbol{\Gamma}_{n_p}(k)\hat{\boldsymbol{\phi}}_{f,L_y,L_u}(k+i-n_p), \ i = 1,\ldots,N-1 \tag{6.70}$$

$$\hat{\boldsymbol{\phi}}_{f,L_y,L_u}(k+j) = \hat{\boldsymbol{\phi}}_{f,L_y,L_u}(1), \quad \text{if } \left| \hat{\phi}_{L_y+1}(k+j) \right| \le \varepsilon \ \text{ or } \ \text{sign}(\hat{\phi}_{L_y+1}(k+j)) \ne \text{sign}(\hat{\phi}_{L_y+1}(1)) \quad j = 1,\ldots,N-1 \tag{6.71}$$

$$\begin{aligned} \Delta \tilde{\boldsymbol{U}}_{N_u}(k) &= \left[\left((I - \Psi_3(k))^{-1} \Psi_4(k) \right)^T \left((I - \Psi_3(k))^{-1} \Psi_4(k) \right) + \lambda I \right]^{-1} \\ &\quad \times \left((I - \Psi_3(k))^{-1} \Psi_4(k) \right)^T \\ &\quad \times \left\{ \tilde{\boldsymbol{Y}}^*_N(k+1) - \tilde{\boldsymbol{Y}}_N(k) - (I - \Psi_3(k))^{-1} \right. \\ &\quad \left. \times \left(\Psi_1(k)\Delta \boldsymbol{Y}_{L_y}(k) + \Psi_2(k)\Delta \boldsymbol{U}_{L_u}(k-1) \right) \right\}, \end{aligned} \tag{6.72}$$

$$u(k) = u(k-1) + g^T \Delta\tilde{U}_{N_u}(k), \tag{6.73}$$

where ε is a positive constant, $\mathbf{P}(-1) > 0$, $\alpha(0) = 0.95$, $\alpha_0 = 0.99$, $\lambda > 0$, $\mu > 0$, and $\eta \in (0,1]$.

Remark 6.6

The selection of other controller parameters is similar to that of Remark 6.3.

Remark 6.7

If the model of the system is a known linear autoregressive integrated moving average (ARIMA) model.

$$A(q^{-1})\Delta y(k) = B(q^{-1})\Delta u(k-1), \tag{6.74}$$

and $L_y = n_y$ and $L_u = n_u$, then the prediction algorithm for PG is no longer needed (since PG is time-invariant), and the above algorithm becomes the classical GPC in this case.

Remark 6.8

If the variation law of the time-varying parameters is known in advance and it is applied to predict these parameters, then the predictive control scheme becomes the one in Ref. [202].

6.4.2 Simulation Results

An example is given to illustrate the effectiveness of FFDL–MFAPC scheme (6.67)–(6.73) for a nonlinear system with time-varying parameters. The comparison with FFDL–MFAC scheme (4.65)–(4.67) shows the prediction impact in the FFDL–MFAPC scheme.

Example 6.5

Discrete-time nonlinear system

$$y(k+1) = \frac{-0.9y(k) + (a(k) + 10)u(k)}{1 + y(k)^2}, \tag{6.75}$$

where $a(k) = 4\text{round}(k/100) + \sin(k/100)$ is a time-varying parameter, which makes the dynamics of the system change greatly.

In the FFDL–MFAPC scheme, the parameters are $L_y = 1$, $L_u = 1$, $N = 5$, $N_u = 1$, and $\lambda = 7$, the initial values of PG are $\hat{\boldsymbol{\phi}}_{f,L_y,L_u}(1) = [-1,10]^T$, and all the elements in Λ are set to a random number in (0, 1). The simulation results for this system with the FFDL–MFAC and FFDL–MFAPC schemes are shown in Figures 6.9 and 6.10, respectively.

Comparing Figure 6.9 with Figure 6.10, we can easily observe that by applying the FFDL–MFAPC scheme, the rising time is decreased and the spikes caused by variation of the model parameter are attenuated. A better control performance is generated by the FFDL–MFAPC scheme.

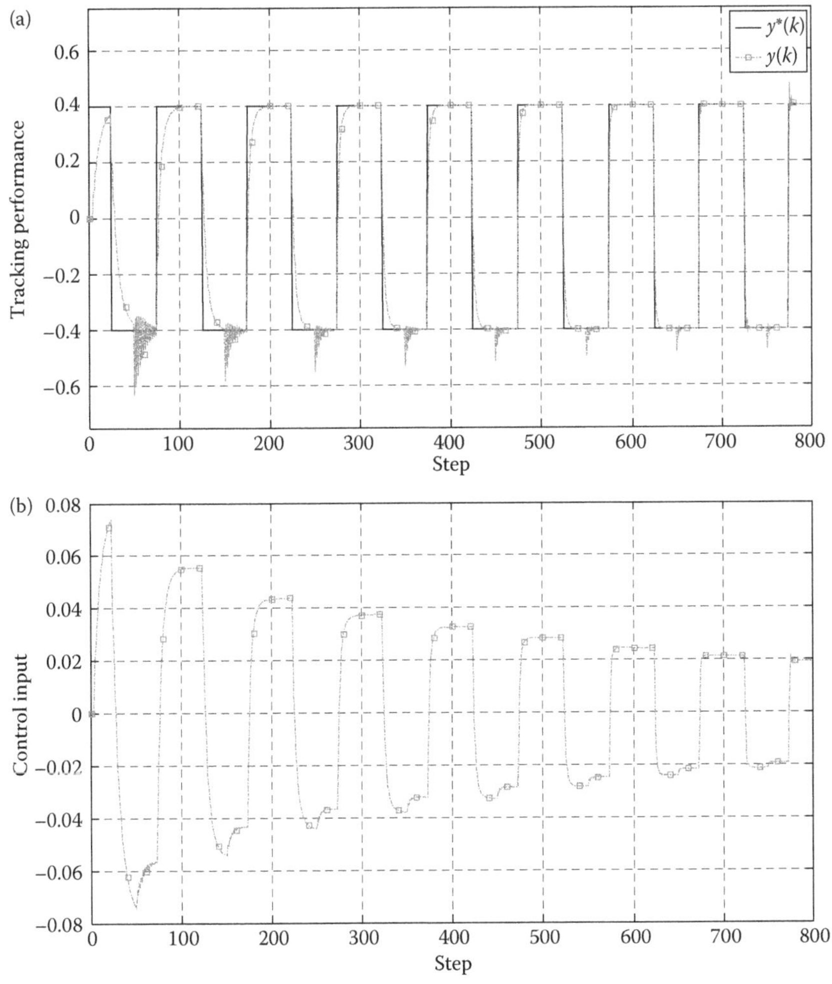

Figure 6.9 Simulation results of FFDL–MFAC scheme in Example 6.5. (a) Tracking performance. (b) Control input.

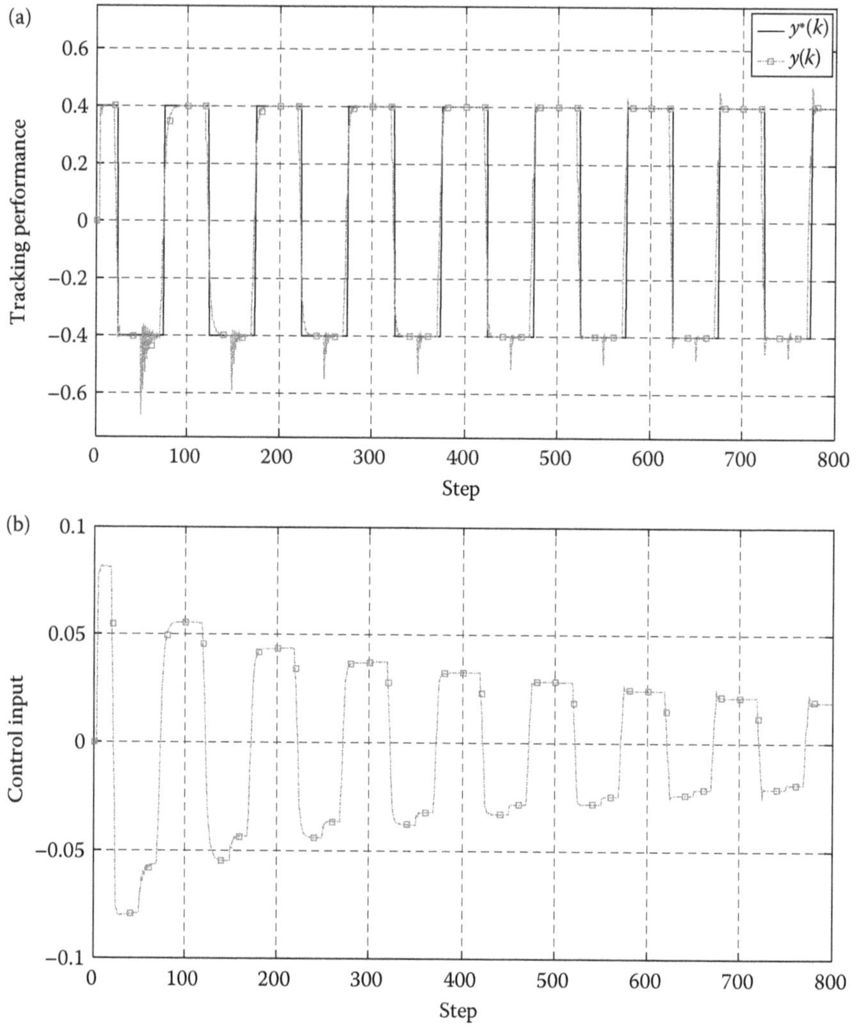

Figure 6.10 Simulation results of FFDL–MFAPC scheme in Example 6.5. (a) Tracking performance. (b) Control input.

6.5 Conclusions

Predictive control is the most widely applied advanced control method in the industrial process excluding the PID control technique; thus, theoretical and practical research on it is of much significance in process control. As we know, the predictive control theory and method for linear time-invariant systems have been developed well, but for the predictive control of nonlinear systems, there is still a lot of work that needs to be done. MFAPC is a kind of data-driven adaptive predictive control method, and its controller design merely depends on the measured I/O data of a closed-loop system without requiring any prior knowledge. Compared with existing MPC methods, the MFAPC scheme does not depend on the system model, and the unmodeled dynamics is not involved in the controller design, so it possesses stronger robustness and may have wider applications. Compared with the MFAC schemes in Chapter 4, the MFAPC schemes generate better control performance, since they utilize the future input and output predictive information of the controlled plant.

Chapter 7

Model-Free Adaptive Iterative Learning Control

7.1 Introduction

In practical applications, many plants repeat the same control tasks over a finite time interval, such as industrial robots, the wafer manufacturing process, the batch-to-batch process, train operation control, and so on. In other words, when a control task is performed repeatedly, the system dynamics will also be repeatable. In fact, the repetition can be viewed as a kind of "experience," which can be used efficiently to improve the tracking control performance. However, the traditional feedback control methods in the time domain, such as PID control, adaptive control, optimal control, and predict control, are not able to learn from the past repeated operations. As a result, although the repeatable plant runs many times over the finite time interval, the tracking error would be the same without any improvement. Instead, iterative learning control (ILC) of repeatable processes is able to refine the control signals at current iteration by utilizing the information of control signals and tracking errors of previous iterative operations. Consequently, the tracking error accuracy is improved as the number of repetitions increases.

ILC was originally proposed for robot manipulators by Arimoto [58] in 1984 by emulating the human learning process. It became one of the hot research topics in control theory and the control engineering field. In Refs. [63,66,203], the newest research results are briefly surveyed. In Refs. [59,62], the recent research results are introduced more comprehensively and systematically. With the increasing development and improvement in theory, ILC is widely applied to many control engineering fields [59,66,67,204–207].

The convergence analysis of ILC is an important theoretical issue. Up to now, three typical controller design methods with corresponding convergence analysis, including contraction mapping based ILC, composite energy function (CEF) based ILC, and optimization based ILC, have been developed. Most ILC methods utilize the contraction mapping and fixed point theory to design a linear iterative learning algorithm, and the pointwise convergence property of tracking error over a finite time interval is achieved in the sense of λ norm with assumptions that the controlled system dynamics is the global *Lipschitz* and the initial point satisfies identical initialization conditions. The controller design in these methods does not require the model information, and thus they can be regarded as a data-driven or model-free control approach. However, the limitation of these methods lies in the fact that the transient performance of the system output along the iteration axis becomes poor sometimes. Consequently, the application of these methods is limited. CEF based ILC relaxes the global *Lipschitz* condition locally by introducing the information of the system states. For the local *Lipschitz* nonlinear systems, the asymptotic convergence of the tracking error along the iteration axis is guaranteed by the *Lyapunov* analysis method under the identical initialization conditions. However, these methods require information of the system dynamics and state. Optimization based ILC is proposed for the known linear systems. The explicit optimal cost function is given and minimized to design the optimal ILC algorithm, and monotonic tracking error convergence along the iteration axis is guaranteed. Compared with contraction mapping based ILC, the CEF based ILC and optimization based ILC overcome the limitation of poor transient performance along the iteration axis. However, they require some *a priori* model information of the controlled systems, such as known linearization in parameter structures or the accurate linear system model, to design the ILC algorithm, and thus lost the ability to deal with the control problem for the unknown nonlinear systems.

In addition, the practical industrial process consists of many production facilities (or processes), linked with each other according to the technological requirement, and the complexity of the industrial process is increasing with the number of manufacturing facilities. Furthermore, the dynamic characteristic of the industrial process varies with the manufacturing conditions and is easily affected by exogenous disturbances. Thus, it is difficult to model an industrial process accurately. And even when the model is obtained, many existing ILC methods, including the CEF based ILC and optimization based ILC, still cannot be applied directly for such models due to the complex structures, high orders, and high nonlinearities.

In this chapter, by exploring the similarities between the MFAC and the ILC, a novel model-free adaptive iterative learning control (MFAILC) design and analysis approach is proposed based on an optimal cost function [179,208–211]. The proposed control approach still retains the desired data-driven model-free feature, and meanwhile possesses the ability to guarantee monotonic convergence of the output tracking error along the iteration axis for a large class of repetitive nonaffine nonlinear systems. The basic idea of the approach is shown as follows: First, the CFDL data model with a simple incremental form is given by introducing the concept of PPD along the iteration axis. Then, the CFDL data model based MFAILC scheme (CFDL–MFAILC) is designed. Theoretical analysis and numerical simulation show that the tracking error converges monotonically to zero along the iteration axis although the initial errors are randomly varying with iterations. The proposed approach can be easily extended to discrete-time MIMO nonlinear systems. In this chapter, only the results for discrete-time SISO nonlinear systems are given due to page limitation.

This chapter is organized as follows. In Section 7.2, the CFDL–MFAILC approach, including the compact form dynamic linearization method in the iteration domain, control system design, convergence analysis, and simulation results, is proposed for a class of discrete-time nonlinear systems. Section 7.3 presents the conclusions.

7.2 CFDL Data Model Based MFAILC

7.2.1 CFDL Data Model in the Iteration Domain

Consider a discrete-time SISO nonlinear system that operates repeatedly in a finite time interval as follows:

$$y(k+1,i) = f(y(k,i),\ldots,y(k-n_y,i),u(k,i),\ldots,u(k-n_u,i)), \tag{7.1}$$

where $u(k,i)$ and $y(k,i)$ are the control input and the system output at time instant k of the ith iteration, $k \in \{0,1,\ldots,T\}$, $i = 1,2,\ldots$. n_y and n_u are two unknown positive integers, and $f(\cdots)$ is an unknown nonlinear function.

Two assumptions are made on system (7.1) before the CFDL data model is elaborated.

Assumption 7.1

The partial derivative of $f(\cdots)$ with respect to the $(n_y + 2)$th variable is continuous.

Assumption 7.2

Suppose that $\forall k \in \{0,1,\ldots,T\}$ and $\forall i = 1,2,\ldots$, and when $|\Delta u(k,i)| \neq 0$, system (7.1) satisfies the generalized *Lipschitz* condition along the iteration axis, that is,

$$|\Delta y(k+1,i)| \le b|\Delta u(k,i)|, \tag{7.2}$$

where $\Delta y(k+1,i) = y(k+1,i) - y(k+1,i-1)$, $\Delta u(k,i) = u(k,i) - u(k,i-1)$; $b > 0$ is a finite positive constant.

From a practical point of view, these assumptions imposed on the plant are reasonable and acceptable. Assumption 7.1 is a typical condition for many control methods which a general nonlinear system should satisfy. Assumption 7.2 imposes an upper-bound limitation on the change rate of the system output driven by changes of the control inputs along the iteration axis. From an energy viewpoint, the energy change inside a system cannot go to infinity if the change of the control input energy is at a finite level. Many practical systems satisfy this assumption, such as the servo control system, temperature control system, pressure control system, traffic control system, and so on.

Theorem 7.1

Consider nonlinear system (7.1) satisfying Assumptions 7.1 and 7.2. If $|\Delta u(k,i)| \neq 0$, then there exists an iteration-dependent time-varying parameter $\phi_c(k,i)$, called PPD, such that system (7.1) can be transformed into the following CFDL data model:

$$\Delta y(k+1,i) = \phi_c(k,i)\Delta u(k,i), \forall k \in \{0,1,\ldots,T\}, \quad i = 1,2,\ldots \tag{7.3}$$

with bounded $\phi_c(k,i)$ for any time k and iteration i.

Proof

From the definition of $\Delta y(k+1,i)$ and (7.1), we have

$$\begin{aligned}
\Delta y(k+1,i) &= f(y(k,i), y(k-1,i),\ldots, y(k-n_y,i), u(k,i), u(k-1,i),\ldots, u(k-n_u,i)) \\
&\quad - f(y(k,i-1), y(k-1,i-1),\ldots, y(k-n_y,i-1), u(k,i-1), \\
&\quad u(k-1,i-1),\ldots, u(k-n_u,i-1)) \\
&= f(y(k,i), y(k-1,i),\ldots, y(k-n_y,i), u(k,i), u(k-1,i),\ldots, u(k-n_u,i)) \\
&\quad - f(y(k,i), y(k-1,i),\ldots, y(k-n_y,i), u(k,i-1), \\
&\quad u(k-1,i),\ldots, u(k-n_u,i)) \\
&\quad + f(y(k,i), y(k-1,i),\ldots, y(k-n_y,i), u(k,i-1), \\
&\quad u(k-1,i),\ldots, u(k-n_u,i)) \\
&\quad - f(y(k,i-1), y(k-1,i-1),\ldots, y(k-n_y,i-1), u(k,i-1), \\
&\quad u(k-1,i-1),\ldots, u(k-n_u,i-1)).
\end{aligned} \tag{7.4}$$

Denote

$$\begin{aligned}\xi(k,i) = {} & f(y(k,i),\ldots,y(k-n_y,i),u(k,i-1),u(k-1,i),\ldots,u(k-n_u,i)) \\ & - f(y(k,i-1),\ldots,y(k-n_y,i-1),u(k,i-1), \\ & u(k-1,i-1),\ldots,u(k-n_u,i-1)).\end{aligned} \tag{7.5}$$

By virtue of Assumption 7.1 and the differential mean value theorem, Equation (7.4) can be rewritten as

$$\Delta y(k+1,i) = \frac{\partial f^*}{\partial u(k,i)}(u(k,i) - u(k,i-1)) + \xi(k,i), \tag{7.6}$$

where $(\partial f^*)/(\partial u(k,i))$ denotes the partial derivative value of $f(\cdots)$ with respect to the $n_y + 2$ variable at a certain point between

$$[y(k,i),\ldots,y(k-n_y,i),u(k,i),u(k-1,i),\ldots,u(k-n_u,i)]^T$$

and

$$[y(k,i),\ldots,y(k-n_y,i),u(k,i-1),u(k-1,i),\ldots,u(k-n_u,i)]^T.$$

For every fixed iteration i and every fixed time k, consider the following equation with a variable $\eta(k,i)$:

$$\xi(k,i) = \eta(k,i)\Delta u(k,i). \tag{7.7}$$

Since $|\Delta u(k,i)| \neq 0$, there must exist a unique solution $\eta^*(k,i)$ to Equation (7.7).

Let $\phi_c(k,i) = (\partial f^*)/(\partial u(k,i)) + \eta^*(k,i)$. Then, Equation (7.6) is rewritten as $\Delta y(k+1,i) = \phi_c(k,i)\Delta u(k,i)$, which is the main conclusion of Theorem 7.1. The secondary concern of the boundedness of $\phi_c(k,i)$ is guaranteed directly by using Assumption 7.2. ■

Remark 7.1

From the proof of Theorem 7.1, we can see that $\phi_c(k,i)$ is related to the input and output signals till the time instant k of the $(i-1)$th and ith iterations. Thus, $\phi_c(k,i)$ is an iteration-related time-varying parameter. On the other hand, $\phi_c(k,i)$ can be considered as a differential signal in some sense and it is bounded for any k and

i. If the sampling period and $\Delta u(k,i)$ are not too large, $\phi_c(k,i)$ may be regarded as a slowly iteration-varying parameter; consequently, we can implement adaptive iterative learning control of the original system by designing a parameter estimator along the iteration axis.

7.2.2 Control System Design

7.2.2.1 Controller Algorithm

Given a desired trajectory $y_d(k)$, $k \in \{0,1,\ldots,T\}$, the control objective is to find a sequence of appropriate control inputs $u(k,i)$ such that the tracking error $e(k+1,i) = y_d(k+1) - y(k+1,i)$ converges to zero as the iteration number i approaches infinity.

Rewrite (7.3) as

$$y(k+1,i) = y(k+1,i-1) + \phi_c(k,i)\Delta u(k,i). \tag{7.8}$$

Consider the cost function of the control input as follows:

$$J(u(k,i)) = |e(k+1,i)|^2 + \lambda |u(k,i) - u(k,i-1)|^2, \tag{7.9}$$

where $\lambda > 0$ is a weighting factor, which is introduced to restrain the changing rate of the control input.

From (7.8) and the definition of $e(k+1,i)$, Equation (7.9) can be rewritten as

$$\begin{aligned} J(u(k,i)) &= |y_d(k+1) - y(k+1,i-1) - \phi_c(k,i)(u(k,i) - u(k,i-1))|^2 \\ &\quad + \lambda |u(k,i) - u(k,i-1)|^2 \\ &= |e(k+1,i-1) - \phi_c(k,i)(u(k,i) - u(k,i-1))|^2 + \lambda |u(k,i) - u(k,i-1)|^2. \end{aligned} \tag{7.10}$$

Using the optimal condition $(1/2)(\partial J/(\partial u(k,i))) = 0$, we have

$$u(k,i) = u(k,i-1) + \frac{\rho \phi_c(k,i)}{\lambda + |\phi_c(k,i)|^2} e(k+1,i-1), \tag{7.11}$$

where the step factor $\rho \in (0,1]$ is added to make the controller algorithm (7.11) more general.

7.2.2.2 PPD Iterative Updating Algorithm

Since $\phi_c(k,i)$ is not available, controller algorithm (7.11) cannot be applied directly. Thus, we present a cost function of PPD estimation as

$$J(\phi_c(k,i)) = \left|\Delta y(k+1,i-1) - \phi_c(k,i)\Delta u(k,i-1)\right|^2 + \mu\left|\phi_c(k,i) - \hat{\phi}_c(k,i-1)\right|^2, \tag{7.12}$$

where $\mu > 0$ is a weighting factor.

Using the optimal condition $(1/2)(\partial J/(\partial \hat{\phi}_c(k,i))) = 0$, we have

$$\hat{\phi}_c(k,i) = \hat{\phi}_c(k,i-1) + \frac{\eta \Delta u(k,i-1)}{\mu + \left|\Delta u(k,i-1)\right|^2} \times (\Delta y(k+1,i-1) - \hat{\phi}_c(k,i-1)\Delta u(k,i-1)), \tag{7.13}$$

where the step factor $\eta \in (0,1]$ is added to make the estimation algorithm (7.13) more general, and $\hat{\phi}_c(k,i)$ is the estimation value of $\phi_c(k,i)$.

On the basis of the PPD estimation algorithm (7.13), controller algorithm (7.11) is rewritten as

$$u(k,i) = u(k,i-1) + \frac{\rho \hat{\phi}_c(k,i)}{\lambda + \left|\hat{\phi}_c(k,i)\right|^2} e(k+1,i-1). \tag{7.14}$$

To cause the parameter estimation algorithm (7.13) to have a strong tracking ability, we present a resetting algorithm as follows:

$$\hat{\phi}_c(k,i) = \hat{\phi}_c(k,1), \quad \text{if } \hat{\phi}_c(k,i) \le \varepsilon \quad \text{or} \quad \Delta u(k,i-1) \le \varepsilon \quad \text{or} \quad \operatorname{sign}(\hat{\phi}_c(k,i)) \ne \operatorname{sign}(\hat{\phi}_c(k,1)), \tag{7.15}$$

where ε is a small positive constant and $\hat{\phi}_c(k,1)$ is the initial value of $\hat{\phi}_c(k,i)$.

7.2.2.3 CFDL–MFAILC Scheme

The CFDL–MFAILC scheme for nonlinear system (7.1) is constructed by integrating controller algorithm (7.14), parameter iterative updating algorithm (7.13), and resetting algorithm (7.15).

Remark 7.2

The CFDL–MFAILC scheme is designed only using I/O data of the plant. Hence, it is a data-driven model-free control approach. It is worth pointing out that the PPD estimation $\hat{\phi}_c(k,i)$ affects the learning gains in controller algorithm (7.14) virtually and can be iteratively calculated by iteratively updating algorithm (7.13) and resetting algorithm (7.15) together, which is quite different from the traditional ILC, where its learning gain is fixed and cannot be tuned automatically and iteratively.

7.2.3 Convergence Analysis

For the rigorous analysis of convergence, another assumption is imposed on the system as follows.

Assumption 7.3

$\forall k \in \{0,1,\ldots,T\}$ and $\forall i = 1,2,\ldots$, the parameter $\phi_c(k,i)$ satisfies that $\phi_c(k,i) > \underline{\varepsilon} > 0$ (or $\phi_c(k,i) < -\underline{\varepsilon} < 0$), where $\underline{\varepsilon}$ is a positive constant. Without loss of generality, we suppose that $\phi_c(k,i) > \underline{\varepsilon} > 0$.

Remark 7.3

Assumption 7.3 means that the system output does not decrease as the corresponding control input increases along the iteration axis, which may be treated as a kind of linear-like characteristic. In fact, many practical systems can satisfy this assumption such as temperature control, pressure control, and so on.

Theorem 7.2

If nonlinear system (7.1), satisfying Assumptions 7.1–7.3, is controlled by the CFDL–MFAILC scheme (7.13)–(7.15), then there exists $\lambda_{\min} > 0$ such that the following three properties hold for any $\lambda > \lambda_{\min}$.

a. $\hat{\phi}_c(k,i)$ is bounded $\forall k \in \{0,1,\ldots,T\}$ and $\forall i = 1,2,\ldots$.
b. The tracking error monotonically converges to zero in a pointwise manner over the finite time interval as i approaches infinity, that is, $\lim_{i\to\infty} |e(k+1,i)| = 0$.
c. $\{u(k,i)\}$ and $\{y(k,i)\}$ are bounded $\forall k \in \{0,1,\ldots,T\}$ and $\forall i = 1,2,\ldots$.

Proof

This proof consists of three steps. The first step is to prove the boundedness of $\hat{\phi}_c(k,i)$. The second step is to prove the pointwise monotonic convergence of tracking error. The third step is to prove the BIBO stability of the system.

Step 1 If $|\hat{\phi}_c(k,i)| \leq \varepsilon$ or $|\Delta u(k,i-1)| \leq \varepsilon$ or $\text{sign}(\hat{\phi}_c(k,i)) \neq \text{sign}(\hat{\phi}_c(k,1))$ then the boundedness of $\hat{\phi}_c(k,i)$ is obvious.

In other cases, define the PPD estimation error as $\tilde{\phi}_c(k,i) = \hat{\phi}_c(k,i) - \phi_c(k,i)$. Subtracting $\phi_c(k,i)$ from both sides of (7.13) yields

$$\begin{aligned}\tilde{\phi}_c(k,i) &= \tilde{\phi}_c(k,i-1) - (\phi_c(k,i) - \phi_c(k,i-1)) \\ &\quad + \frac{\eta\Delta u(k,i-1)}{\mu + |\Delta u(k,i-1)|^2} \times (\Delta y(k,i-1) - \hat{\phi}_c(k,i-1)\Delta u(k,i-1)).\end{aligned} \tag{7.16}$$

Let $\Delta\phi_c(k,i) = \phi_c(k,i) - \phi_c(k,i-1)$. Substituting the CFDL data model (7.3) into (7.16) yields

$$\begin{aligned}\tilde{\phi}_c(k,i) &= \tilde{\phi}_c(k,i-1) - \Delta\phi_c(k,i) \\ &\quad + \frac{\eta\Delta u(k,i-1)}{\mu + |\Delta u(k,i-1)|^2} \times (\phi_c(k,i-1)\Delta u(k,i-1) - \hat{\phi}_c(k,i-1)\Delta u(k,i-1)) \\ &= \tilde{\phi}_c(k,i-1) - \frac{\eta|\Delta u(k,i-1)|^2}{\mu + |\Delta u(k,i-1)|^2}\tilde{\phi}_c(k,i-1) - \Delta\phi_c(k,i) \\ &= \left(1 - \frac{\eta|\Delta u(k,i-1)|^2}{\mu + |\Delta u(k,i-1)|^2}\right)\tilde{\phi}_c(k,i-1) - \Delta\phi_c(k,i).\end{aligned} \tag{7.17}$$

Note that, for $0 < \eta \leq 1$ and $\mu > 0$, the function $(\eta|\Delta u(k,i-1)|^2)/(\mu + |\Delta u(k,i-1)|^2)$ is monotonically increasing with respect to $|\Delta u(k,i-1)|^2$ and its minimum value is $\eta\varepsilon^2/(\mu + \varepsilon^2)$. Thus, there exists a positive constant d_1 such that

$$0 < \left|\left(1 - \frac{\eta|\Delta u(k,i-1)|^2}{\mu + |\Delta u(k,i-1)|^2}\right)\right| \leq 1 - \frac{\eta\varepsilon^2}{\mu + \varepsilon^2} = d_1 < 1. \tag{7.18}$$

From Theorem 7.1, we can see that $|\phi_c(k,i)|$ is bounded by a constant $\bar{b}$, which leads to $|\phi_c(k,i) - \phi_c(k,i-1)| \leq 2\bar{b}$. Taking the absolute value on both sides of (7.17) and using (7.18) yield

$$\begin{aligned}|\tilde{\phi}_c(k,i)| &= \left|1 - \frac{\eta|\Delta u(k,i-1)|^2}{\mu + |\Delta u(k,i-1)|^2}\right| |\tilde{\phi}_c(k,i-1)| + |\Delta\phi_c(k,i)| \\ &\leq d_1 |\tilde{\phi}_c(k,i-1)| + 2\bar{b} \\ &\vdots \\ &\leq d_1^{i-1} |\tilde{\phi}_c(k,1)| + \frac{2\bar{b}}{1-d_1}. \end{aligned} \tag{7.19}$$

Thus, $\tilde{\phi}_c(k,i)$ is bounded. Then, $\forall k \in \{0,1,\ldots,T\}$ and $\forall i = 1,2, \ldots,$ and the boundedness of $\hat{\phi}_c(k,i)$ is also guaranteed since $\phi_c(k,i)$ is bounded.

Step 2 Using the CFDL data model (7.3), the tracking error is rewritten as follows:

$$\begin{aligned} e(k+1,i) &= y_d(k+1) - y(k+1,i) = y_d(k+1) - y(k+1,i-1) - \phi_c(k,i)\Delta u(k,i) \\ &= e(k+1,i-1) - \phi_c(k,i)\Delta u(k,i). \end{aligned} \tag{7.20}$$

Substituting controller algorithm (7.14) into (7.20) yields

$$e(k+1,i) = \left(1 - \phi_c(k,i)\frac{\rho\hat{\phi}_c(k,i)}{\lambda + |\hat{\phi}_c(k,i)|^2}\right)e(k+1,i-1). \tag{7.21}$$

Let $\lambda_{\min} = (\bar{b}^2/4)$. Since $\alpha^2 + \beta^2 \geq 2\alpha\beta$, there exists a positive constant M_1 such that the following inequality holds for $\lambda > \lambda_{\min}$:

$$0 < M_1 \leq \frac{\phi_c(k,i)\hat{\phi}_c(k,i)}{\lambda + |\hat{\phi}_c(k,i)|^2} \leq \frac{\bar{b}\,\hat{\phi}_c(k,i)}{\lambda + |\hat{\phi}_c(k,i)|^2} \leq \frac{\bar{b}\,\hat{\phi}_c(k,i)}{2\sqrt{\lambda}\,\hat{\phi}_c(k,i)} < \frac{\bar{b}}{2\sqrt{\lambda_{\min}}} = 1. \tag{7.22}$$

According to (7.22), $\rho \in (0,1]$, and $\lambda > \lambda_{\min}$, there exists a positive constant $d_2 < 1$ such that

$$\left|1 - \frac{\rho\phi_c(k,i)\hat{\phi}_c(k,i)}{\lambda + |\hat{\phi}_c(k,i)|^2}\right| = 1 - \frac{\rho\phi_c(k,i)\hat{\phi}_c(k,i)}{\lambda + |\hat{\phi}_c(k,i)|^2} \leq 1 - \rho M_1 \triangleq d_2 < 1. \tag{7.23}$$

Taking the absolute value on both sides of (7.21) and using (7.23) yield

$$
\begin{aligned}
|e(k+1,i)| &= \left|1 - \frac{\rho\phi_c(k,i)\hat{\phi}_c(k,i)}{\lambda + |\hat{\phi}_c(k,i)|^2}\right| |e(k+1,i-1)| \\
&\le d_2 |e(k+1,i-1)| \le \cdots \le d_2^{i-1} |e(k+1,1)|. \qquad (7.24)
\end{aligned}
$$

Inequality (7.24) implies that $e(k+1,i)$ converges to zero in a pointwise manner over the finite time interval as i approaches infinity.

Step 3 Since $y_d(k)$ is iteration-invariant, the convergence of $e(k,i)$ implies that $y(k,i)$ is also bounded.

From controller algorithm (7.14), we have

$$
\Delta u(k,i) = \frac{\rho\hat{\phi}_c(k,i)}{\lambda + |\hat{\phi}_c(k,i)|^2} e(k+1,i-1). \qquad (7.25)
$$

Using $(\sqrt{\lambda})^2 + |\hat{\phi}_c(k,i)|^2 \ge 2\sqrt{\lambda}\,\hat{\phi}_c(k,i)$ and $\lambda > \lambda_{\min}$, and taking the absolute value on both sides of (7.25) yield

$$
\begin{aligned}
|\Delta u(k,i)| &= \left|\frac{\rho\hat{\phi}_c(k,i)e(k+1,i-1)}{\lambda + |\hat{\phi}_c(k,i)|^2}\right| \le \left|\frac{\rho\hat{\phi}_c(k,i)}{2\sqrt{\lambda}\,\hat{\phi}_c(k,i)}\right| |e(k+1,i-1)| \\
&\le \left|\frac{\rho}{2\sqrt{\lambda_{\min}}}\right| |e(k+1,i-1)| = M_2 |e(k+1,i-1)|, \qquad (7.26)
\end{aligned}
$$

where M_2 is a bounded positive constant.

From (7.24) and (7.26), we have

$$
\begin{aligned}
|u(k,i)| &= |u(k,i) - u(k,1) + u(k,1)| \le |u(k,i) - u(k,1)| + |u(k,1)| \\
&= |u(k,i) - u(k,i-1) + u(k,i-1) \cdots - u(k,2) + u(k,2) - u(k,1)| + |u(k,1)| \\
&\le |\Delta u(k,i)| + |\Delta u(k,i-1)| + \cdots + |\Delta u(k,2)| + |u(k,1)| \\
&\le M_2 |e(k+1,i-1)| + M_2 |e(k+1,i-2)| + \cdots + M_2 |e(k+1,1)| + |u(k,1)| \\
&\le M_2 \frac{1}{1-d_2} |e(k+1,1)| + |u(k,1)|.
\end{aligned}
\qquad (7.27)
$$

This implies that $u(k,i)$ is bounded $\forall k \in \{0,1,\ldots,T\}$ and $\forall i = 1,2,\ldots$. ■

From the proof of Theorem 7.1, we can see that the CFDL–MFAILC scheme is suitable for a class of repetitive nonaffine nonlinear systems satisfying the generalized *Lipschitz* condition. The proposed scheme only utilizes the online measurement I/O data of the controlled system and does not require any information on the system dynamic model. The proposed scheme can achieve pointwise convergence over a finite time interval monotonically along the iteration axis, and overcome the limitations under identical initial conditions. In other words, it not only retains the property of monotonic convergence of optimization based ILC for the known linear time-invariant system but also keeps the ability of the original ILC to deal with the unknown repetitive nonlinear system. Furthermore, it also retains the data-driven and model-free feature of the original ILC.

7.2.4 Simulation Results

In this section, a numerical simulation is given to verify the correctness and effectiveness of the CFDL–MFAILC scheme for a discrete-time nonlinear system. It should be noted that the model information, including system structure (linear or nonlinear), orders, and relative degree, is not involved in the design of the CFDL–MFAILC scheme. The model in the following example is only used to generate the I/O data of the controlled plant.

Example 7.1

Discrete-time nonlinear system

$$y(k+1)=\begin{cases}\dfrac{y(k)}{1+y(k)^2}+u(k)^3, & 0\le k\le 50\\[2ex] \dfrac{y(k)y(k-1)y(k-2)u(k-1)(y(k-2)-1)+a(k)u(k)}{1+y(k-1)^2+y(k-2)^2}, & 50\le k\le 100\end{cases}\tag{7.28}$$

where $a(k) = 1 + \text{round}(k/50)$ is a time-varying parameter, $k = \{0,1,\ldots,100\}$.

The controlled system is a nonaffine nonlinear system with time-varying structure, time-varying order, and time-varying parameter.

The desired trajectory is given as

$$y_d(k+1)=\begin{cases}0.5\times(-1)^{\text{round}(k/10)}, & 0\le k\le 30\\ 0.5\sin(k\pi/10)+0.3\cos(k\pi/10), & 30<k\le 70\\ 0.5\times(-1)^{\text{round}(k/10)}, & 70<k\le 100.\end{cases}\tag{7.29}$$

The parameters in the CFDL–MFAILC scheme (7.13)–(7.15) are set to $\rho = 1$, $\lambda = 1$, $\eta = 1$, and $\mu = 1$. The control input of the first iteration is set to 0. The initial state value $y(0,i)$ is randomly varying in the interval [−0.05, 0.05] when iteration i evolves.

Figure 7.1 shows the profile of the initial value $y(0,i)$ over 100 iterations. Figure 7.2 shows the profile of the maximum learning error $e_{\max}(i) = \max\limits_{k\in\{1,\ldots,100\}} |e(k,i)|$. The effectiveness of the proposed scheme can be seen from Figures 7.1 and 7.2. Despite

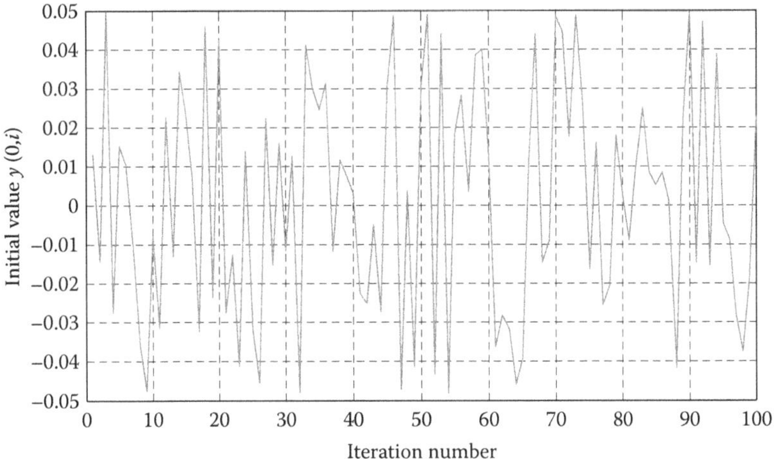

Figure 7.1 Profile of the random initial value in Example 7.1.

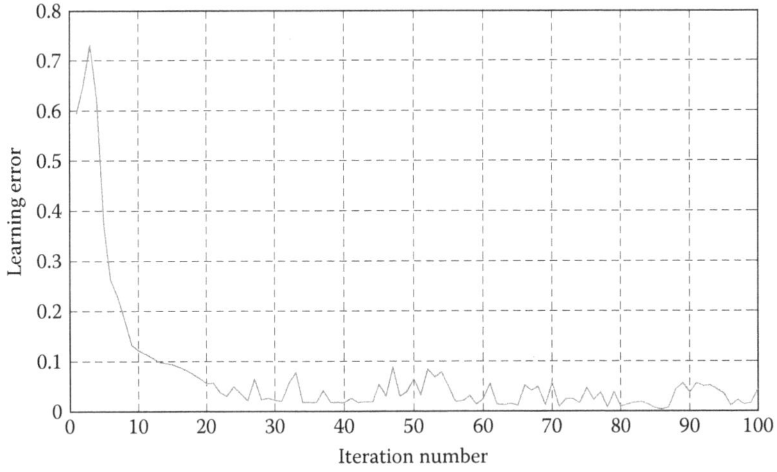

Figure 7.2 Profile of the maximum learning error $e_{max}(i)$ in Example 7.1.

the random initial values along the iteration axis, the tracking error converges asymptotically to zero along the iteration axis.

7.3 Conclusions

Motivated by the similarity in controller structure between MFAC in time domain and ILC in iteration domain, a new CFDL–MFAILC scheme is proposed in this chapter. Compared with the contraction mapping based ILC, the proposed MFAILC for unknown discrete-time nonlinear systems can guarantee monotonic tracking error convergence. Compared with the optimization based ILC, developed for known linear time-invariant systems, the MFAILC does not require the dynamic model of the controlled plant, and it is designed depending only on the I/O measurement data. Furthermore, the proposed approach can perform well when the initial state value is varying along the iteration axis. Both the theoretical analysis and simulations further verify the effectiveness of the proposed approach.

It is worth pointing out that the PFDL data model based MFAILC and the FFDL data model based MFAILC schemes can be easily obtained by using the similar procedure in Section 7.2 and they are omitted here due to page limition.

Chapter 8

Model-Free Adaptive Control for Complex Connected Systems and Modularized Controller Design

8.1 Introduction

With the development of computer and communication technology, practical engineering systems in many fields, such as the chemical industry, metallurgy, machinery, electricity, and transportation, have undergone significant changes. The scale has become larger and the structure has also become more complicated. A complex system usually consists of many subsystems, which may be widely distributed in space with complex connection structure. Consequently, it is difficult to obtain an accurate first-principles or the identified input output model of such a complex system, as well as satisfactory control performance by utilizing MBC methods, which renders traditional estimation and control theory difficult to be applied efficiently. With the help of modern information technology, mass data from the production process can be easily collected, stored, and processed. Thus, it is desired for researchers and engineers to study the data driven control method by effectively using these data.

There are two kinds of traditional control methods for complex systems: centralized control methods and decentralized control methods. For large-scale systems, the presupposition of centrality often fails to hold either due to the lack of centralized information or due to the lack of centralized computing capability. The positional separation and complicated links of such practical systems provide impetus for a decentralized scheme. Thus, for reasons of economics and reliability, decentralized decision making, distributed computation, and hierarchical control have become a trend [212–214].

For decentralized control methods, the local controller is also needed to design for each subsystem correspondingly. However, it is difficult to design an effective controller for whether centralized or decentralized systems, since accurate system models are difficult to build due to the fact that subsystems are distributed widely with complex connection structures [215–217]. Therefore, seeking a control method that does not rely on the model of the controlled plant becomes a new research direction [123,125]. It should be pointed out that most of complex systems are composed of some basic blocks, which are connected in series, parallel, and feedback form. Designing controllers for such interconnected systems with satisfactory performance is one of recent research focuses [218,219]. When a system is strongly coupled and cannot be decomposed into the aforementioned three basic connection structures, its control problem becomes more challenging. In this chapter, four types of interconnected systems are considered, and the corresponding MFAC methods are presented. In addition, both the DDC methods and the MBC methods have their own advantages and disadvantages. Therefore, MFAC based modularized controller design schemes, a complementarily cooperating mechanism between MFAC method and other control methods, are also discussed.

This chapter is organized as follows. In Section 8.2, data-driven model-free adaptive control algorithms, for systems in series connection, parallel connection, feedback connection, and complex interconnected systems, are studied, and their effectiveness is verified by simulations. In Section 8.3, complementary mechanisms based on different control algorithms are discussed and two types of modularized controller design schemes are proposed. The effectiveness of the proposed methods is verified by simulations. The conclusions are presented in Section 8.4.

8.2 MFAC for Complex Connected Systems

When a system consists of a few subsystems connected in a certain structure, it can be regarded as a new augmented large system. Since the MFAC is a DDC method, that is, the controller design does not depend on the model and the scale of the controlled plant, it should have the ability to deal with the control problem for this new large system. This may be referred as the centralized estimation and centralized control scheme hereafter. For the scheme, however, some helpful information, such as the known connection structure, for improving the control performance may be lost if we crudely apply the MFAC method to the augmented large system. Based on this observation, a series of novel MFAC schemes for complex systems, called the decentralized estimation and centralized control type MFAC methods, are proposed, and their correctness and effectiveness are also demonstrated with numerical simulations.

8.2.1 Series Connection

If subsystems of a complex system are connected in series, the control performance can be improved by utilizing the decentralized estimation and centralized control-type MFAC method. The main idea of this method is that, first, the dynamic linearization data model, taking the PFDL data model as an example, of each subsystem is built along the operation points; next, the PG of each subsystem is estimated by using local I/O data individually, and then the estimated PG vector in the PFDL data model of the complex system is constructed by virtue of the series connection structure. Finally, MFAC scheme is designed for the complex system using the integrated data model.

Consider series connection of the following two SISO discrete-time nonlinear subsystems as shown in Figure 8.1.

$$\text{P1}: y_1(k+1) = f_1(y_1(k),\ldots,y_1(k-n_{y_1}),u_1(k),\ldots,u_1(k-n_{u_1})), \tag{8.1}$$

$$\text{P2}: y_2(k+1) = f_2(y_2(k),\ldots,y_2(k-n_{y_2}),u_2(k),\ldots,u_2(k-n_{u_2})), \tag{8.2}$$

where $y_i(k) \in R$ and $u_i(k) \in R$ are the system output and the control input of the ith subsystem at time k, respectively; $f_i(\ldots)$ represents the dynamics of the ith subsystem; and n_{y_i} and n_{u_i} are two unknown integers, $i = 1,2$.

According to the analysis in Chapter 3, both the subsystems P1 and P2 can be equivalently transformed into one of the three kinds of dynamic linearization data models, including the CFDL data model, the PFDL data model, and the FFDL data model. In this section, the decentralized estimation and centralized control-type MFAC method is presented by taking the PFDL data model as an example due to page limitation.

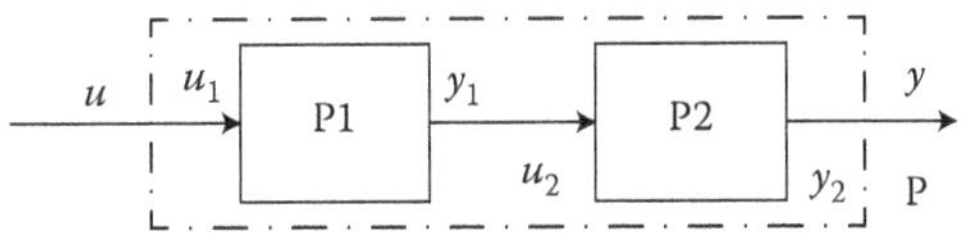

Figure 8.1 Block diagram of a system in series connection.

According to Theorem 3.3, the PFDL data models of two subsystems P1 and P2 are given as follows:

$$\mathrm{P1}: \Delta y_1(k+1) = \boldsymbol{\phi}^T_{1,p,L_1}(k)\Delta \boldsymbol{U}_{1,L_1}(k),$$

$$\mathrm{P2}: \Delta y_2(k+1) = \boldsymbol{\phi}^T_{2,p,L_2}(k)\Delta \boldsymbol{U}_{2,L_2}(k),$$

where $\Delta \boldsymbol{U}_{i,L_i}(k) = [\Delta u_i(k),\ldots,\Delta u_i(k-L_i+1)]^T \in R^{L_i}$, L_i is the control input linearization length constant of the ith subsystem; and $\boldsymbol{\phi}_{i,p,L_i}(k) = [\phi_{i1}(k),\ldots, \phi_{iL_i}(k)]^T \in R^{L_i}$ is the unknown PG of the ith subsystem, $i = 1,2$.

Since the two subsystems are connected in series, it means that $u = u_1$, $u_2 = y_1$, and $y = y_2$. Therefore, the equivalent PFDL data model of the augmented system can be described as follows:

$$\begin{aligned}
\Delta y(k+1) &= \boldsymbol{\phi}^T_{2,p,L_2}(k)\Delta \boldsymbol{U}_{2,L_2}(k) = \boldsymbol{\phi}^T_{2,p,L_2}(k)\begin{bmatrix} \Delta y_1(k) \\ \vdots \\ \Delta y_1(k-L_2+1)\end{bmatrix} \\
&= \boldsymbol{\phi}^T_{2,p,L_2}(k)\begin{bmatrix} \boldsymbol{\phi}^T_{1,p,L_1}(k-1)\Delta \boldsymbol{U}_{1,L_1}(k-1) \\ \vdots \\ \boldsymbol{\phi}^T_{1,p,L_1}(k-L_2)\Delta \boldsymbol{U}_{1,L_1}(k-L_2)\end{bmatrix} \\
&= \boldsymbol{\phi}^T_{2,p,L_2}(k)\begin{bmatrix} \phi_{11}(k-1)\Delta u(k-1) + \cdots + \phi_{1L_1}(k-L_1)\Delta u(k-L_1) \\ \vdots \\ \phi_{11}(k-L_2)\Delta u(k-L_2) + \cdots + \phi_{1L_1}(k-L_1-L_2+1)\Delta u(k-L_1-L_2+1)\end{bmatrix} \\
&= \phi_{21}(k)(\phi_{11}(k-1)\Delta u(k-1) + \cdots + \phi_{1L_1}(k-L_1)\Delta u(k-L_1)) \\
&\quad + \phi_{22}(k)(\phi_{11}(k-2)\Delta u(k-2) + \cdots + \phi_{1L_1}(k-L_1-1)\Delta u(k-L_1-1)) \\
&\quad + \cdots \\
&\quad + \phi_{2L_2}(k)(\phi_{11}(k-L_2)\Delta u(k-L_2) + \cdots + \phi_{1L_1}(k-L_1-L_2+1)\Delta u(k-L_1-L_2+1)).
\end{aligned} \tag{8.3}$$

Define two $L_1 + L_2 - 1$-dimensional vectors as follows:

$$\Delta \boldsymbol{U}_S(k-1) = \left[\Delta u(k-1),\ldots, \Delta u(k-L_1-L_2+1)\right]^T,$$

$$\boldsymbol{\phi}_S(k) = \left[\phi_1(k),\ldots, \phi_{L_1+L_2-1}(k)\right]^T,$$

where

$$\phi_i(k) = \begin{cases} \sum_{j=1}^{i} \phi_{1(i-j+1)}(k-j)\phi_{2j}(k), & 1 \le i \le L_1 \\ \sum_{j=i-L_1+1}^{L_2} \phi_{1(i-j+1)}(k-j)\phi_{2j}(k), & L_1 < i \le L_1 + L_2 - 1. \end{cases}$$

Then, the PFDL data model (8.3) can be rewritten in a compact form:

$$\Delta y(k+1) = \boldsymbol{\phi}_S^T(k)\Delta \boldsymbol{U}_S(k-1), \tag{8.4}$$

where $\boldsymbol{\phi}_S(k)$ is the PG of the augmented system.

Remark 8.1

If the system consists of N subsystems in series, its PFDL data model can be derived based on integrating local PFDL data models of all subsystems with similar derivation.

Using PFDL model (8.4), the PFDL–MFAC algorithm for the large augmented system is constructed:

$$u(k) = u(k-1) + \frac{\rho\hat{\phi}_1(k+1)}{\lambda + |\hat{\phi}_1(k+1)|^2} \times \left((y^*(k+2) - y(k) - \hat{\boldsymbol{\phi}}_S^T(k)\Delta \boldsymbol{U}_S(k-1) - \sum_{i=2}^{L_1+L_2-1} \hat{\phi}_i(k+1)\Delta u(k-i+1) \right), \tag{8.5}$$

where $\rho \in (0,1]$ is a step factor; $\lambda > 0$ is a weighting factor; $\hat{\boldsymbol{\phi}}_S(k)$ and $\hat{\phi}_i(k+1), i = 2,\ldots,L_1 + L_2 - 1$ are the estimation and prediction of $\boldsymbol{\phi}_S(k)$ and $\phi_i(k+1)$, $i = 2,\ldots,L_1 + L_2 - 1$, respectively.

The PG of each subsystem at current time k can be estimated by using the PG estimation algorithm (4.26) and resetting algorithm (4.27) in Chapter 4. Since the $\hat{\phi}_i(k+1), i = 1,\ldots,L_1 + L_2 - 1$ in (8.5) is unavailable at current time k, it leads to a noncausal formulation of (8.5), which means that it is not ready to be used in practice. In such circumstances, we need to adopt the prediction algorithm to estimate $\phi_i(k+1)$ at time k. In fact, any prediction algorithm for time-varying parameter can be used here, such as the multilayer hierarchical forecasting method [79].

Based on the estimation algorithm (4.26), resetting algorithm (4.27), prediction algorithm in Refs. [78,79], and controller algorithm (8.5), the MFAC for the systems in series connection of two subsystems can be realized.

Remark 8.2

It is worth pointing out that the system connected in series can be considered as a new unknown system, and its PFDL data model could be built directly. Then, the MFAC scheme can be designed based on this PFDL data model, whose PG is estimated by using I/O data of the series system. However, since this method is a centralized estimation and centralized control scheme, which does not make full use of the structure information, it may lead to a big bias of estimation and influence the control performance. Comparatively, applying the decentralized estimation and centralized control scheme, which utilizes the structure information, it may lead to better estimation and control performance.

Remark 8.3

In particular, when $L_1 = L_2 = 1$, the PFDL data model (8.4) becomes $\Delta y(k+1) = \phi_{21}(k)\phi_{11}(k-1)\Delta u(k-1)$. The corresponding control law is given as follows:

$$u(k) = u(k-1) + \frac{\rho\hat{\phi}_{21}(k+1)\hat{\phi}_{11}(k)(y^*(k+2) - y(k) - \hat{\phi}_{21}(k)\hat{\phi}_{11}(k-1)\Delta u(k-1))}{\lambda + |\hat{\phi}_{21}(k+1)\hat{\phi}_{11}(k)|^2}. \tag{8.6}$$

In general, the MFAC scheme for a large augmented system with N subsystems in series can be developed by using a similar designing procedure.

8.2.2 Parallel Connection

If subsystems of a complex system are connected in a parallel, then the main idea of the decentralized estimation and centralized control type MFAC method is that, first, the dynamic linearization data model for each subsystem is built; next, the PG of each subsystem is estimated using each local I/O data individualy, and then the dynamic linearization data model of the augmented system is formulated by virtue of the parallel structure. Finally, the MFAC scheme is designed for the augmented system using the integrated data model.

Consider parallel connection of two SISO discrete-time nonlinear subsystems (8.1) and (8.2), as shown in Figure 8.2.

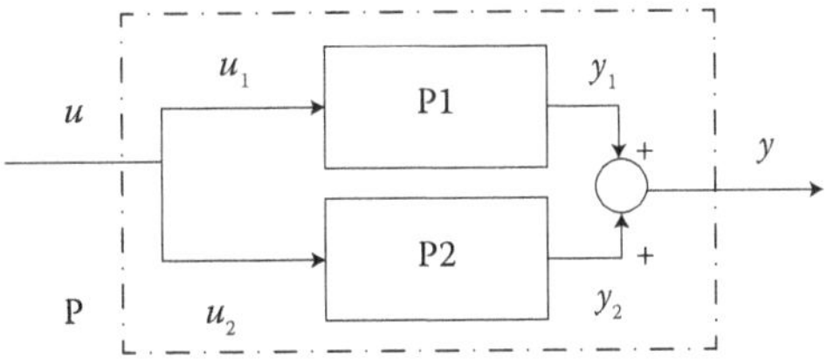

Figure 8.2 Block diagram of a system in parallel connection.

Since the two subsystems are connected in parallel, it is obvious that the relationships $u = u_1 = u_2$ and $y = y_1 + y_2$ hold. Thus, the PFDL data model of the system in parallel is

$$\begin{aligned}\Delta y(k+1) &= \Delta y_1(k+1) + \Delta y_2(k+1)\\ &= \boldsymbol{\phi}^T_{1,p,L_1}(k)\Delta \boldsymbol{U}_{1,L_1}(k) + \boldsymbol{\phi}^T_{2,p,L_2}(k)\Delta \boldsymbol{U}_{2,L_2}(k).\end{aligned} \tag{8.7}$$

Without loss of generality, we assume that $L_1 \le L_2 = L$, and define two L-dimensional vectors as follows:

$$\Delta \boldsymbol{U}_P(k) = [\Delta u(k), \ldots, \Delta u(k-L+1)]^T,$$

$$\boldsymbol{\phi}_P(k) = \left[\phi_1(k), \ldots, \phi_L(k)\right]^T,$$

where

$$\phi_i(k) = \begin{cases}\phi_{1i}(k) + \phi_{2i}(k), & 1 \le i \le L_1\\ \phi_{2i}(k), & L_1 < i \le L_2.\end{cases}$$

Then, PFDL data model (8.7) is rewritten in a compact form:

$$\Delta y(k+1) = \boldsymbol{\phi}^T_P(k)\Delta \boldsymbol{U}_P(k), \tag{8.8}$$

where $\boldsymbol{\phi}_P(k)$ is the PG vector of the large system.

Remark 8.4

If the system consists of N subsystems in parallel, its PFDL data model can be similarly derived based on integrating local PFDL data models of all subsystems.

Using the PFDL data model (8.8), the PFDL–MFAC algorithm for the large augmented system is constructed

$$u(k) = u(k-1) + \frac{\rho\hat{\phi}_1(k)}{\lambda + |\hat{\phi}_1(k)|^2}\left(y^*(k+1) - y(k) - \sum_{i=2}^{L}\hat{\phi}_i(k)\Delta u(k-i+1)\right), \tag{8.9}$$

where $\rho \in (0,1]$ is a step factor, $\lambda > 0$ is a weighting factor, $\hat{\phi}_i(k)$ is the estimation of $\phi_i(k)$ at time instant k, $i = 1,\ldots,L$.

Based on the estimation algorithm (4.26), resetting algorithm (4.27), and controller algorithm (8.9), the MFAC for the system in parallel connection of two subsystems can be realized.

Remark 8.5

In particular, when $L_1 = L_2 = 1$, the PFDL data model (8.8) becomes $\Delta y(k+1) = (\phi_{11}(k) + \phi_{21}(k))\Delta u(k)$. The corresponding control law is given as follows:

$$u(k) = u(k-1) + \frac{\rho(\hat{\phi}_{11}(k) + \hat{\phi}_{21}(k))}{\lambda + \left|\hat{\phi}_{11}(k) + \hat{\phi}_{21}(k)\right|^2}(y^*(k+1) - y(k)). \tag{8.10}$$

In general, the MFAC scheme for systems in parallel connection of N subsystems can be developed by using a similar design procedure.

8.2.3 Feedback Connection

If subsystems of a complex system are connected in a feedback style, the outline of the corresponding decentralized estimation and centralized control type MFAC method is that, first, the dynamic linearization data model of each subsystem is built; next, the PG of subsystem is estimated using local I/O data individually, and then the dynamic linearization data model for the augmented system is formulated by virtue of the feedback structure. Finally, the MFAC scheme is designed for the augmented system using the integrated data model.

Consider feedback connection of two SISO discrete-time nonlinear subsystems (8.1) and (8.2), shown in Figure 8.3.

According to the connection structure, we have $u_1 = u - y_2$, $y = y_1$, $u_2 = y_1$. Thus, the PFDL data model of the system in feedback connection is

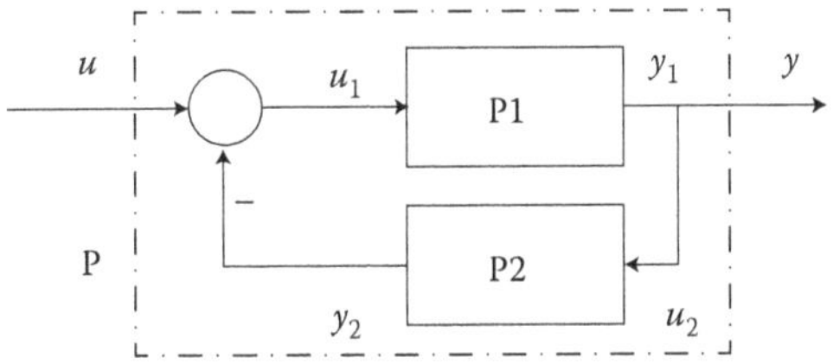

Figure 8.3 Block diagram of system in feedback connection.

$$\Delta y(k+1) = \boldsymbol{\phi}^T_{1,p,L_1}(k)\begin{bmatrix}\Delta u_1(k)\\ \vdots \\ \Delta u_1(k-L_1+1)\end{bmatrix}$$
$$= \boldsymbol{\phi}^T_{1,p,L_1}(k)\left(\begin{bmatrix}\Delta u(k)\\ \vdots \\ \Delta u(k-L_1+1)\end{bmatrix} - \begin{bmatrix}\Delta y_2(k)\\ \vdots \\ \Delta y_2(k-L_1+1)\end{bmatrix}\right)$$
$$= \boldsymbol{\phi}^T_{1,p,L_1}(k)\left(\begin{bmatrix}\Delta u(k)\\ \vdots \\ \Delta u(k-L_1+1)\end{bmatrix} - \begin{bmatrix}\boldsymbol{\phi}^T_{2,p,L_2}(k-1)\begin{bmatrix}\Delta y(k-1)\\ \vdots \\ \Delta y(k-L_2)\end{bmatrix}\\ \vdots \\ \boldsymbol{\phi}^T_{2,p,L_2}(k-L_1)\begin{bmatrix}\Delta y(k-L_1)\\ \vdots \\ \Delta y(k-L_1-L_2+1)\end{bmatrix}\end{bmatrix}\right). \tag{8.11}$$

Define two $(L_1 + L_1 + L_2 - 1)$-dimensional vectors as follows:

$$\Delta \boldsymbol{H}_F(k) = \left[\Delta \boldsymbol{U}^T_{L_1}(k) \quad \Delta \boldsymbol{Y}^T_{L_1+L_2-1}(k-1)\right]^T,$$

$$\boldsymbol{\phi}_F(k) = [\phi_1(k), \ldots, \phi_{L_1+L_1+L_2-1}(k)]^T = [\boldsymbol{\phi}^T_U(k) \quad \boldsymbol{\phi}^T_Y(k)]^T,$$

where

$$\Delta \boldsymbol{U}_{L_1}(k) = [\Delta u_1(k), \ldots, \Delta u_1(k-L_1+1)]^T,$$

$$\Delta \boldsymbol{Y}_{L_1+L_2-1}(k-1)=[\Delta y(k-1),\ldots,\Delta y(k-L_1-L_2+1)]^T,$$

$$\boldsymbol{\phi}_U^T(k)=\boldsymbol{\phi}_{L_1}^T(k)$$

$$\boldsymbol{\phi}_Y^T(k)=-\boldsymbol{\phi}_{1,p,L_1}^T(k)\begin{bmatrix}\phi_{21}(k-1)\,\phi_{22}(k-2) & \cdots & \phi_{2L_2}(k-L_2)\\ \phi_{21}(k-2)\,\phi_{2}(k-3) & \cdots & \phi_{2L_2}(k-L_2-1)\\ \vdots & & \\ \phi_{21}(k-L_1) & \cdots & \phi_{2L_2}(k-L_1-L_2+1)\end{bmatrix}.$$

Then, the PFDL data model (8.11) implies

$$\Delta y(k+1)=\boldsymbol{\phi}_F^T(k)\Delta \boldsymbol{H}_F(k). \tag{8.12}$$

Using PFDL data model (8.12) of the system in feedback connection, the PFDL–MFAC algorithm is given as

$$u(k)=u(k-1)+\frac{\rho\hat{\phi}_1(k)}{\lambda+\hat{\phi}_1(k)^2}\Bigg(y^*(k+1)-y(k)$$
$$-\sum_{i=2}^{L_1}\hat{\phi}_i(k)\Delta u(k-i+1)-\sum_{i=L_1+1}^{L_1+L_1+L_2-1}\hat{\phi}_i(k)\Delta y(k-i+L_1)\Bigg). \tag{8.13}$$

After the PG of each subsystem at current time k is estimated by using PG estimation algorithm (4.26) and resetting algorithm (4.27) in Chapter 4, the MFAC scheme for the system in feedback connection of two subsystems can be realized based on the controller algorithm (8.13).

Remark 8.6

In particular, when $L_1=L_2=1$, the PFDL data model becomes $\Delta y(k+1)=\phi_{11}(k)\Delta u(k)-\phi_{11}(k)\phi_{21}(k-1)\Delta y(k-1)$. The corresponding control law is as follows:

$$u(k)=u(k-1)+\frac{\rho\hat{\phi}_{11}(k)}{\lambda+\hat{\phi}_{11}(k)^2}\left(y^*(k+1)-y(k)-\hat{\phi}_{11}(k)\hat{\phi}_{21}(k-1)\Delta y(k-1)\right). \tag{8.14}$$

8.2.4 Complex Interconnection

When a system is strongly coupled and cannot be decomposed into one or combination of the aforementioned three basic connection structures, its control problem becomes more challenging. In this section, under the condition that the dynamic interactions among subsystems are measurable, the complex interconnected system is decomposed into N independent subsystems. The dynamic linearization data models are built for these subsystems individually, and then the MFAC scheme based on these data models is designed for the complex interconnected system.

Consider a complex interconnected system with N subsystems and the ith subsystem is described as

$$\begin{aligned} y_i(k+1) &= f_i(y_i(k),\ldots,y_i(k-n_{y_i}),u_i(k),\ldots,u_i(k-n_{u_i}),z_{1i}(k),\ldots,\\ &\quad z_{1i}(k-n_{z_{1i}}),\ldots,z_{ji}(k),\ldots,z_{ji}(k-n_{ji})\ldots,z_{Ni}(k),\ldots,z_{Ni}(k-n_{z_{Ni}}))\\ &\quad i \neq j, i=1,\ldots,N. \end{aligned} \tag{8.15}$$

where $u_i(k) \in R$ and $y_i(k) \in R$ are the control input and the system output of the ith subsystem at time k, respectively; $z_{ji}(k) \in R$, $j \neq i$ is the measurable interaction of the ith subsystem imposed by the jth subsystem; $f_i(\ldots)$ is an unknown nonlinear function describing the ith nonlinear system; and $n_{y_i}, n_{u_i}, n_{z_{ji}} \in Z^+$ are unknown integers.

If the interactions among subsystems are measurable, the complex interconnected system (8.15) can be decomposed into N independent subsystems, as depicted in Figure 8.4. The subsystem $f_i(\cdots)$ can be viewed as a MISO system, whose inputs are $u_i(k)$ and the other $N-1$ measurable interactions z_{ji}, $j \neq i$.

Define the augmented input of the ith subsystem as

$$\boldsymbol{u}_i(k) = \left[u_i(k), z_{1i}(k),\ldots,z_{ji}(k),\ldots,z_{Ni}(k)\right]^T, \quad i \neq j, i, j = 1,\ldots,N. \tag{8.16}$$

If subsystem (8.15) satisfies Assumptions 3.7′ and 3.8′, it can be transformed into the following CFDL data model according to Corollary 3.1:

$$y_i(k+1) = y_i(k) + \boldsymbol{\phi}_{c,i}^T(k)\Delta \boldsymbol{u}_i(k), \tag{8.17}$$

where $\boldsymbol{\phi}_{c,i}(k) = \left[\phi_i(k), \varphi_{1i}(k)\cdots\varphi_{ji}(k),\ldots,\varphi_{Ni}(k)\right]^T$ is the PG of the ith subsystem, $j \neq i$, $i = 1,\ldots,N$, $\phi_i(k)$ is the PPD of $f_i(\ldots)$ with respect to the control input $u_i(k)$, $\varphi_{ji}(k)$ is the PPD of $f_i(\ldots)$ with respect to measurable interaction $z_{ji}(k)$, and $\Delta \boldsymbol{u}_i(k) = \boldsymbol{u}_i(k) - \boldsymbol{u}_i(k-1)$.

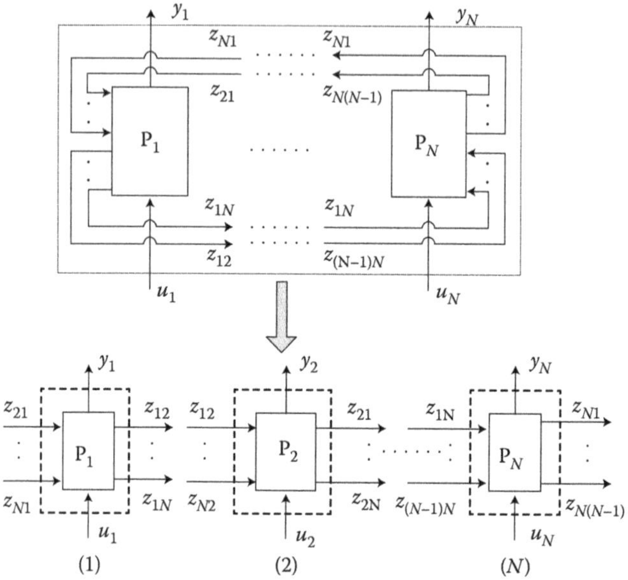

Figure 8.4 Decomposition of a complex interconnected system.

The essence of the above procedure is to transform a strongly coupled MIMO nonlinear system into N decoupled MISO subsystems with measurable interactions. Therefore, the decentralized estimation and decentralized control-type MFAC scheme can be designed for the complex interconnected system based on the CFDL data models of N decoupled subsystems.

Using the control scheme (5.14)–(5.16) in the MISO case, the MFAC scheme for each subsystem is as follows:

$$u_i(k) = u_i(k-1) + \frac{\rho\hat{\phi}_i(k)\left(y_i^*(k+1) - y_i(k) - \sum_{j=1, j\neq i}^{N} \hat{\varphi}_{ji}(k)\Delta z_{ji}(k)\right)}{\lambda_i + \hat{\phi}_i^2(k)}, \tag{8.18}$$

$$\hat{\boldsymbol{\phi}}_{c,i}(k) = \hat{\boldsymbol{\phi}}_{c,i}(k-1) + \frac{\eta\Delta\boldsymbol{u}_i(k-1)\left(y_i(k) - y_i(k-1) - \hat{\boldsymbol{\phi}}_{c,i}^T(k-1)\Delta\boldsymbol{u}_i(k-1)\right)}{\mu + \left\|\Delta\boldsymbol{u}_i(k-1)\right\|^2}, \tag{8.19}$$

$$\hat{\phi}_i(k) = \hat{\phi}_i(1), \quad \text{if } |\hat{\phi}_i(k)| \le \varepsilon \quad \text{or} \quad \text{sign}(\hat{\phi}_i(k)) \neq \text{sign}(\hat{\phi}_i(1)), \tag{8.20}$$

where, $\lambda > 0$, $\mu > 0$, $\rho \in (0,1]$, $\eta \in (0,2]$, and ε is a positive constant. Equation (8.20) is the parameter resetting algorithm.

8.2.5 Simulation Results

In this section, two numerical simulations are presented to verify the correctness and effectiveness of the proposed decentralized estimation and centralized control type MFAC schemes and decentralized estimation and decentralized control-type MFAC schemes for systems in series connection, parallel connection, feedback connection, and complex interconnected system, respectively. It is worth pointing out that the models of the following systems are merely used to generate the input–output data but are not involved in controller design.

8.2.5.1 Series, Parallel, and Feedback Connection

Example 8.1

Consider a system consisting of the following two nonlinear subsystems in series, parallel, and feedback connections:

$$\begin{aligned} &\text{P1: } y_1(k+1) = 0.5y_1(k)/(1+y_1^2(k)) + 1.2u_1^2(k) + 0.4u_1^2(k-1), \\ &\text{P2: } y_2(k+1) = 0.1u_2^2(k)/(1+0.25y_2^2(k)). \end{aligned} \tag{8.21}$$

The desired output trajectory is

$$y^*(k) = 1 \tag{8.22}$$

For three systems in series connection, parallel connection, and feedback connection, simulation comparisons between the proposed scheme and the CFDL–MFAC scheme (4.7)–(4.9) are presented. For the CFDL–MFAC scheme (4.7)–(4.9), the interconnected system is treated as an unknown controlled plant integrally. The decentralized estimation and centralized control type MFAC method utilizes projection algorithm (4.7) and resetting algorithm (4.8) to estimate the PG of each subsystem for computing the PG of the complex interconnection system, and adopts controller algorithms (8.6), (8.10), and (8.14), respectively. For simulations, the weighting factors in the CFDL–MFAC scheme are set to 0.8, 5, and 0.25, respectively; the other parameters are set to $\hat{\phi}_c(0) = 1, \eta = 1, \mu = 1,$ and $\rho = 1$, the weighting factors in the decentralized estimation and centralized control-type MFAC scheme are set to 0.8, 5, and 0.25, respectively; the other parameters are set $\hat{\phi}_{11}(0) = \hat{\phi}_{21}(0) = 1, \eta_1 = \eta_2 = 1, \mu_1 = \mu_2 = 1,$ and $\rho = 1$.

Simulation comparisons between the two schemes under three connection structures are shown in Figures 8.5–8.7, respectively. The simulation results show that the control performance of the two control schemes in three cases is acceptable. Compared with the CFDL–MFAC, the new proposed MFAC scheme yields a better control performance with a shorter settling time, and smoother control input, since the PG of each subsystem is estimated separately and the connection information is fully utilized.

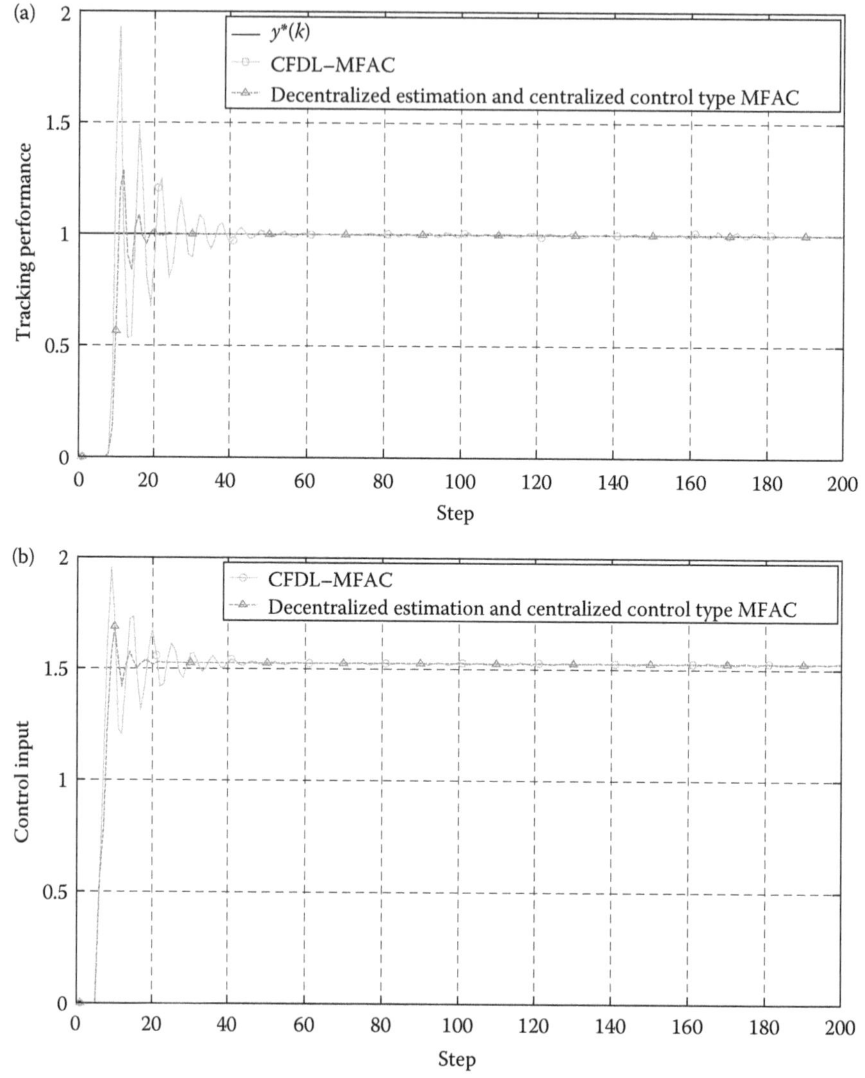

Figure 8.5 Simulation comparison between CFDL–MFAC and decentralized estimation and centralized control type MFAC for a system in series connection. (a) Tracking performance. (b) Control input.

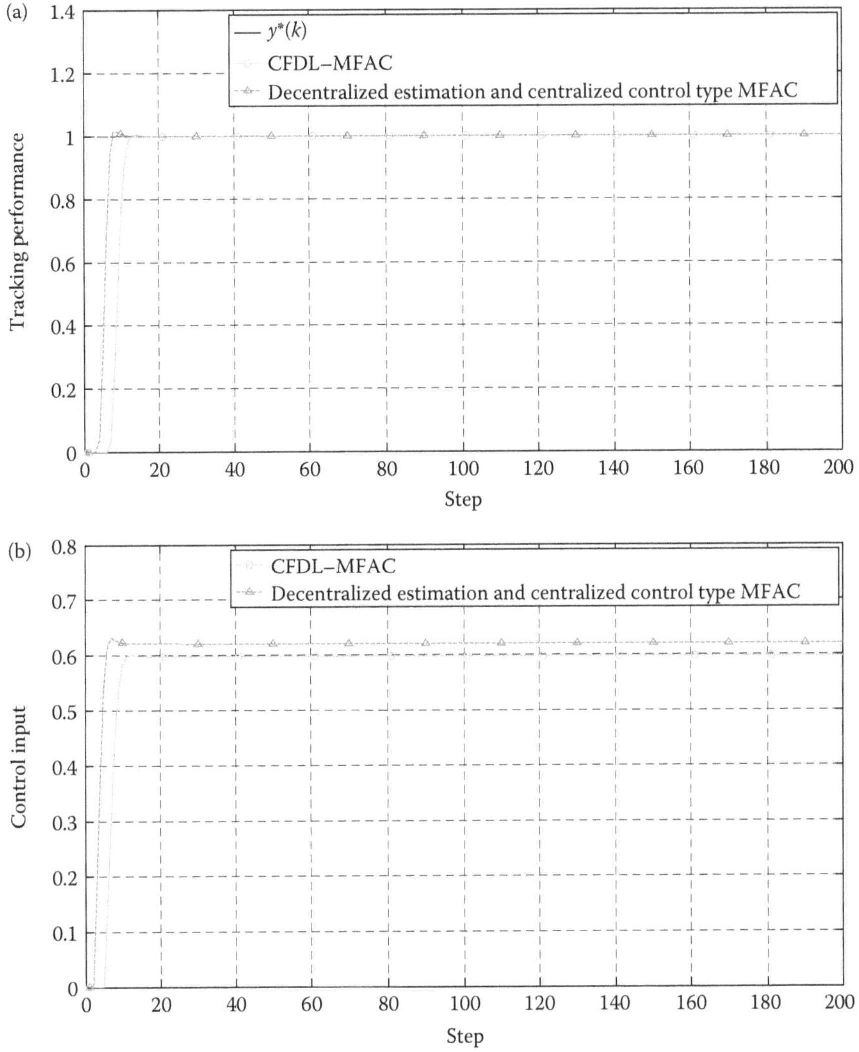

Figure 8.6 Simulation comparison between CFDL–MFAC and decentralized estimation and centralized control type MFAC for a system in parallel connection. (a) Tracking performance. (b) Control input.

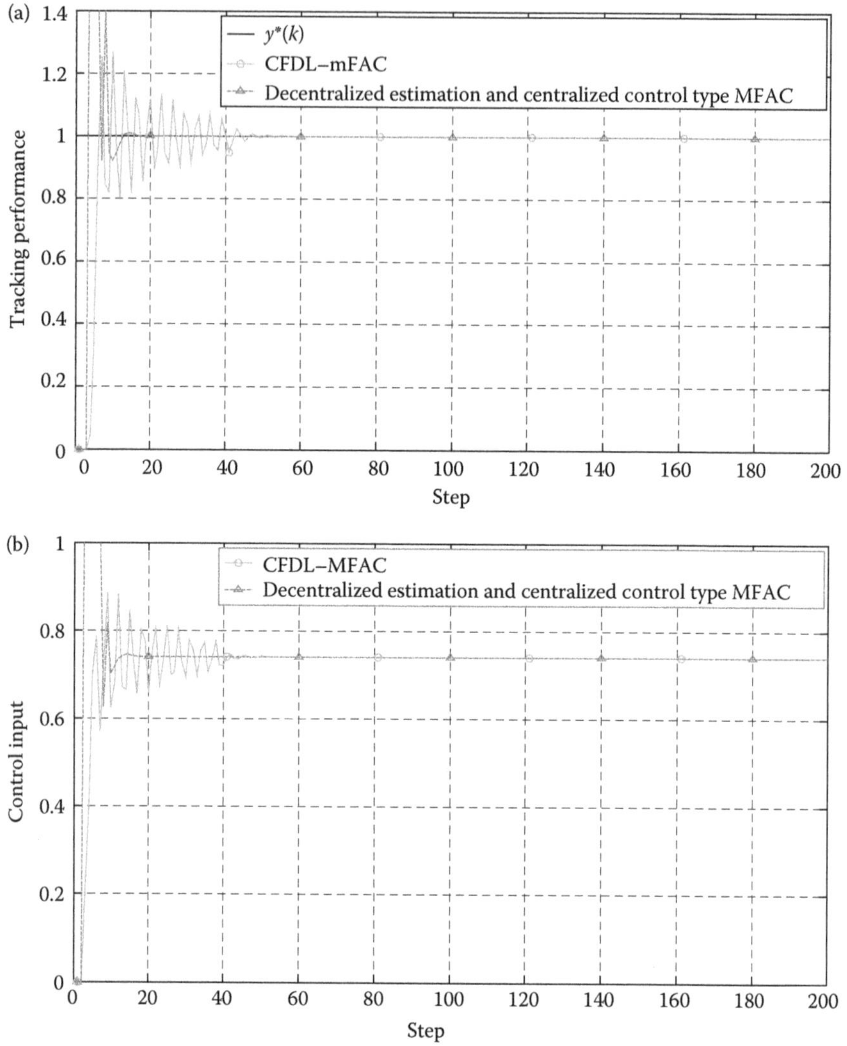

Figure 8.7 Simulation comparison between CFDL–MFAC and decentralized estimation and centralized control type MFAC for a system in feedback connection. (a) Tracking performance. (b) Control input.

8.2.5.2 Complex Interconnection

Example 8.2

Consider the following complex interconnected nonlinear system:

$$
\begin{aligned}
y_1(k+1) &= y_1(k)/\left(1+y_1^3(k)\right) + u_1(k) + 0.1z_{21}(k) + 0.2z_{31}(k),\\
z_{12}(k+1) &= 0.1y_1(k)\left(u_1(k)-u_1(k-1)\right) + 0.2u_1(k),\\
z_{13}(k+1) &= -0.1y_1(k)\left(u_1(k)-u_1(k-1)\right) + 0.1u_1(k),\\
y_2(k+1) &= 0.5y_2(k)y_2(k-1)/\left(1+y_2^2(k)+y_2^2(k-1)+y_2^2(k-2)\right)\\
&\quad + u_2(k) + 0.1u_2(k-1)z_{12}^2(k) + 0.1y_2(k)/\left(1+z_{32}^2(k)\right),\\
z_{21}(k+1) &= 0.1u_2(k-1)y_2^2(k) + 0.05y_2(k) + 0.1u_2(k),\\
z_{23}(k+1) &= 0.5u_2(k-1)/\left(3+y_2^2(k)\right) + 0.2u_2(k),\\
y_3(k+1) &= 1.2y_3(k)y_3(k-1)/(1+y_3^2(k)+y_3^2(k-1))\\
&\quad + 0.4u_3(k-1)e^{z_{23}(k)} + 0.3\sin(0.5\,(y_3(k)+y_3(k-1)))\\
&\quad \times \cos\,(0.5\;y_3(k)+y_3(k-1))\cos\,(z_{13}(k)) + 1.2u_3(k),\\
z_{31}(k+1) &= 0.1y_3(k)\,(u_3(k-1)-u_3\;(k-2))\\
&\quad + 0.03y_3(k-2)\,(u_3(k)-u_3(k-1)),\\
z_{32}(k+1) &= 0.1y_3(k)y_3(k-1)\,(u_3(k)-u_3(k-1))/(0.1+y_3^2(k)+y_3^2(k-1)),
\end{aligned}
\tag{8.23}
$$

where $z_{ji}(k) \in R$ is the measurable interaction of the ith subsystem imposed by the jth subsystem, $i,j = 1,2,3$, $i \neq j$.

The desired trajectories are given as follows:

$$
y_1^*(k) = \begin{cases} 2, & 1 \le k < 333 \\ 0.5, & 333 \le k < 667 \\ 1, & 667 \le k < 1000 \end{cases};\quad y_2^*(k) = 1.5;\, y_3^*(k) = \text{round}\left(\frac{k}{350}\right).
$$

It is a strongly coupled nonlinear system with three inputs and three outputs. The mathematical description of the controlled system is so complex that one cannot imagine how an MBC method can deal with its control problem even when the precise model is known. Here, a simulation comparison between the CFDL–MFAC for MIMO nonlinear systems in Section 5.2 and the decentralized estimation and decentralized control-type MFAC (8.18)–(8.20) is carried out. The parameters of CFDL–MFAC are set to $\rho = 0.9$, $\lambda = 0.05$, $\eta = 0.1$, and $\mu = 1$; the parameters of the proposed MFAC scheme are set to $\rho_1 = \rho_2 = \rho_3 = 0.9$, $\lambda_1 = \lambda_2 = \lambda_3 = 0.05$, $\eta_1 = \eta_2 = \eta_3 = 0.1$, and $\mu_1 = \mu_2 = \mu_3 = 1$.

The comparison results are shown in Figures 8.8–8.10. From the simulation results, we can observe that both schemes give a satisfactory control performance for such a complex MIMO system. It is worth pointing out that the decentralized estimation and decentralized control-type MFAC method give much better control performance with smaller overshoot and smoother control input since it takes the measurable interactions among three subsystems into account.

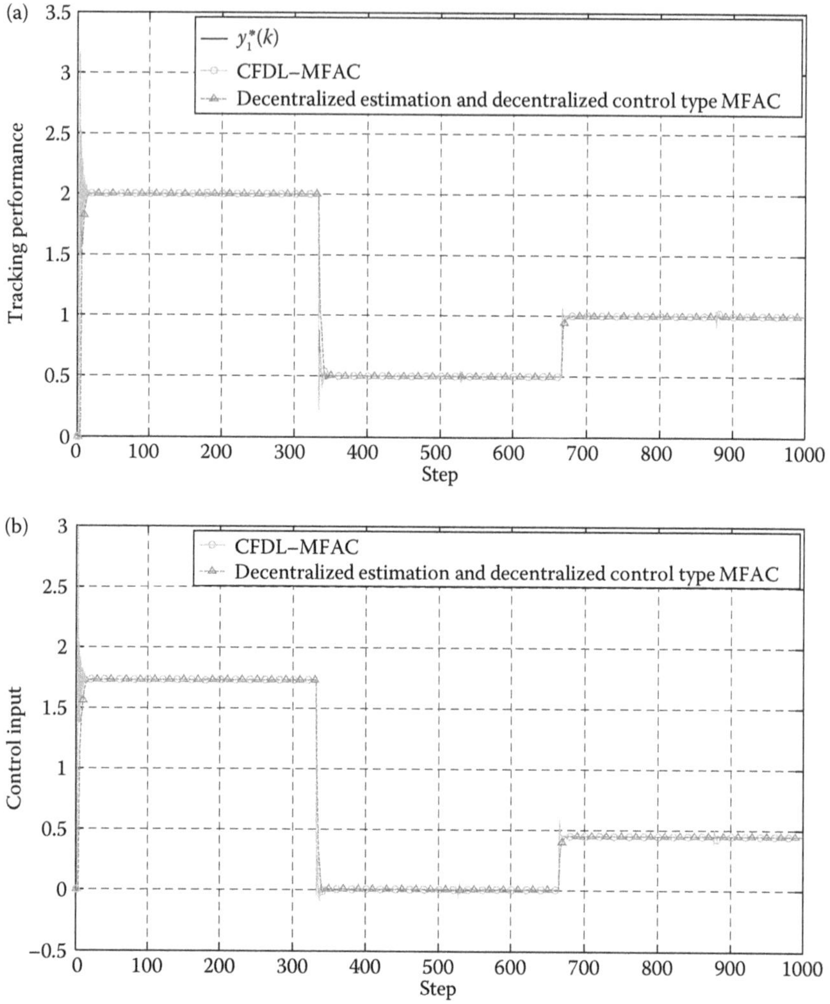

Figure 8.8 Simulation results of subsystem P1. (a) Tracking performance. (b) Control input.

8.3 Modularized Controller Design

Each control method, no matter whether it is model-based or data-driven, does have its own advantages and disadvantages and cannot be replaced entirely by others. MBC methods have been applied in many practical fields, and the engineers in practical fields know these equipments and devices well, although their control performance may be questionable sometimes. Completely removing or replacing them may lead to enormous waste of resources. Moreover, it is difficult for engineers to

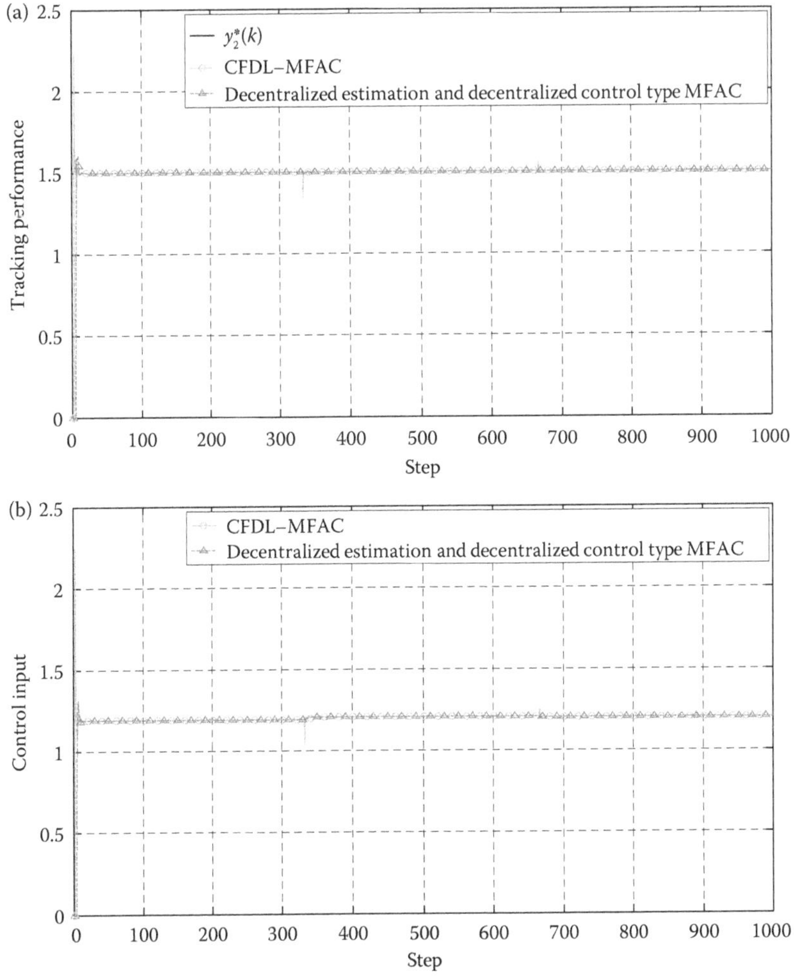

Figure 8.9 Simulation results of subsystem P2. (a) Tracking performance. (b) Control input.

accept a new method quickly due to doubt and unfamiliarity. Thus, it is of great significance to develop a complementarily modularized controller by incorporating MFAC with other control methods [125,220,221].

8.3.1 Estimation-Type Control System Design

Consider a general discrete-time SISO nonlinear system:

$$y(k+1) = f(y(k),\ldots,y(k-n_y),u(k),\ldots,u(k-n_u)), \tag{8.24}$$

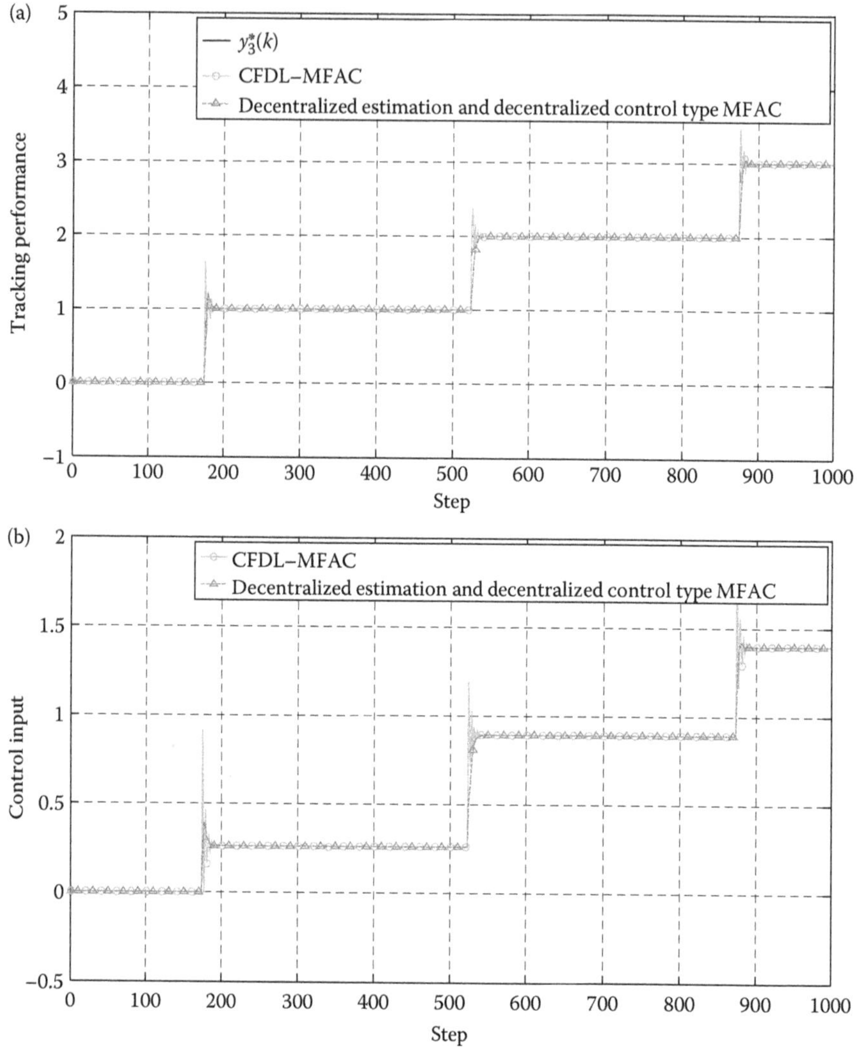

Figure 8.10 Simulation results of subsystem P3. (a) Tracking performance. (b) Control input.

where $u(k) \in R$ and $y(k) \in R$ are the control input and the system output at time k, n_y and n_u are two integers, and $f(\ldots)$ is an unknown nonlinear function.

In practice, control engineers often design the control system using a linear time-invariant model with a lower order to approximate the controlled plant. For example, the following first-order or second-order linear model is used for designing the adaptive control law for system (8.24):

$$y_m(k+1) = \left[y(k), u(k)\right]\begin{bmatrix}\theta_1\\ \theta_2\end{bmatrix}, \tag{8.25}$$

$$y_m(k+1) = \left[y(k), y(k-1), u(k), u(k-1)\right]\begin{bmatrix}\theta_1\\ \theta_2\\ \theta_3\\ \theta_4\end{bmatrix}. \tag{8.26}$$

Parameters $[\theta_1, \theta_2]^T$ in model (8.25) or parameters $[\theta_1, \theta_2, \theta_3, \theta_4]^T$ in model (8.26) are estimated by using the projection algorithm or the least-squares algorithm, and then the following adaptive control laws are derived according to the certainty equivalence principle:

$$u(k) = \frac{1}{\hat{\theta}_2(k)}(y^*(k+1) - \hat{\theta}_1(k)y(k)), \tag{8.27}$$

or

$$u(k) = \frac{1}{\hat{\theta}_3(k)}(y^*(k-1) - \hat{\theta}_1(k)y(k) - \hat{\theta}_2(k)y(k-1) - \hat{\theta}_4(k)u(k-1)), \tag{8.28}$$

where $\hat{\theta}_i(k)$ is the estimation of $\theta_i(k)$, $i = 1,2$ or $i = 1,2,3,4$.

It is obvious that there must exist unmodeled dynamics in (8.25) or (8.26) for its approximation of (8.24). Hence, the control performance of the above-mentioned adaptive control system may not be satisfactory due to the possible influence from the unmodeled dynamics in practice. Actually, the accurate relationship between plant (8.24) and approximate model (8.25) or (8.26) is as follows:

$$y(k+1) = y_m(k+1) + NL, \tag{8.29}$$

where NL represents the unmodeled dynamics

$$NL = f(y(k),\ldots,y(k-n_y),u(k),\ldots,u(k-n_u)) - \left[y(k), u(k)\right]\begin{bmatrix}\theta_1\\ \theta_2\end{bmatrix}, \tag{8.30}$$

or

$$NL = f\,(y(k),\ldots,y(k-n_y),u(k),\ldots,u(k-n_u)) - [y(k), y(k-1), u(k), u(k-1)]\begin{bmatrix}\theta_1\\ \theta_2\\ \theta_3\\ \theta_4\end{bmatrix}. \tag{8.31}$$

If NL is known or can be estimated, then the following modified control law can be obtained:

$$u(k) = \frac{1}{\hat{\theta}_2(k)}(y^*(k+1) - \hat{\theta}_1(k)y(k) - NL), \tag{8.32}$$

or

$$u(k) = \frac{1}{\hat{\theta}_3(k)}(y^*(k+1) - \hat{\theta}_1(k)y(k) - \hat{\theta}_2(k)y(k-1) - \hat{\theta}_4(k)u\,(k-1) - NL). \tag{8.33}$$

Taking the unmodeled dynamics into consideration, the adaptive control law (8.32) or (8.33) should give a better control performance when it is used in practice compared with the original ones. However, the NL used in (8.32) or (8.33) includes not only the unmodeled dynamics, but also the possible error of the parameter estimation, and thus its dynamics may be very complex. By now, there are many works on designing a certain mechanism to compensate the effect from the unmodeled dynamics and parameter estimation error, but it still needs to develop simple and effective methods.

According to the above analysis and discussion, an estimation-type control system design scheme based on MFAC, shown in Figure 8.11, is introduced. This scheme consists of controller algorithm (8.32), the projection algorithm or least-squares algorithm for parameter estimation, and the following unmodeled dynamics estimation algorithm:

$$\begin{aligned}\hat{\phi}_c(k) &= \hat{\phi}_c(k-1) + \frac{\eta\Delta u(k-1)}{\mu + \Delta u(k-1)^2}\left(\Delta e(k) - \hat{\phi}_c(k-1)\Delta u(k-1)\right),\\ \hat{NL}\,(k+1) &= e(k) + \hat{\phi}_c(k)\Delta u\,(k).\end{aligned} \tag{8.34}$$

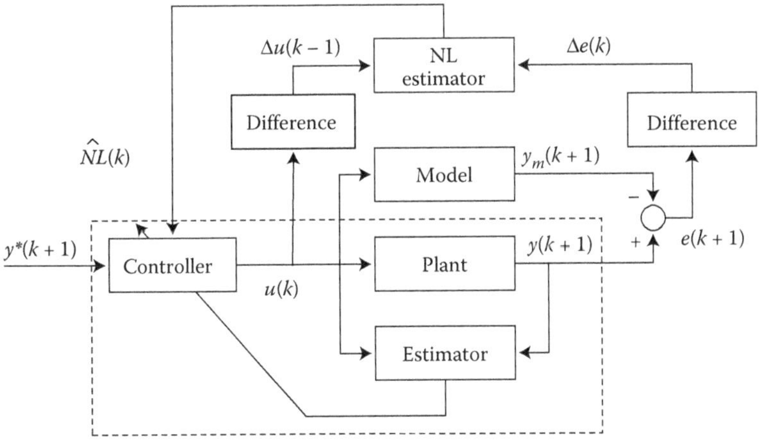

Figure 8.11 Block diagram of estimation-type MFAC control system.

The external loop of MFAC is used to overcome the effect caused by the unmodeled dynamics or parameter estimation error and to improve the quality of the control system. It is easy for us to understand the idea of the modularized design from Figure 8.11; that is, adding the external loop to the existing internal feedback loop does not affect the original control system if the existing internal feedback loop works well, and the external loop will help the internal loop in improving the system performance if there exists tracking error. The external and internal loops cowork complementarily. It should be noted that MFAC in this control scheme plays an assistant role, that is, it estimates or predicts the unmodeled dynamics and then compensates the effect from the unmodeled dynamics and parameter estimation error. It becomes the original adaptive control system when the MFAC external loop is disconnected.

8.3.2 Embedded-Type Control System Design

8.3.2.1 Embedded-Type Control System Design for Nonrepetitive Systems

Though MBC methods are perfect in theory, their control performance may not be acceptable when they are used in practice due to the existence of unmodeled dynamics and external uncertainties. If a nearly accurate mathematical model of plant is available, it is not a wise choice to ignore the model. Thus, to improve control performance of MBC methods, such as adaptive control, optimal control, and so on, MFAC can be embedded into the existing MBC systems to achieve modularized design, which intends to exploit the advantages from both the MFAC and MBC methods by making them work together.

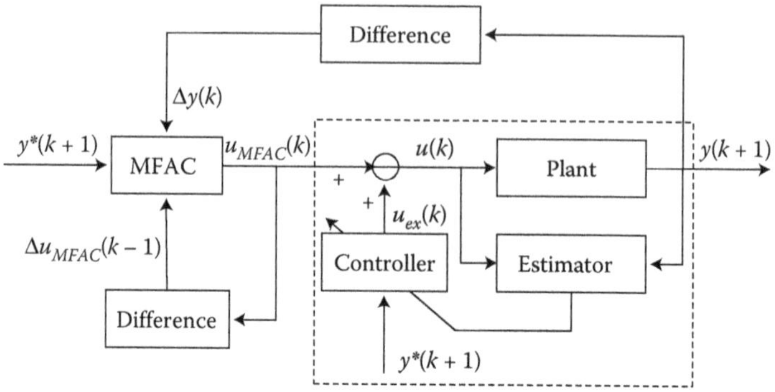

Figure 8.12 Block diagram of the embedded-type MFAC control system.

The modularized design scheme of the embedded-type control system based on MFAC is shown in Figure 8.12. The block in the dashed square is the original control system based on MBC methods, such as adaptive control.

The idea behind this kind of modularized design methods can be well interpreted from Figure 8.12. The embedded MFAC external loop does not affect the original existing feedback control system. The system becomes the original adaptive control system when the external loop of MFAC is disconnected. On the contrary, only the MFAC module takes effect when the internal loop is disconnected. If the control performance of the adaptive control system becomes perfect, it means the output of the controlled plant tracks the desired one, and then the MFAC external module is disconnected. In other words, the existing control system based on adaptive control can be regarded as a new augmented controlled plant, and then the MFAC method is applied to the augmented plant.

The actual control input is the sum of control signals generated by the MBC method and the MFAC method, that is,

$$u(k) = u_{ex}(k) + u_{MFAC}(k), \tag{8.35}$$

where $u_{ex}(k)$ is the control input generated by the MBC method and $u_{MFAC}(k)$ is the control input generated by MFAC.

Furthermore, if the controlled plant consists of several subsystems, then the following three control schemes can be applied according to the type of connection.

Scenario 1

If the plant can be decomposed into P1 and P2 in series as depicted in Figure 8.13, and the control performance of this system in series is not acceptable; the MFAC module can be embedded at the node between P1 and P2. Compared with the

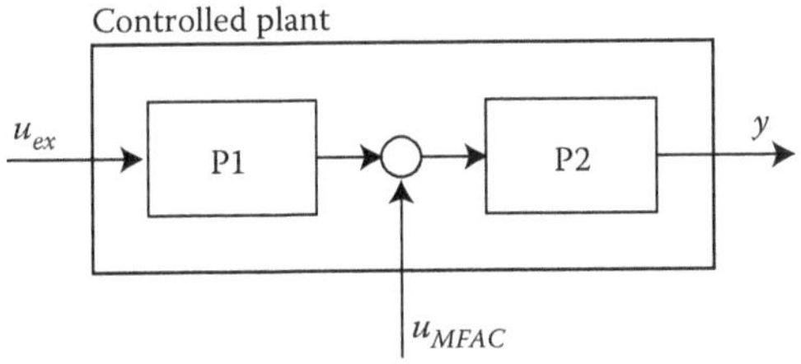

Figure 8.13 System in series connection.

aforementioned embedded-type method shown in Figure 8.12, the MFAC module of this scheme affects P2 more directly if the embedded mechanism is allowed to be inserted here. Thus, the performance of P2 can be improved.

Scenario 2

If the plant can be decomposed into P1 and P2 in parallel, and the control input signal generated from the embedded MFAC module can be embedded in the position 1 or position 2 as depicted in Figure 8.14. Compared with the embedded method shown in Figure 8.12, the performance of each subsystem can be improved directly by properly selecting the inserting position.

Scenario 3

If the plant can be decomposed into P1 and P2 in the feedback connection, the control signal generated from the embedded MFAC module can be inserted at the

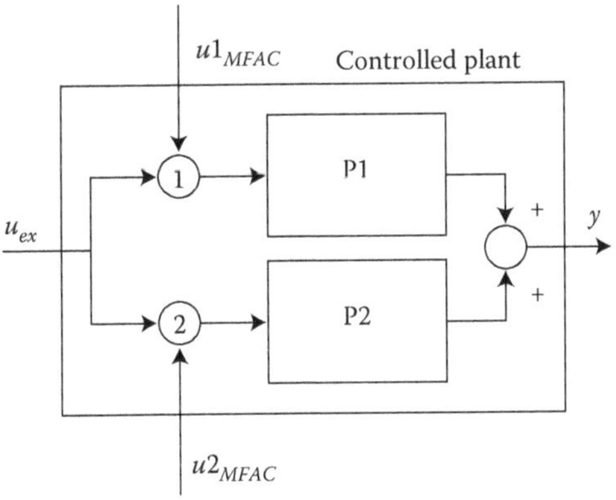

Figure 8.14 System in parallel connection.

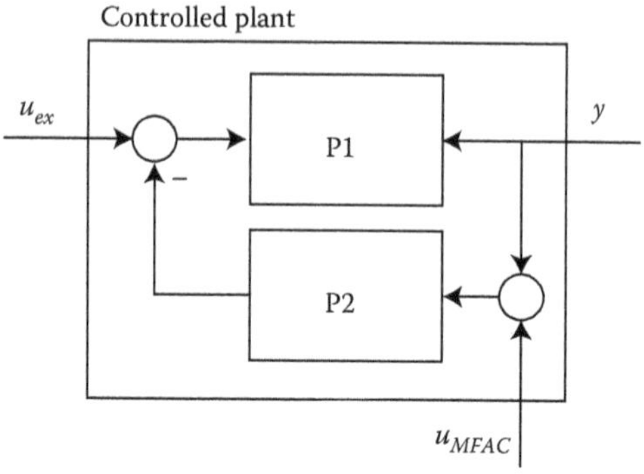

Figure 8.15 System in feedback connection.

node depicted in Figure 8.15. Compared with the embedded-type method shown at Figure 8.12, the performance of subsystem P2 can be improved.

In one word, the embedded position for the MFAC module is largely dependent on the structure and characteristics of the controlled system. No matter at which node the MFAC signal is inserted, the existing control system can always be treated as an augmented controlled plant of the embedded-type MFAC control scheme.

8.3.2.2 Modularized Controller Design Scheme for Repetitive Systems

Feedback, as the most common form of automatic control methods, is designed to satisfy certain specifications according to the desires of control engineers. Usually, the control input signal of a feedback controller is generated by utilizing the error between the system's actual behavior (output) and the desired behavior (output reference). Feedback control is generally considered as a control design method in the time domain. Theoretically speaking, asymptotic convergence of a feedback control system can only be obtained when the operation duration is long enough. In other words, perfect tracking cannot be achieved by the feedback controller if the operation is within a finite time interval. Feedback control methods, however, have been widely applied in many practical systems. For repetitive operation processes, such as robotic arms on the assembly line, batch process, chemical injection molding, and so on, control input signal by using feedback control methods is generated only according to the output error(s) in the current operation process, instead of learning from previous operation process(es); thus the control performance cannot be improved no matter how many times the process repeats. To make full use of the repetitive operation pattern information of the system, integration design of feedback control and the feedforward ILC is proposed. With this mechanism, the feedback control is responsible for stabilizing the system and dealing with the external disturbance, while the feedforward learning loop targets the high-precision tracking. Control performance improvement of this integrated scheme comes from the design mechanism in which advantages of each control method are complementarily exploited. Thus, this type of control system is of great significance in application.

In recent years, research on the integration of feedback control and ILC has attracted increasing attention of scholars. The discussion on this topic is often carried out under the assumption that the system is strictly repetitive. In such circumstance, adding ILC mechanism can improve control performance compared with the one using the feedback controller alone. The parameters of repetitive systems and the desired trajectory, however, may vary at different iterations due to the change of the external environment; thus, the control performance of ILC may not be as satisfactory as expected due to the nonrepetitive factors in practical systems. MFAC is designed merely by utilizing the I/O data of the system without any prior knowledge of its physical model, and can realize the adaptive control for the unknown nonlinear system. On the other hand, MFAC is a feedback control method, and it does not have the ability to learn from the experiences of the previous operation when the system excutes the same control task repetitively; that is, the control performance of any two consecutive operations are the same without any improvement.

On the basis of the above observation, an iterative learning-enhanced MFAC algorithm is proposed for the following general discrete-time nonlinear system with m inputs and m outputs in this section.

$$\boldsymbol{y}_n(k+1) = \boldsymbol{f}(\boldsymbol{u}_n(k), \boldsymbol{y}_n(k), \xi(k)), \tag{8.36}$$

where n denotes the iteration number, $\boldsymbol{f}(\ldots)$ is a function with appropriate dimension, and $\boldsymbol{y}_n(k)$ and $\boldsymbol{u}_n(k)$ are the system output and the control input at time k of the nth iteration. $\xi(k)$ is a bounded external disturbance.

The iterative learning-enhanced MFAC control scheme is constructed as follows:

$$\boldsymbol{u}_n(k) = \boldsymbol{u}_n^f(k) + \boldsymbol{u}_n^b(k), \tag{8.37}$$

$$\boldsymbol{u}_n^f(k) = \boldsymbol{u}_{n-1}^f(k) + \beta \boldsymbol{e}_{n-1}(k+1), \tag{8.38}$$

$$\boldsymbol{u}_n^b(k) = \boldsymbol{u}_n^b(k-1) + \frac{\rho \hat{\boldsymbol{\Phi}}_c^T(k)(\boldsymbol{y}^*(k+1) - \boldsymbol{y}_n(k))}{\lambda + \|\hat{\boldsymbol{\Phi}}_c(k)\|^2}, \tag{8.39}$$

$$\hat{\boldsymbol{\Phi}}_c(k) = \hat{\boldsymbol{\Phi}}_c(k-1) + \frac{\eta(\Delta \boldsymbol{y}_n(k) - \hat{\boldsymbol{\Phi}}_c(k-1)\Delta \boldsymbol{u}_n^b(k-1))\Delta \boldsymbol{u}_n^b(k-1)^T}{\mu + \left\|\Delta \boldsymbol{u}_n^b(k-1)\right\|^2}, \tag{8.40}$$

$$\hat{\phi}_{ii}(k) = \hat{\phi}_{ii}(1),$$

$$\text{if} \, |\hat{\phi}_{ii}(k)| < b_2 \;\text{ or }\; |\hat{\phi}_{ii}(k)| > \alpha b_2 \;\text{ or }\; \text{sign}\,(\hat{\phi}_{ii}(k)) \neq \text{sign}\,(\hat{\phi}_{ii}(1)),\; i = 1,\ldots,m$$

$$\hat{\phi}_{ij}(k) = \hat{\phi}_{ij}(1),$$

$$\text{if} \, |\hat{\phi}_{ij}(k)| > b_1 \;\text{ or }\; \text{sign}\,(\hat{\phi}_{ij}(k)) \neq \text{sign}\,(\hat{\phi}_{ij}(1)),\; i,j = 1,\ldots,m,\; i \neq j \tag{8.41}$$

where $\Delta \boldsymbol{u}_n^b(k) = \boldsymbol{u}_n^b(k) - \boldsymbol{u}_n^b(k-1)$, $\Delta \boldsymbol{y}_n(k) = \boldsymbol{y}_n(k) - \boldsymbol{y}_n(k-1)$, $\lambda > 0$, $\mu > 0$, $\eta \in (0,2)$, $\rho \in (0,1]$, β is the learning gain, ε is a sufficiently small positive constant, $\hat{\phi}_{ij}(1)$ is the initial value of $\hat{\phi}_{ij}(k)$, $\boldsymbol{e}_n(k) = \boldsymbol{y}_d(k) - \boldsymbol{y}_n(k)$ denotes the tracking error at the nth iteration, $\boldsymbol{u}_n^f$ represents the control input signal from feedforward ILC, and $\boldsymbol{u}_n^b$ is the control input signal generated by the original MFAC. There are two parts in this algorithm: ILC as a feedforward controller and MFAC as a feedback controller.

It can be seen that the feedback controller and the feedforward controller work individually. In practice, a simple way to implement the embedded controller design for repetitive systems is by adding ILC algorithm (8.38) as the external loop controller to an existing MFAC scheme (8.39)–(8.41). ILC, as a feedforward controller with learning function, targets to improve the control performance and can realize perfect tracking over a finite time interval. Meanwhile, the MFAC scheme is in charge of stabilizing the system by the feedback mechanism. In this modularized control design, ILC is embedded directly into the MFAC controller without any change of the existing MFAC facility, as shown in Figure 8.16.

The convergence analysis of MFAC with learning loop has been proposed in Ref. [222].

For the proposed control method, there is no need to modify the existing control device or system, but only to embed the ILC controller as the external loop to the MFAC controller in a modularized style to make full use of the available repetitive operation pattern that the controlled system possesses. The other modularized controller design methods can also be found in Refs. [67,204,223].

8.3.3 Simulations

In this section, three numerical simulations are given to illustrate the effectiveness of the modularized controller design approach. All these models are only used as an I/O data generator of the controlled plants.

Example 8.3: Estimation-type MFAC

The discrete-time nonlinear system

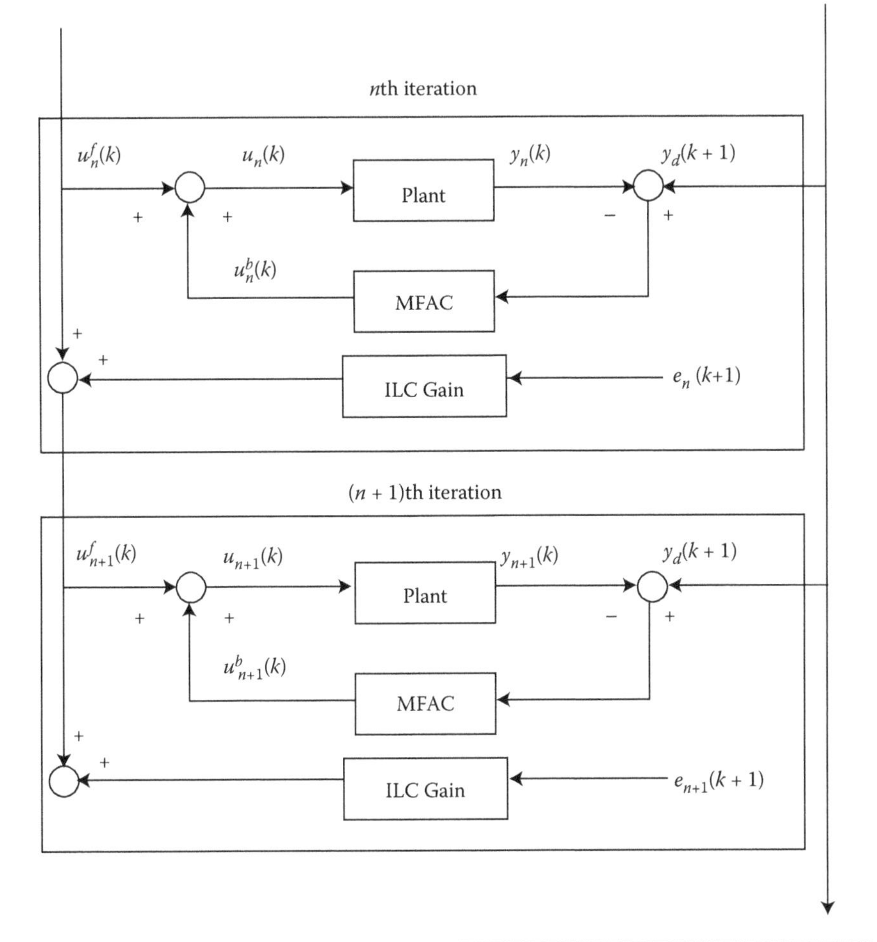

Figure 8.16 Block diagram of a learning-enhanced MFAC control system.

$$y(k+1) = \frac{y(k)}{(1+y^2(k))} + a(t)u^3(k) + 0.2y(k-1), \tag{8.42}$$

where $a(k) = 0.1 + 0.1^{\text{round}(k/100)}$, $k = 1,2,\ldots,1000$.

The desired trajectory is given as

$$y^*(k) = 5\sin\left(\frac{k\pi}{500}\right) \tag{8.43}$$

The simulation results obtained by applying adaptive controller (8.28) and estimation-type MFAC scheme (8.33)–(8.34) are shown in Figure 8.17. The simulation implies that the estimation-type MFAC gives better tracking performance and smoother control input with the help of estimation and compensation of the unmodeled dynamics.

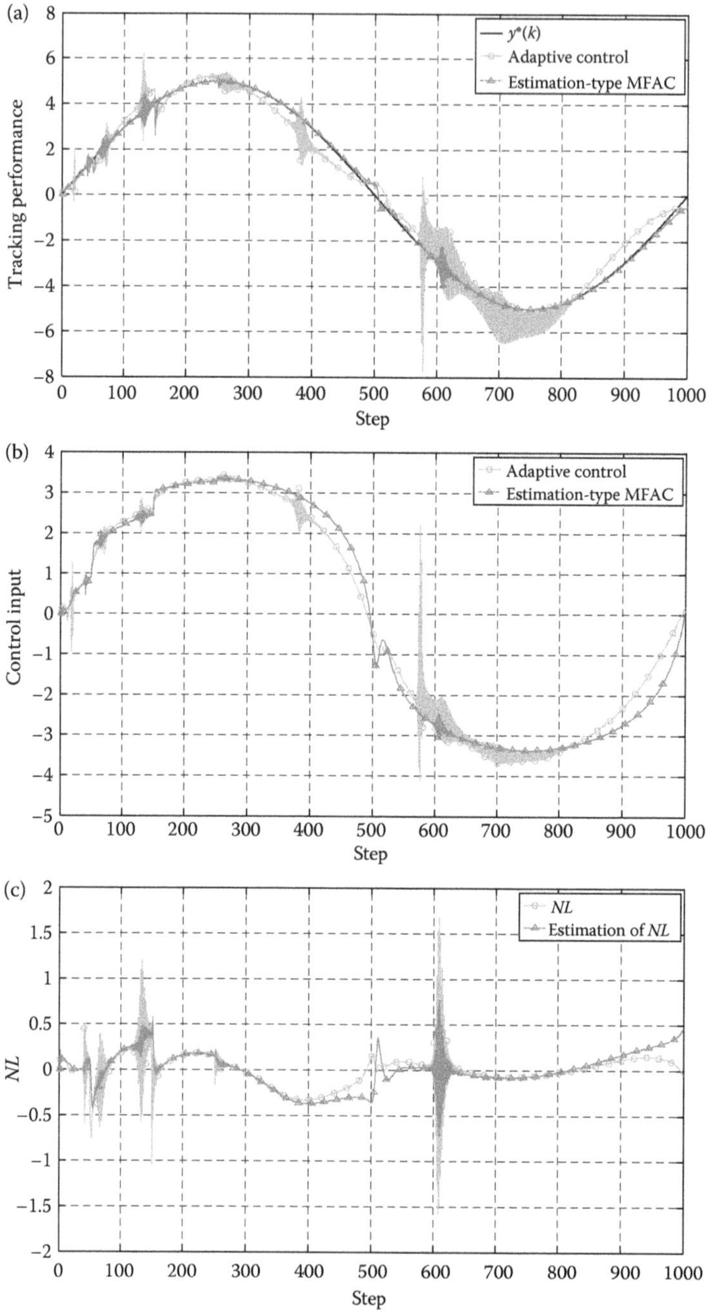

Figure 8.17 Simulation comparison between adaptive control and estimation-type MFAC. (a) Tracking performance. (b) Control input. (c) *NL*.

Example 8.4: Embedded-type MFAC

The plant and the desired trajectory are the same as in Example 8.3. The control performances obtained by applying adaptive controller (8.28) and embedded-type MFAC are shown in Figure 8.18. In the embedded-type control scheme, the internal loop adopts adaptive control method (8.28) and the external loop uses the PFDL–MFAC scheme with $L=2$, $\rho_1=\rho_2=0.3$, $\lambda=0.1$, $\eta=0.1$, and $\mu=2$. From this simulation, one can see that when embedded-type MFAC is applied, the control performance obtained by applying the embedded-type MFAC method is significantly improved.

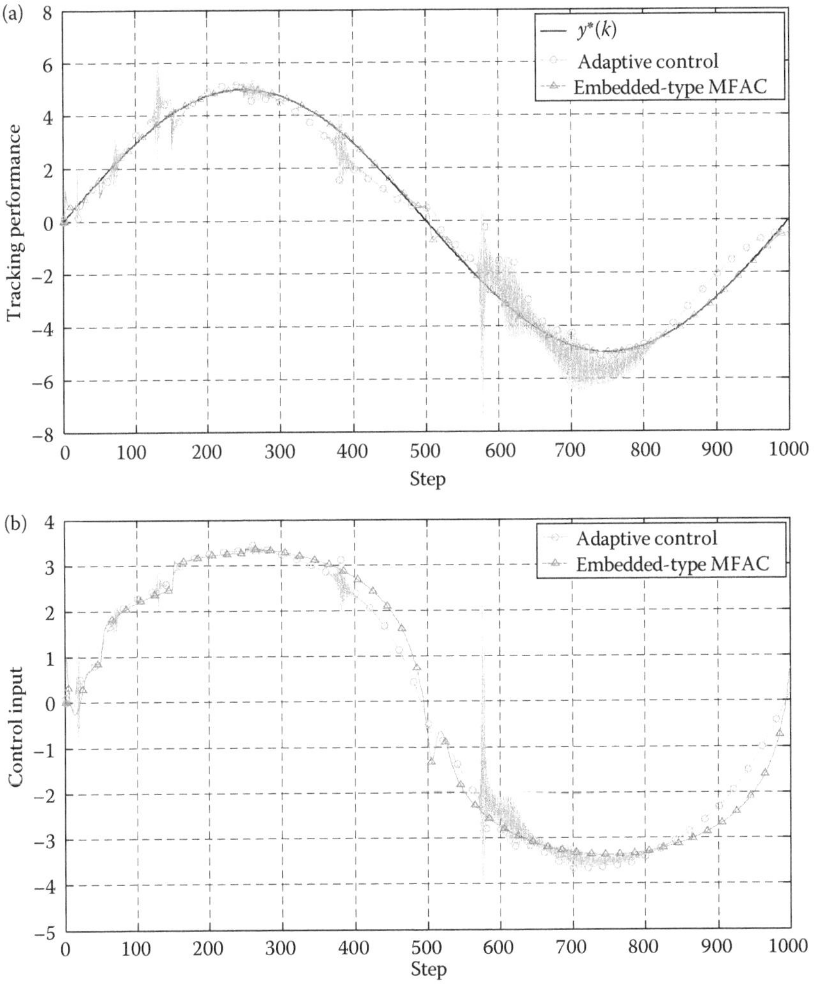

Figure 8.18 Simulation comparison between adaptive control and embedded-type MFAC. (a) Tracking performance. (b) Control input.

Example 8.5: Embedded-type MFAC for repetitive system

The discrete-time nonlinear system

$$\begin{cases} x_i(k) = 1.5u_i(k) - 1.5u_i^2(k) + 0.5u_i^3(k), \\ y_i(k+1) = 0.6y_i(k) - 0.1y_i(k-1) + 1.2x_i(k) - 0.1x_i(k-1) + v(k), \\ \quad k = 1,\ldots,200. \end{cases} \tag{8.44}$$

where $v(k)$ is an iteration-independent Gaussian white noise with standard deviation 0.05.

The desired trajectory is given as

$$y^*(k) = \begin{cases} 0.5, & 0 \le k \le 50; \\ 1.0, & 50 \le k \le 100; \\ 2.0, & 100 \le k \le 150; \\ 1.5, & 150 \le k \le 200. \end{cases} \tag{8.45}$$

First, applying PFDL–MFAC (8.39)–(8.41) with parameters $\rho = 0.9$, $\lambda = 4.6$, $\eta = 1$, $\mu = 1$, and $\varepsilon = 0.0001$, the control performance is acceptable as shown in Figure 8.19. Owing to the existence of noise, perfect tracking is not achieved. The control performance of MFAC with an iterative learning loop is shown in Figure 8.20. Embedded-type MFAC (8.37)–(8.41) uses the same MFAC parameters as above and a learning gain $\beta = 0.55$ is adopted. The simulation results show that the tracking error converges to zero as the iteration number increases. It should be

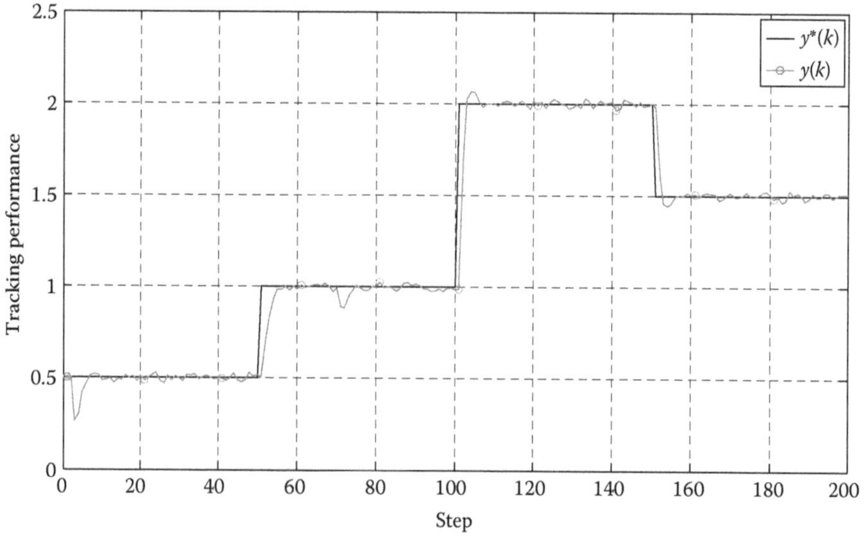

Figure 8.19 Control performance of MFAC without learning mechanism.

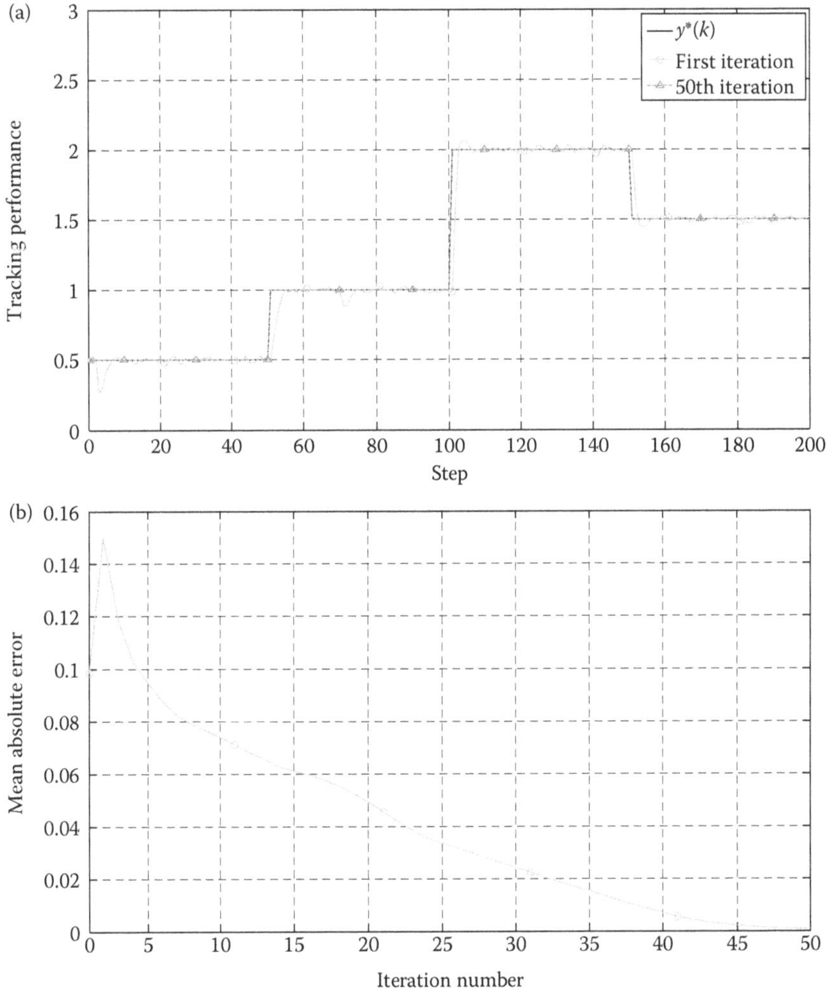

Figure 8.20 Control performance of MFAC with an external learning loop. (a) Control performance at the first and 50th iteration. (b) Mean absolute error.

noted that the tracking error at the first iteration, though ILC does not work, is small owing to the existence of MFAC.

8.4 Conclusions

For the MFAC methods, only the I/O data of the controlled plant are required to design the controller, rather than the mathematical model of the plant. Obviously, the complex system can be viewed as a new large augmented system. From this point of view, the MFAC methods can be applied directly to such augmented systems since they do not require the system model. If the control performance is not satisfactory by directly using the MFAC for this kind of systems, the MFAC methods proposed for complex interconnected systems in this chapter can be applied to improve the control performance. Using the presented methods, PG information of the PFDL data model of each subsystem can be estimated by using the collected I/O data from each subsystem, and then the MFAC scheme, that can make use of the available structure connection information or the measurable interactions, can be designed. In addition, based on the idea of complementation, estimation-type MFAC and embedded-type MFAC are also presented for improving the control performance. This design approach gives MFAC the ability to cowork with the classical feedback control or the feedforward ILC.

All the results in this chapter can be easily extended to the cases of the CFDL and FFDL data models for the SISO and MIMO nonlinear systems.

Chapter 9

Robustness of Model-Free Adaptive Control

9.1 Introduction

In practical applications, the plants may contain various uncertainties and are affected by different disturbances as well. In addition, there might be data dropouts if the control system is implemented using networks. Thus, the robustness of the closed-loop control system should be considered in controller design for the plants with uncertainties, disturbances, and data dropouts, that is, the designed closed-loop control systems should have the ability to remain stable.

Adaptive control mainly focuses on designing a controller for the controlled system with known structure and unknown slowly time-varying or time-invariant parameters. The robustness of adaptive control refers to the ability to stabilize the system with unmodeled dynamics, disturbances, and so on. To enhance the robustness of an adaptive control system, the following methods are widely applied in adaptation mechanism: the dead zone method [224,225], the normalization method [226,227], the parameter projection method [228,229], the σ modification [230,231], the averaging method [132,232], and so on. Although there are many relatively mature methods for model-based robust adaptive control, it is still difficult to find an effective robust adaptive control method when we face a practical control task.

The MFAC approach uses only the online measured I/O data to design the controller without involving a plant model; as a result, the unmodeled dynamics and traditional robustness issue does not exist under the framework of the MFAC. On the other hand, from the aspect of data, the measured I/O data of the system are always contaminated by noise. In addition, data dropouts caused by sensor fault, actuator fault, or network fault are inevitable. Thus, it is of great significance to study the robustness of the data-driven MFAC in the presence of measurement noise and data dropouts [14,126,233,234,321,322].

In this chapter, the robustness of the MFAC scheme for the unknown nonlinear system with measurement noise or data dropouts is analyzed. This chapter is organized as follows. In Section 9.2, the robustness of the MFAC scheme for the system with measurement noise is analyzed. In Section 9.3, the robustness of the MFAC scheme for the controlled system suffering from data dropouts is analyzed first and then a robust MFAC scheme with data compensation is presented. Section 9.4 presents the conclusions.

9.2 MFAC in the Presence of Output Measurement Noise

In many practical systems, both input and output measurements of the controlled system are always contaminated by noise. As shown in the previous chapters, the analysis and design of the data-driven MFAC system only depend on the measured I/O data of the controlled plant without any system information, such as model, structure, and parameters. An intuitive question is: how much of the stability and tracking performance of a closed-loop MFAC system subjected to noise remains? In this section, the robustness of the CFDL–MFAC scheme for a class of discrete-time SISO nonlinear systems in the presence of measurement noise is discussed. It is worth pointing out that the result for the unknown SISO nonlinear system can be easily extended to the unknown MISO and MIMO nonlinear systems.

9.2.1 Robust Stability Analysis

Consider a class of discrete-time SISO nonlinear systems as follow:

$$y(k+1) = f(y(k),\ldots,y(k-n_y),u(k),\ldots,u(k-n_u)), \tag{9.1}$$

where $y(k) \in R$ and $u(k) \in R$ are the system output and the control input at time k, respectively. n_y and n_u are two unknown integers, and $f(\cdots)$: $R^{n_u+n_y+2} \mapsto R$ is an unknown nonlinear function.

The diagram of the MFAC system with measurement noise is shown in Figure 9.1, where $|w(k)| < b_w$ is a bounded measurement noise, b_w is a positive constant, and $y_m(k) = y(k) + w(k)$ is the output measured.

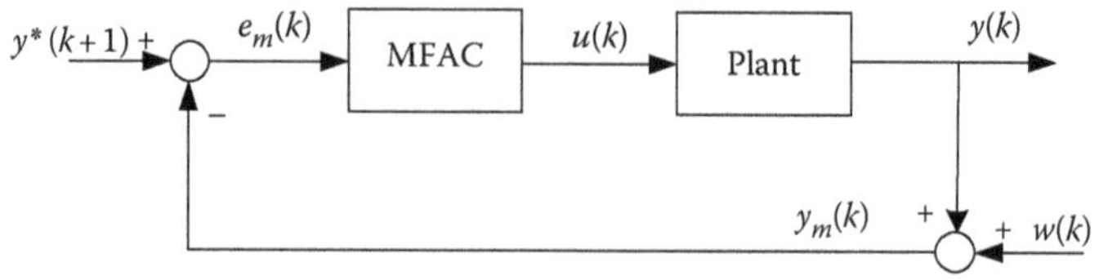

Figure 9.1 MFAC system with measurement noise.

For system (9.1) with measurement noise, the CFDL–MFAC scheme is constructed as follows:

$$\hat{\phi}_c(k) = \hat{\phi}_c(k-1) + \frac{\eta \Delta u(k-1)}{\mu + |\Delta u(k-1)|^2}(\Delta y_m(k) - \hat{\phi}_c(k-1)\Delta u(k-1)), \quad (9.2)$$

$$\hat{\phi}_c(k) = \hat{\phi}_c(1), \quad \text{if } |\hat{\phi}_c(k)| \leq \varepsilon \text{ or } |\Delta u(k-1)| \leq \varepsilon \text{ or } \operatorname{sign}(\hat{\phi}_c(k)) \neq \operatorname{sign}(\hat{\phi}_c(1)), \quad (9.3)$$

$$u(k) = u(k-1) + \frac{\rho\, \hat{\phi}_c(k)}{\lambda + |\hat{\phi}_c(k)|^2}[y^*(k+1) - y_m(k)], \quad (9.4)$$

where $\Delta y_m(k) \triangleq y_m(k) - y_m(k-1)$, $\lambda > 0, \mu > 0, \rho \in (0,1], \eta \in (0,1]$, $\hat{\phi}_c(1)$ is the initial value of $\hat{\phi}_c(k)$, and ε is a sufficiently small positive constant.

The following lemma will be used for robust stability analysis of the MFAC scheme.

Lemma 9.1

Estimation algorithm (9.2) and resetting algorithm (9.3) for nonlinear system (9.1), satisfying Assumptions 3.1, 3.2, 4.1, and 4.2, give that $\hat{\phi}_c(k)$ is bounded by a constant $\bar{b}_1$.

Proof

If $|\hat{\phi}_c(k)| \leq \varepsilon$ or $|\Delta u(k-1)| \leq \varepsilon$ or $\operatorname{sign}(\hat{\phi}_c(k)) \neq \operatorname{sign}(\hat{\phi}_c(1))$, then the boundedness of $\hat{\phi}_c(k)$ is guaranteed by resetting algorithm (9.3).

In the other case, defining the PPD estimation error as $\tilde{\phi}_c(k) \triangleq \hat{\phi}_c(k) - \phi_c(k)$ and subtracting $\phi_c(k)$ from both sides of (9.2), we have

$$\begin{aligned}\tilde{\phi}_c(k) &= \tilde{\phi}_c(k-1) - \phi_c(k) + \phi_c(k-1) \\ &\quad + \frac{\eta \Delta u(k-1)}{\mu + |\Delta u(k-1)|^2}(\Delta y_m(k) - \hat{\phi}_c(k-1)\Delta u(k-1)).\end{aligned} \quad (9.5)$$

According to the definition of $y_m(k)$ and Theorem 3.1, we have

$$\begin{aligned}\Delta y_m(k) &= \Delta y(k) + \Delta w(k)\\ &= \phi_c(k-1)\Delta u(k-1) + \Delta w(k),\end{aligned} \tag{9.6}$$

where $\Delta w(k) \triangleq w(k) - w(k-1)$.
Substituting (9.6) into (9.5) yields

$$\begin{aligned}\tilde{\phi}_c(k) &= \left(1 - \frac{\eta|\Delta u(k-1)|^2}{\mu + |\Delta u(k-1)|^2}\right)\tilde{\phi}_c(k-1)\\ &+ \frac{\eta\Delta u(k-1)}{\mu + |\Delta u(k-1)|^2}\Delta w(k) - \phi_c(k) + \phi_c(k-1).\end{aligned} \tag{9.7}$$

Since $\mu > 0$, $\eta \in (0,1]$, and function $\eta x/(\mu + x)$ is monotonically increasing with respect to argument x, the following inequalities hold:

$$0 < \frac{\eta\varepsilon^2}{\mu + \varepsilon^2} \le \frac{\eta|\Delta u(k-1)|^2}{\mu + |\Delta u(k-1)|^2} < \eta \le 1, \tag{9.8}$$

$$\frac{\eta|\Delta u(k-1)|}{\mu + |\Delta u(k-1)|^2} = \frac{\eta}{\dfrac{\mu}{|\Delta u(k-1)|} + |\Delta u(k-1)|} \le \frac{\eta}{2\sqrt{\mu}}. \tag{9.9}$$

Theorem 3.1 shows that there exists a positive constant $\bar{b}$ such that $|\phi_c(k)| \le \bar{b}$. Using $|\phi_c(k)| \le \bar{b}$ and (9.9), the following inequality can be easily obtained:

$$\left|\frac{\eta\Delta u(k-1)}{\mu + |\Delta u(k-1)|^2}\right||\Delta w(k)| + |\phi_c(k) - \phi_c(k-1)| \le \frac{\eta}{2\sqrt{\mu}}2b_w + 2\bar{b} = \frac{\eta b_w}{\sqrt{\mu}} + 2\bar{b}. \tag{9.10}$$

Let $c = \eta b_w/\sqrt{\mu} + 2\bar{b}$ and $\delta = 1 - (\eta\varepsilon^2/(\mu + \varepsilon^2))$. Taking the absolute value on both sides of (9.7) and using (9.8) and (9.10) yield

$$
\begin{aligned}
\left|\tilde{\phi}_c(k)\right| &\leq \left|1 - \frac{\eta \Delta u^2(k-1)}{\mu + \Delta u^2(k-1)^2}\right| \left|\tilde{\phi}_c(k-1)\right| \\
&\quad + \left|\frac{\eta \Delta u(k-1)}{\mu + \Delta u^2(k-1)}\right| \left|\Delta w(k)\right| + \left|\phi_c(k) - \phi_c(k-1)\right| \\
&\leq (1-\delta)\left|\tilde{\phi}_c(k-1)\right| + c \leq (1-\delta)^2 \left|\tilde{\phi}_c(k-2)\right| + c(1-\delta) + c \\
&\leq \cdots \leq (1-\delta)^{k-1}\left|\tilde{\phi}_c(1)\right| + \frac{c}{1-(1-\delta)}.
\end{aligned}
\tag{9.11}
$$

Inequality (9.11) implies that $\tilde{\phi}_c(k)$ is bounded. Thus, $\hat{\phi}_c(k)$ is also bounded since $\phi_c(k)$ is bounded. ■

Defining the tracking error as $e(k) = y^*(k) - y(k)$, and using Lemma 9.1, the following theorem can be given.

Theorem 9.1

Nonlinear system (9.1), satisfying Assumptions 3.1, 3.2, 4.1, and 4.2, is controlled by the CFDL–MFAC scheme (9.2)–(9.4); if $y^*(k+1) = y^*$ and $|w(k)| < b_w$, then there exists a constant $\lambda_{\min} > 0$, such that the following property holds for any $\lambda > \lambda_{\min}$:

$$
\lim_{k \to \infty} \left|e(k)\right| \leq \frac{d_2 b_w}{d_1},
\tag{9.12}
$$

where $d_1 = \rho \underline{\varepsilon} \varepsilon / (\lambda + \bar{b}_1^2)$ and $d_2 = \rho \bar{b} / (2\sqrt{\lambda})$.

Proof

Define the tracking error of measured output as $e_m(k) = y^* - y_m(k)$. From the definition of the tracking error, we have

$$
e_m(k) = e(k) - w(k).
\tag{9.13}
$$

Using (9.4) and (9.13) yields

$$
u(k) = u(k-1) + \frac{\rho \hat{\phi}_c(k)}{\lambda + \left|\hat{\phi}_c(k)\right|^2} e_m(k) = u(k-1) + \frac{\rho \hat{\phi}_c(k)}{\lambda + \left|\hat{\phi}_c(k)\right|^2} (e(k) - w(k)).
\tag{9.14}
$$

Since

$$y^* - y(k+1) = y^* - y(k) - \phi_c(k)\Delta u(k), \tag{9.15}$$

$\underline{\varepsilon} < \phi_c(k) \le \bar{b}$, $\varepsilon < \hat{\phi}_c(k) \le \bar{b}_1$, and $\rho \in (0,1]$, the following inequality holds for any $\lambda > \bar{b}^2/4$:

$$0 < \frac{\rho\varepsilon\underline{\varepsilon}}{\lambda + \bar{b}_1^2} \le \vartheta(k) \le \frac{\rho\bar{b}\hat{\phi}_c(k)}{2\sqrt{\lambda}\hat{\phi}_c(k)} = \frac{\rho\bar{b}}{2\sqrt{\lambda}} < 1,$$

where $\vartheta(k) = \dfrac{\rho\phi_c(k)\hat{\phi}_c(k)}{\lambda + \hat{\phi}_c^2(k)}$.

Substituting (9.14) into (9.15) yields

$$e(k+1) = (1 - \vartheta(k))e(k) + \vartheta(k)w(k). \tag{9.16}$$

Let $d_1 = \rho\varepsilon\underline{\varepsilon}/(\lambda + \bar{b}_1^2)$ and $d_2 = \rho\bar{b}/(2\sqrt{\lambda})$. Taking the absolute value on both sides of (9.16) yields

$$\begin{aligned}|e(k+1)| &\le |(1 - \vartheta(k))||e(k)| + |\vartheta(k)||w(k)|.\\ &\le (1-d_1)|e(k)| + d_2 b_w\\ &\le (1-d_1)^2|e(k-1)| + (1-d_1)d_2 b_w + d_2 b_w\\ &\vdots\\ &\le (1-d_1)^k|e(1)| + \frac{d_2 b_w}{d_1},\end{aligned} \tag{9.17}$$

and thus $\lim_{k\to\infty}|e(k)| \le (d_2 b_w/d_1)$. ■

Remark 9.1

Theorem 9.1 shows that the MFAC scheme (9.2)–(9.4) guarantees the boundedness of the tracking error of the closed-loop system as long as the measurement noise is bounded. The upper bound of the tracking error is dependent on the upper bound of the measurement noise $w(k)$. If the system is not contaminated by measurement noise, that is, $w(k) = 0$, then the tracking error converges to zero. Thus, the MFAC system is robust and stable.

9.2.2 Simulations

In this section, a numerical simulation is given to show the robustness of the CFDL–MFAC scheme (9.2)–(9.4) for a nonlinear system with measurement noise.

Example 9.1

Discrete-time SISO nonlinear system

$$y(k+1) = \begin{cases} \dfrac{y(k)}{1+y^2(k)} + u^3(k), & k \le 500 \\ \dfrac{y(k)y(k-1)y(k-2)u(k-1)\big(y(k-2)-1\big) + a(k)u(k)}{1+y^2(k-1)+y^2(k-2)}, & k > 500 \end{cases} \tag{9.18}$$

where $a(k) = 1 + \text{round}(k/500)$ is a time-varying parameter.

The desired output signal is

$$y^*(k+1) = \begin{cases} 1, & 1 \le k \le 500 \\ -1, & 500 < k \le 1000. \end{cases}$$

MFAC scheme (9.2)–(9.4) is applied to nonlinear system (9.18). The initial values of the system are $u(1) = u(2) = 0$, $y(2) = 1$, and $y(1) = -1$. The controller parameters are set to $\hat{\phi}_c(1) = 2$, $\rho = 1$, $\lambda = 10$, $\eta = 1$, $\mu = 1$, and $\varepsilon = 10^{-5}$. The measurement noise is given as follows:

$$w(k) = \begin{cases} 0.1 + 0.05\text{rand}(1), & \text{if } k \le 250 \\ -0.1 + 0.05\text{rand}(1), & \text{if } 250 < k \le 500 \\ 0.1\sin(2\pi k/250) + 0.05\text{rand}(1), & \text{if } 500 < k \le 750 \\ 0.1\cos(2\pi k/250) + 0.05\text{rand}(1), & \text{if } 750 < k \le 1000. \end{cases}$$

The measurement noise, tracking performance, and control input of the closed-loop system are shown in Figure 9.2a, b, and c, respectively. The simulation results show that CFDL–MFAC scheme (9.2)–(9.4) guarantees robust stability of the closed-loop system despite the measurement noise.

9.3 MFAC in the Presence of Data Dropouts

With the development of control science, computer networks, and communication techniques, networked control systems (NCS) have been widely used [235,236]. Compared with traditional point-to-point connections, the use of the

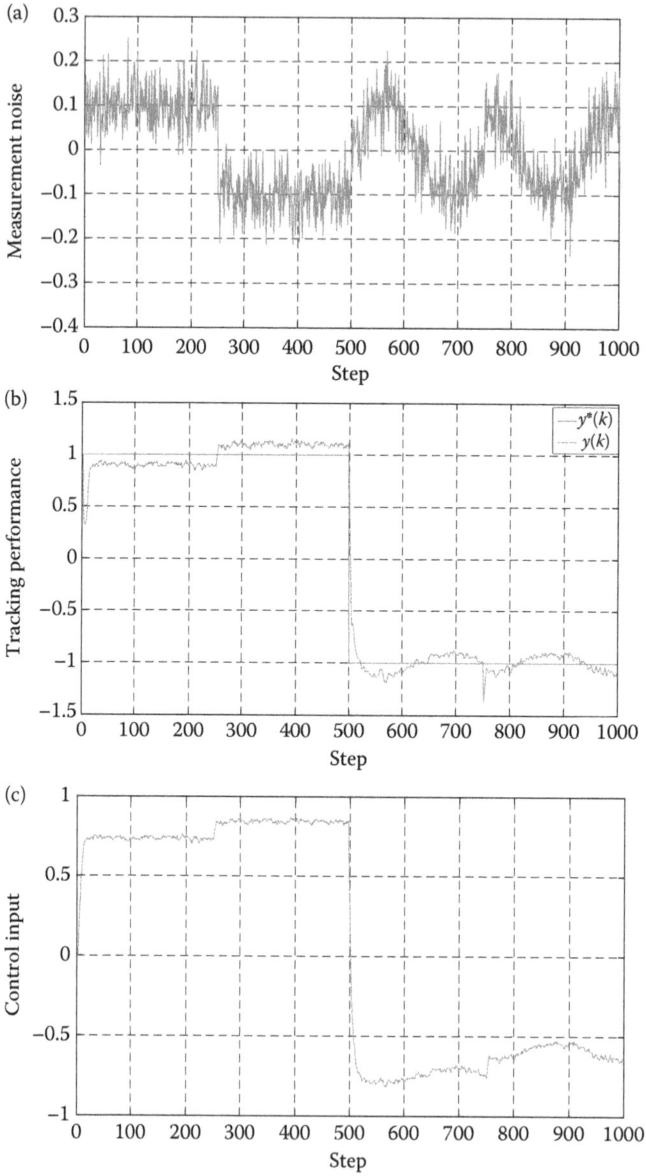

Figure 9.2 Control performance of the CFDL–MFAC scheme with measurement noise. (a) Measurement noise $w(k)$. (b) Tracking performance. (c) Control input.

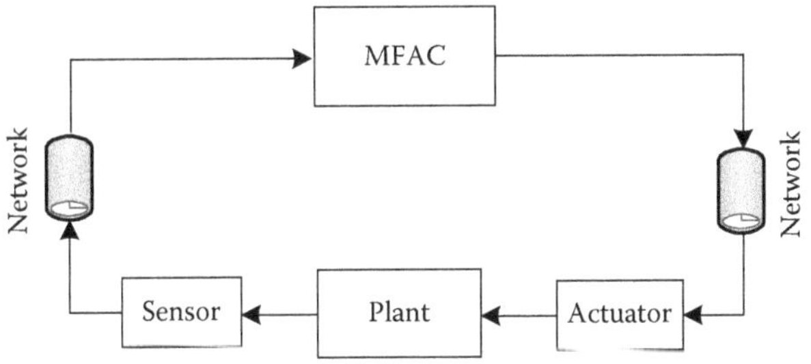

Figure 9.3 Model-free adaptive control system with data dropouts.

communication channels can reduce the costs of cables and implementations, and simplify the installation and maintenance of the whole system. However, network congestion and connection failure, which will lead to the occurrence of data dropouts, are inevitable in networked systems. Data dropout is an important issue in NCS, and the analysis and design of an NCS with data dropouts has become the basic issue in NCS. On the basis of the above consideration, the robustness of the MFAC system is discussed in this section.

As shown in Figure 9.3, when the MFAC approach is used in an NCS, the system output data and the control input data of the controlled system are transmitted from the sensor to the controller and from the controller to the actuator via a network, respectively. Thus, the control input data and system output data may be dropped out due to network fault and sensor fault. Without loss of generality, we only consider the robustness of the MFAC system with only output data dropouts in this section. The result can be easily extended to the case of control input data dropouts.

9.3.1 Robust Stability Analysis

Consider a class of discrete-time SISO nonlinear systems as follows:

$$y(k+1) = f(y(k),\dots,y(k-n_y),u(k),\dots,u(k-n_u)). \tag{9.19}$$

According to Theorem 3.1, if $\Delta u(k) \neq 0$, nonlinear system (9.19) can be transformed into the following CFDL data model:

$$y(k+1) = y(k) + \phi_c(k)\Delta u(k), \tag{9.20}$$

and the corresponding CFDL–MFAC scheme (4.7)–(4.9) is designed.

From control scheme (4.7)–(4.9), one can see that the system output data $y(k)$ is required for updating the PPD estimated value $\hat{\phi}_c(k)$ and the control input signal $u(k)$. Thus, control scheme (4.7)–(4.9) needs to be modified under the case of output data dropouts.

It is supposed that the controller can detect whether the output $y(k)$ is dropped or not. Define the following two variables:

$$\beta(k) = \begin{cases} 1, & \text{if } y(k) \text{ is not dropped} \\ 0, & \text{if } y(k) \text{ is dropped} \end{cases}, \quad \bar{y}(k) = \begin{cases} y(k), & \text{if } \beta(k) = 1 \\ \bar{y}(k-1), & \text{if } \beta(k) = 0 \end{cases}$$

and $P\{\beta(k) = 1\} = \bar{\beta}$, where $P\{\cdot\}$ denotes probability.

The CFDL–MFAC scheme for data dropouts is constructed as

$$\hat{\phi}_c(k) = \hat{\phi}_c(k-1) + \beta(k)\left[\frac{\eta \Delta u(k-1)}{\mu + |\Delta u(k-1)|^2}(\bar{y}(k) - \bar{y}(k-1) - \hat{\phi}_c(k-1)\Delta u(k-1))\right], \tag{9.21}$$

$$\hat{\phi}_c(k) = \hat{\phi}_c(1), \quad \text{if } \beta(k-1) = 1 \quad \text{and} \quad \left(|\Delta u(k-1)| \le \varepsilon \quad \text{or} \quad |\hat{\phi}_c(k)| \le \varepsilon \quad \text{or} \quad \text{sign}(\hat{\phi}_c(k)) \ne \text{sign}(\hat{\phi}_c(1))\right), \tag{9.22}$$

$$u(k) = u(k-1) + \beta(k)\left[\frac{\rho \hat{\phi}_c(k)}{\lambda + |\hat{\phi}_c(k)|^2}(y^*(k+1) - \bar{y}(k))\right], \tag{9.23}$$

where $\eta \in (0,1]$, $\rho \in (0,1]$, $\lambda > 0$, $\mu > 0$, and ε is a sufficiently small positive constant.

Theorem 9.2

If nonlinear system (9.19), satisfying Assumptions 3.1, 3.2, 4.1, and 4.2, is controlled by the CFDL–MFAC scheme (9.21)–(9.23) with $y^*(k+1) = y^*$, then the tracking error converges to zero as long as not all the output data are dropped.

Proof

The proof of the boundedness of $\hat{\phi}_c(k)$ is similar to that of Lemma 10.1.

Define the tracking error as $e(k) = y^* - y(k)$, and let k_i be the time instants without data dropouts, $i = 1,2,\ldots$.

Since not all the output data are dropped, $\bar{\beta} \ne 0$ holds. When k is within k_{i-1} and k_i, controller algorithm (9.23) gives

$$u(k) = \begin{cases} u(k-1) + \dfrac{\rho \hat{\phi}_c(k)e(k)}{\lambda + |\hat{\phi}_c(k)|^2}, & \text{if } k = k_{i-1} \text{ or } k = k_i \\ u(k_{i-1}), & \text{if } k_{i-1} < k < k_i. \end{cases} \tag{9.24}$$

Thus, the following equation holds as $e(k) \neq 0$:

$$\Delta u(k) = \begin{cases} \dfrac{\rho \hat{\phi}_c(k)e(k)}{\lambda + |\hat{\phi}_c(k)|^2} \neq 0, & \text{if } k = k_{i-1} \text{ or } k = k_i \\ 0, & \text{if } k_{i-1} < k < k_i. \end{cases} \tag{9.25}$$

The above equation implies that $u(k_i - 1) = u(k_i - 2) = \cdots = u(k_{i-1}) \neq u(k_{i-1} - 1)$. Using Theorem 3.2, this gives

$$y(k) = y(k_{i-1}) + \phi_c(k-1)(u(k-1) - u(k_{i-1} - 1)), \quad \text{if } k_{i-1} < k \leq k_i. \tag{9.26}$$

Substituting (9.26) into the tracking error equation and taking the absolute value yield

$$\begin{aligned} |e(k)| &= |y^* - y(k)| \\ &= |y^* - y(k_{i-1}) - \phi_c(k-1)(u(k-1) - u(k_{i-1} - 1))| \\ &= |y^* - y(k_{i-1}) - \phi_c(k-1)(u(k_{i-1}) - u(k_{i-1} - 1))| \\ &= \left| e(k_{i-1}) - \phi_c(k-1)\frac{\rho \hat{\phi}_c(k_{i-1})e(k_{i-1})}{\lambda + |\hat{\phi}_c(k_{i-1})|^2} \right| \\ &\leq \left| \left(1 - \frac{\rho\phi_c(k-1)\hat{\phi}_c(k_{i-1})}{\lambda + |\hat{\phi}_c(k_{i-1})|^2} \right) \right| |e(k_{i-1})|, \end{aligned} \tag{9.27}$$

where $k_{i-1} < k \leq k_i$, $i = 1,2,\ldots$.

From the boundedness of $\phi_c(k)$ and $\hat{\phi}_c(k)$ and resetting algorithm (9.22), there exist two positive constants $\bar{b}$ and $\bar{b}_1$ such that $\underline{\varepsilon} < \phi_c(k) \leq \bar{b}$ and $\varepsilon < \hat{\phi}_c(k) \leq \bar{b}_1$ hold for any k. Since $\rho \in (0,1]$ and $\lambda > \bar{b}^2/4$, the following inequality holds:

$$0 < \frac{\rho\underline{\varepsilon}\varepsilon}{\lambda + \bar{b}_1^2} \leq \frac{\rho\phi_c(k-1)\hat{\phi}_c(k_{i-1})}{\lambda + |\hat{\phi}_c(k_{i-1})|^2} \leq \frac{\rho\bar{b}\,\hat{\phi}_c(k_{i-1})}{2\sqrt{\lambda}\,\hat{\phi}_c(k_{i-1})} = \frac{\rho\bar{b}}{2\sqrt{\lambda}} < 1.$$

Let $d_1 = \rho\underline{\varepsilon}\varepsilon/(\lambda + \bar{b}_1^2) < 1$. Equation (9.27) can be rewritten as

$$|e(k)| \leq (1 - d_1)|e(k_{i-1})|, \tag{9.28}$$

where $k_{i-1} < k \leq k_i$, $i = 1,2,\ldots$.

This inequality implies that the control scheme guarantees tracking error convergence as long as not all the output data are dropped. ■

9.3.2 MFAC Scheme with Data-Dropped Compensation

In NCS, there are many algorithms used to compensate data dropout, such as weighted average algorithm [237], adaptive filtering approach [238], and predictive approach using the past data [239,240]. All these algorithms can be used to design the MFAC scheme with data-dropped compensation. In this section, a data-dropped compensation algorithm is proposed based on the CFDL data model, and then used to design the MFAC scheme to attenuate the effect on the control performance caused by the data dropouts.

According to Theorem 3.1, discrete-time SISO nonlinear system (9.19) is transformed into the following CFDL data model:

$$y(k) = y(k-1) + \phi_c(k-1)\Delta u(k-1). \tag{9.29}$$

Define the following variables:

$$\beta(k) = \begin{cases} 1, & \text{if } y(k) \text{ is not dropped} \\ 0, & \text{if } y(k) \text{ is dropped} \end{cases}, \quad \bar{\bar{y}}(k) = \begin{cases} y(k), & \text{if } \beta(k) = 1 \\ \hat{y}(k), & \text{if } \beta(k) = 0 \end{cases},$$

and $P\{\beta(k) = 1\} = \bar{\beta}$, where $\hat{y}(k)$ is the estimation of the output $y(k)$, and $P\{\cdot\}$ denotes probability.

If the output data $y(k)$ is dropped, then $y(k-1)$, $\hat{\phi}_c(k-1)$, and $\Delta u(k-1)$ can be used to estimate the value of $y(k)$ according to (9.29). That is,

$$\hat{y}(k) = \bar{\bar{y}}(k-1) + \hat{\phi}_c(k-1)\Delta u(k-1), \tag{9.30}$$

and consequently, the CFDL–MFAC with data-dropped compensation is given as follows:

$$\hat{\phi}_c(k) = \hat{\phi}_c(k-1) + \beta(k)\left[\frac{\eta\Delta u(k-1)}{\mu + |\Delta u(k-1)|^2}(\bar{\bar{y}}(k) - \bar{\bar{y}}(k-1) - \hat{\phi}_c(k-1)\Delta u(k-1))\right], \tag{9.31}$$

$$\hat{\phi}_c(k) = \hat{\phi}_c(1), \quad \text{if } |\hat{\phi}_c(k)| \leq \varepsilon \text{ or } |\Delta u(k-1)| \leq \varepsilon \quad \text{or} \quad \text{sign}(\hat{\phi}_c(k)) \neq \text{sign}(\hat{\phi}_c(1)), \tag{9.32}$$

$$u(k) = u(k-1) + \frac{\rho\hat{\phi}_c(k)}{\lambda + |\hat{\phi}_c(k)|^2}(y^*(k+1) - \bar{\bar{y}}(k)), \tag{9.33}$$

where $\mu > 0$, $\eta \in (0,1]$, $\rho \in (0,1]$, $\lambda > 0$, and ε is a small positive constant.

The CFDL–MFAC scheme (9.31)–(9.33) for the nonlinear system with data dropouts has the following results.

Theorem 9.3

If nonlinear system (9.29) satisfying Assumptions 3.1, 3.2, 4.1, and 4.2, is controlled by the CFDL–MFAC scheme with data-dropped compensation (9.31)–(9.33) for $y^*(k+1) = y^*$, then the tracking error converges to zero as long as not all the output data are dropped.

Proof

The proof of the boundedness of $\hat{\phi}_c(k)$ is similar to that of Lemma 9.1.

Define k_i as the time instant without data dropouts, $i = 1,2,\ldots$. Since not all the output data are dropped, $\bar{\beta} \neq 0$ holds. According to (9.30), at time instant $k_{i-1}+1$ with data dropout, the following equation holds:

$$\bar{\bar{y}}(k_{i-1}+1) = y(k_{i-1}) + \hat{\phi}_c(k_{i-1})\Delta u(k_{i-1}).$$

At time instant $k_{i-1}+2$ with data dropout, we have

$$\begin{aligned}\bar{\bar{y}}(k_{i-1}+2) &= \bar{\bar{y}}(k_{i-1}+1) + \hat{\phi}_c(k_{i-1}+1)\Delta u(k_{i-1}+1)\\ &= y(k_{i-1}) + \hat{\phi}_c(k_{i-1})\Delta u(k_{i-1}) + \hat{\phi}_c(k_{i-1}+1)\Delta u(k_{i-1}+1).\end{aligned}$$

Analogously, at time instant $k_i - 1$ with data dropout, it gives

$$\begin{aligned}\bar{\bar{y}}(k_i-1) &= \bar{\bar{y}}(k_i-2) + \hat{\phi}_c(k_i-2)\Delta u(k_i-2)\\ &= \bar{\bar{y}}(k_i-3) + \hat{\phi}_c(k_i-3)\Delta u(k_i-3) + \hat{\phi}_c(k_i-2)\Delta u(k_i-2)\\ &\vdots\\ &= y(k_{i-1}) + \hat{\phi}_c(k_{i-1})\Delta u(k_{i-1}) + \cdots + \hat{\phi}_c(k_i-3)\Delta u(k_i-3) + \hat{\phi}_c(k_i-2)\Delta u(k_i-2).\end{aligned}$$

Thus, the following Equation (9.34) holds for any time instants $k_{i-1} < k < k_i$:

$$\begin{aligned}\bar{\bar{y}}(k) &= \bar{\bar{y}}(k-1) + \hat{\phi}_c(k-1)\Delta u(k-1)\\ &= \bar{\bar{y}}(k-2) + \hat{\phi}_c(k-2)\Delta u(k-2) + \hat{\phi}_c(k-1)\Delta u(k-1)\\ &\vdots\\ &= y(k_{i-1}) + \hat{\phi}_c(k_{i-1})\Delta u(k_{i-1}) + \cdots + \hat{\phi}_c(k-2)\Delta u(k-2) + \hat{\phi}_c(k-1)\Delta u(k-1).\end{aligned} \tag{9.34}$$

Defining the tracking error as $e(k) = y^* - y(k)$ and using Equation (9.34) and controller algorithm (9.33), we have

$$u(k) = \begin{cases} u(k-1) + \dfrac{\rho\hat{\phi}_c(k)e(k)}{\lambda + |\hat{\phi}_c(k)|^2}, & \text{if } k = k_{i-1} \text{ or } k = k_i \\ u(k-1) + \dfrac{\rho\hat{\phi}_c(k)}{\lambda + |\hat{\phi}_c(k)|^2} & \\ \times (e(k_{i-1}) - \hat{\phi}_c(k_{i-1})\Delta u(k_{i-1}) - \cdots - \hat{\phi}_c(k-1)\Delta u(k-1)), & \text{if } k_{i-1} < k < k_i. \end{cases} \tag{9.35}$$

Let $c(k) = 1 - \dfrac{\rho\hat{\phi}_c(k)^2}{\lambda + |\hat{\phi}_c(k)|^2}$. Then (9.35) gives

$$\Delta u(k_{i-1}) = \frac{\rho\hat{\phi}_c(k_{i-1})}{\lambda + |\hat{\phi}_c(k_{i-1})|^2}e(k_{i-1}),$$

$$\begin{aligned}\Delta u(k_{i-1}+1) &= \frac{\rho\hat{\phi}_c(k_{i-1}+1)}{\lambda + |\hat{\phi}_c(k_{i-1}+1)|^2}(e(k_{i-1}) - \hat{\phi}_c(k_{i-1})\Delta u(k_{i-1}))\\ &= \frac{\rho\hat{\phi}_c(k_{i-1}+1)}{\lambda + |\hat{\phi}_c(k_{i-1}+1)|^2}\left(1 - \frac{\rho\hat{\phi}_c(k_{i-1})^2}{\lambda + |\hat{\phi}_c(k_{i-1})|^2}\right)e(k_{i-1})\\ &= \frac{\rho\hat{\phi}_c(k_{i-1}+1)c(k_{i-1})}{\lambda + |\hat{\phi}_c(k_{i-1}+1)|^2}e(k_{i-1}).\end{aligned}$$

$$\Delta u(k_{i-1}+2) = \frac{\rho\hat{\phi}_c(k_{i-1}+2)}{\lambda+|\hat{\phi}_c(k_{i-1}+2)|^2}(e(k_{i-1})-\hat{\phi}_c(k_{i-1})\Delta u(k_{i-1})-\hat{\phi}_c(k_{i-1}+1)\Delta u(k_{i-1}+1))$$

$$= \frac{\rho\hat{\phi}_c(k_{i-1}+2)}{\lambda+|\hat{\phi}_c(k_{i-1}+2)|^2}\left(c(k_{i-1})e(k_{i-1})-\frac{\rho\hat{\phi}_c(k_{i-1}+1)^2c(k_{i-1})}{\lambda+|\hat{\phi}_c(k_{i-1}+1)|^2}e(k_{i-1})\right)$$

$$= \frac{\rho\hat{\phi}_c(k_{i-1}+2)c(k_{i-1}+1)c(k_{i-1})}{\lambda+|\hat{\phi}_c(k_{i-1}+2)|^2}e(k_{i-1}). \tag{9.36}$$

Similarly, the following equation holds for any time instants $k_{i-1} < k < k_i$ with data dropout:

$$\Delta u(k) = \frac{\rho\hat{\phi}_c(k)}{\lambda+|\hat{\phi}_c(k)|^2}(e(k_{i-1})-\hat{\phi}_c(k_{i-1})\Delta u(k_{i-1})-\cdots-\hat{\phi}_c(k-1)\Delta u(k-1))$$

$$= \frac{\rho\hat{\phi}_c(k)\prod_{i=k_{i-1}}^{k-1}c(i)}{\lambda+|\hat{\phi}_c(k)|^2}e(k_{i-1}). \tag{9.37}$$

If $e(k_{i-1}) \neq 0$, then $\Delta u(k) \neq 0$ from (9.37), where $k_{i-1} < k < k_i$. Thus, according to Theorem 3.1, we have

$$\begin{aligned} y(k) &= y(k-1)+\phi_c(k-1)\Delta u(k-1) \\ &= y(k-2)+\phi_c(k-2)\Delta u(k-2)+\phi_c(k-1)\Delta u(k-1) \\ &\vdots \\ &= y(k_{i-1})+\phi_c(k_{i-1})\Delta u(k_{i-1})+\cdots+\phi_c(k-2)\Delta u(k-2)+\phi_c(k-1)\Delta u(k-1). \end{aligned} \tag{9.38}$$

Substituting (9.37) and (9.38) into the tracking error equation yields

$$e(k) = e(k_{i-1}) - \phi_c(k_{i-1})\Delta u(k_{i-1}) - \cdots - \phi_c(k-2)\Delta u(k-2) - \phi_c(k-1)\Delta u(k-1)$$

$$= e(k_{i-1})\left(1 - \frac{\rho\phi_c(k_{i-1})\,\hat{\phi}_c(k_{i-1})}{\lambda + \left|\hat{\phi}_c(k_{i-1})\right|^2} - \cdots - \frac{\rho\phi_c(k-2)\,\hat{\phi}_c(k-2)\prod_{i=k_{i-1}}^{k-2} c(i)}{\lambda + \left|\hat{\phi}_c(k-2)\right|^2} - \frac{\rho\phi_c(k-1)\,\hat{\phi}_c(k-1)\prod_{i=k_{i-1}}^{k-1} c(i)}{\lambda + \left|\hat{\phi}_c(k-1)\right|^2}\right). \tag{9.39}$$

The boundedness of $\hat{\phi}_c(k)$ and the resetting algorithm imply that there exists a positive constant $\bar{b}_1$ such that $\varepsilon < \hat{\phi}_c(k) \le \bar{b}_1$ holds for any time instant k. Note that the function $\rho x/(\lambda + x)$ is a monotonically increasing function with respect to argument x, then its minimum and maximum values can be determined by the following inequality:

$$0 < \underline{c} = \frac{\rho\varepsilon^2}{\lambda + \varepsilon^2} < c(k) = \frac{\rho\,\hat{\phi}_c(k)^2}{\lambda + \left|\hat{\phi}_c(k)\right|^2} \le \frac{\rho\bar{b}_1^2}{\lambda + \bar{b}_1^2} = \bar{c}.$$

From the boundedness of $\phi_c(k)$ and $\hat{\phi}_c(k)$, we know that there exist two constants d_1 and d_2 such that

$$0 < d_1 = \frac{\rho\bar{\varepsilon}\varepsilon}{\lambda + \bar{b}_1^2} \le \frac{\rho\phi_c(k)\,\hat{\phi}_c(k)}{\lambda + \left|\hat{\phi}_c(k)\right|^2} \le \frac{\rho\bar{b}\,\hat{\phi}_c(k)}{\lambda + \left|\hat{\phi}_c(k)\right|^2} \le \frac{\rho\bar{b}}{2\sqrt{\lambda}} = d_2.$$

From the definition of d_2 and $\bar{c}$, it implies

$$d_2 + \bar{c} = \frac{\rho\bar{b}}{2\sqrt{\lambda}} + \frac{\rho\bar{b}_1^2}{\lambda + \bar{b}_1^2} \le \frac{\rho\bar{b}}{2\sqrt{\lambda}} + \frac{\rho\bar{b}_1^2}{2\sqrt{\lambda}\bar{b}_1} = \frac{\rho\bar{b}}{2\sqrt{\lambda}} + \frac{\rho\bar{b}_1}{2\sqrt{\lambda}}.$$

Selecting $\rho \in (0,1]$ and using $\lambda > \left(\bar{b} + \bar{b}_1\right)^2/4$, we have

$$d_1 + \underline{c}d_1 + \cdots + \underline{c}^{k-k_{i-1}}d_1$$

$$\leq \frac{\rho\phi_c(k_{i-1})\,\hat{\phi}_c(k_{i-1})}{\lambda + \left|\hat{\phi}_c(k_{i-1})\right|^2} + \cdots + \frac{\rho\phi_c(k-2)\,\hat{\phi}_c(k-2)\prod_{i=k_{i-1}}^{k-2}c(i)}{\lambda + \left|\hat{\phi}_c(k-2)\right|^2}$$

$$+ \cdots + \frac{\rho\phi_c(k-1)\,\hat{\phi}_c(k-1)\prod_{i=k_{i-1}}^{k-1}c(i)}{\lambda + \left|\hat{\phi}_c(k-1)\right|^2}$$

$$\leq d_2 + d_2\bar{c} + \cdots + d_2\bar{c}^{k-k_{i-1}} < \frac{d_2}{1-\bar{c}} < 1. \tag{9.40}$$

According to (9.40), there exists a positive constant $d'(k-k_{i-1})$ such that

$$\left|1 - \frac{\rho\phi_c(k_{i-1})\,\hat{\phi}_c(k_{i-1})}{\lambda + \left|\hat{\phi}_c(k_{i-1})\right|^2} - \cdots - \frac{\rho\phi_c(k-2)\,\hat{\phi}_c(k-2)\prod_{i=k_{i-1}}^{k-2}c(i)}{\lambda + \left|\hat{\phi}_c(k-2)\right|^2}\right.$$

$$\left. - \frac{\rho\phi_c(k-1)\,\hat{\phi}_c(k-1)\prod_{i=k_{i-1}}^{k-1}c(i)}{\lambda + \left|\hat{\phi}_c(k-1)\right|^2}\right|$$

$$\leq \left|1 - \left(d_1 + \underline{c}d_1 + \cdots + \underline{c}^{k-k_{i-1}}d_1\right)\right| \triangleq d'(k-k_{i-1}) < 1. \tag{9.41}$$

Taking the absolute value on both sides of (9.39) and using (9.41) yields

$$|e(k)| \leq \left|1 - (d_1 + \underline{c}d_1 + \cdots + \underline{c}^{k-k_{i-1}}d_1)\right|\left|e(k_{i-1})\right| = d'(k-k_{i-1})\left|e(k_{i-1})\right|. \tag{9.42}$$

Inequality (9.42) implies that the control scheme guarantees tracking error convergence as long as not all the output data are dropped. ■

Remark 9.2

From the proof of Theorem 9.2, the tracking error convergence rate for MFAC scheme (9.21)–(9.23) is as follows:

$$1 - d_1 = 1 - \frac{\rho\bar{\varepsilon}\varepsilon}{\lambda + \bar{b}_1^2}.$$

From the proof of Theorem 9.3, we can see that the tracking error convergence rate for the improved MFAC scheme (9.31)–(9.33) is as follows:

$$d'(k - k_{i-1}) = \left|1 - d_1 + \underline{c}d_1 + \cdots + \underline{c}^{k-k_{i-1}}d_1\right| = \left|1 - \frac{\rho\bar{\varepsilon}\varepsilon}{\lambda + \bar{b}_1^2}(1 + \underline{c} + \cdots + \underline{c}^{k-k_{i-1}})\right|.$$

Obviously, for a given weighting factor λ, the improved MFAC scheme (9.31)–(9.33) has a faster convergence rate than that of the MFAC scheme (9.21)–(9.23).

9.3.3 Simulations

In this section, a numerical simulation is given to illustrate the difference in the control performance and the robustness between the MFAC scheme without data-dropped compensation (9.21)–(9.23) and the MFAC scheme with data-dropped compensation (9.31)–(9.33).

Example 9.2

The nonlinear system with the initial values and the desired output signal is the same as that in Example 9.1. The parameters of the two CFDL–MFAC schemes are set as $\hat{\phi}_c(1) = 2$, $\rho = 1$, $\lambda = 10$, $\eta = 1$, $\mu = 1$, and $\varepsilon = 10^{-5}$.

The simulation results between the two MFAC schemes with 80% data dropout rate and 90% data dropout rate are shown in Figure 9.4, respectively. The simulation results show that both control schemes can yield the same steady-state performance even if the output data dropout rate is higher. It is worth pointing out that the MFAC scheme with data-dropped compensation can attenuate the effect on the control performance by data dropouts and improve the transit response and the convergence rate of the tracking error as well.

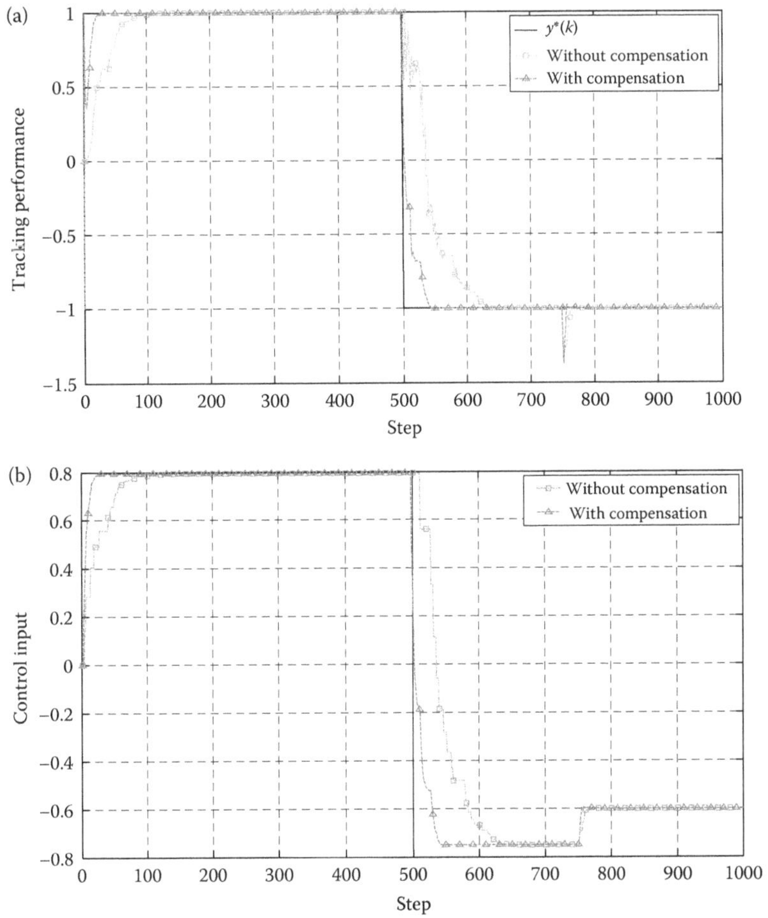

Figure 9.4 Simulation comparison between the two MFAC schemes with data dropouts. (a) Tracking performance with 80% data dropout rate. (b) Control input with 80% data dropout rate. (c) Tracking performance with 90% data dropout rate. (d) Control input with 90% data dropout rate.

9.4 Conclusions

In this chapter, the robust stability of the CFDL–MFAC scheme for a class of nonlinear systems with measurement noise and data dropouts is analyzed. The theoretical analysis and simulation results show that the CFDL–MFAC scheme guarantees boundedness of the tracking error as long as the measurement noise is bounded, and convergence of the tracking error if not all the output data are lost. On the basis of the analysis of CFDL–MFAC with data dropouts, a CFDL–MFAC scheme with data-dropped compensation is presented for improving the convergence rate of the tracking error.

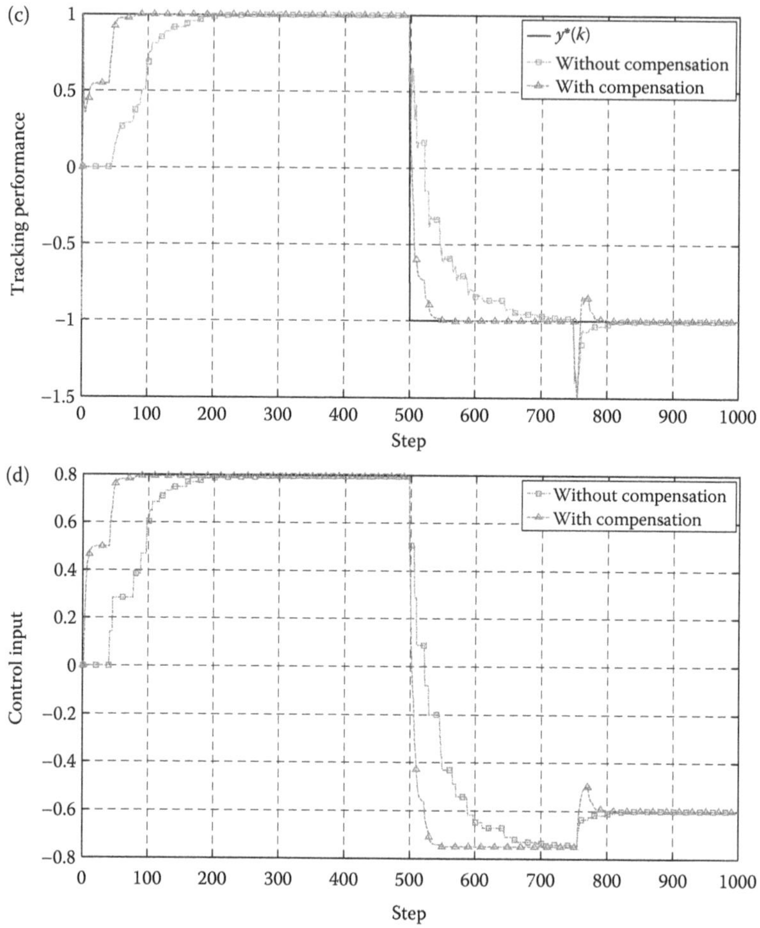

Figure 9.4 (continued) Simulation comparison between the two MFAC schemes with data dropouts. (a) Tracking performance with 80% data dropout rate. (b) Control input with 80% data dropout rate. (c) Tracking performance with 90% data dropout rate. (d) Control input with 90% data dropout rate.

It is worth pointing out that, MFAC approach uses only the online measured I/O data to design the controller without involving a plant model, and the performance of controlled system is not affected by the system model, just depending on I/O data. Since the uncertainties of data mainly contain noise signal or incompleteness in many practical systems, the framework of robustness for MFAC in this chapter can provide reference to other data-driven control methods.

Chapter 10

Symmetric Similarity for Control System Design

10.1 Introduction

Most of the studies on the control system design were generally conducted from the viewpoint of "mathematical analysis." In other words, the analysis for the controlled plant, the controller design, and the performance of the closed-loop system were explored separately. Taking adaptive control as an example, there are usually three steps in the design and analysis of an adaptive control system. First, the parameter estimator is designed and analyzed using the measured data of the controlled system based on a mathematical model whose structure is given. Then, the model-based controller is developed in which the information on parameter estimation is included. Finally, the properties of the closed-loop system are analyzed based on the error dynamics theoretically. Instead, there are only a few works that explore the control system analysis and synthesis directly using the control system structure as a whole.

As we all know, symmetric similarity structures exist widely in many animal and plant species in nature, and they are formed by a "self-optimizing" process with evolvement and natural selection in a long time [23,24,241]. For example, it seems that the left and right parts of a plant leaf are symmetric, but they are only similar in structure. This fact regarding "optimal structure systems" in nature reveals that systems with symmetric similarity structures must have some incomparable advantages, compared with other systems without symmetric similarity structures. In other words, systems with symmetric similarity structures are "superior." Can we learn from nature to design artificial control systems with symmetric similarity structures as "superior systems"? Apparently, the exploration of this problem, considered as a "self-optimizing" evolution progress of the control theory itself, is of significant interest both in theory and practice. Up to now, however, there is little work focusing on symmetric similarity in the control system design.

In addition, all the species on the earth evolve following the natural laws, including competition and hybridization, and as a result, create a colorful world. On the basis of the similarities existing in the real world, humans are able to analyze, design, and solve practical problems by taking advantage of the similarities between different systems. Using the similarity is an important way for us to discover and understand the world. The symmetric similarity in structure is regarded as one of the intrinsic characteristics of humans as an intelligent organism.

Since the last century, many control theories and methodologies have been developed, and they have significantly influenced our lives in various aspects. Although these theories and methods were proposed and developed independently, some of them are intrinsically similar to each other. If we can clearly define and take advantage of the similarity, a new viewpoint would be provided for us to study and solve the control problems encountered in practice. As a result, new "superior" controllers may be constructed.

It is, however, full of challenges to study the symmetric similarity in control systems, and the similarities between different control theories. In particular, many more mathematical tools are required to be developed further to assist in the definition and analysis of similarity. In Ref. [241] and the references therein, initial works on linear systems, as well as some complex systems, with symmetric similarity structures were reported. In Ref. [23], the basic idea and principle of adaptive control system design with symmetric similarity structure was introduced originally, and some preliminary works based on this idea were in Refs. [124,242–246]. Similar results can also be found in Refs. [247–252]. Recently, model-free adaptive iterative learning control, which is similar to the existing MFAC, was developed in Refs. [208,253]. Some adaptive iterative learning control schemes based on the similarity analysis between adaptive control and iterative learning control were proposed in Refs. [254–261].

In this chapter, the design and analysis of control systems with symmetric similarity structures are explored by taking discrete-time adaptive control as an example.

First, the design methods of control systems with symmetric similarity structures are presented for adaptive control and MFAC, respectively. For both of them, the structures of the controller and the estimator are having symmetric similarity, and the design procedures of the controller and the estimator are also having symmetric similarity.

Second, the similarities between MFAC and model-free adaptive iterative learning control are explored, both of them are analyzed using the contraction mapping method. The two control methods are similar in many points, such as the design procedure and structure of the controller and the estimator, the convergence property, the analysis tool, and so on. The design and analysis of the two control methods have been presented in Chapters 4 and 7, respectively. In this chapter, we only provide a comparison table for readers to explore the similarities between them. Readers interested in these two control approaches can read Chapters 4 and 7 for comparison.

Finally, this chapter also discusses the similarities between adaptive control and iterative learning control, which are two completely different control methods under different control backgrounds. As a direct result, a new superior control system, adaptive iterative learning control, is proposed. Compared with traditional adaptive control, the proposed adaptive iterative learning control can achieve perfect tracking over a finite time interval for nonlinear systems with unknown time-varying parameters. Compared with iterative learning control, the proposed approach can track the iteration-varying reference trajectory with random initial states, and thus relaxes the strictly repeatable condition, which is essential for ILC to guarantee convergence.

This chapter is organized as follows. Section 10.2 gives the concepts, the design principle, and the symmetric similarity framework of adaptive control. Section 10.3 shows the similarities between MFAC and MFAILC based on the contents in Chapters 4 and 7. The similarities between adaptive control and ILC are explored in Section 10.4. Section 10.5 presents the conclusions.

10.2 Symmetric Similarity for Adaptive Control Design

In general, an adaptive control system consists of two main parts: an estimator and a controller. Since it is difficult to extend the well-known concepts in linear systems, such as zero-pole assignment, characteristic polynomial, and so on, to general discrete-time nonlinear systems, adaptive control design methods for nonlinear systems are mainly based on the minimized prediction error. In this section, the design of adaptive control systems with symmetric similarity is also developed by utilizing the minimized prediction error.

Consider a class of discrete-time SISO nonlinear systems:

$$y(k+1) = f(y(k),\ldots,y(k-n_y),u(k),\ldots,u(k-n_u),\boldsymbol{\theta}(k)), \tag{10.1}$$

where $y(k) \in R$, $u(k) \in R$, and $\boldsymbol{\theta}(k)$ are the system output, the control input, and an unknown slowly time-varying parameter at time instant k, respectively. n_y and n_u are two integers, and $f(\ldots)$ is a nonlinear scalar function.

The main steps of adaptive control design for system (10.1) are to update the estimation $\hat{\boldsymbol{\theta}}(k)$ of the unknown parameter by minimizing a cost function of parameter estimation at first, and then to calculate the control input by minimizing a cost function of control input $u(k)$ based on the certainty equivalence principle and the estimation $\hat{\boldsymbol{\theta}}(k)$.

It can be seen that the roles of the slowly time-varying parameter $\boldsymbol{\theta}(k)$ and the control input $u(k)$ in the adaptive control system design are of equal importance. First, both their updating laws are obtained by minimizing a specified cost function, correspondingly. Second, the constraints considered both for controller algorithm design and parameter estimation algorithm design are similar. In particular, it is a common requirement that the variation of the control input between two consecutive time instants, $\Delta u(k)$, should not be very large to reduce actuator abrasion and energy consumption. Similarly, to design the estimation algorithm of $\boldsymbol{\theta}(k)$, $\Delta\hat{\boldsymbol{\theta}}(k)$ is also required to be not very large in order to enhance the robustness against abrupt variation caused by disturbance or sensor failure. Otherwise, the abrupt change of $\hat{\boldsymbol{\theta}}(k)$ may lead to an unstable system, where the estimated parameter is used. Next, for predictive control, it needs to predict and adjust the future values of the control input $u(k)$ when computing the control input signal by using the receding horizon control technology. In the same way, the parameter estimation $\hat{\boldsymbol{\theta}}(k+j), j=1,\ldots,n$ is also needed to be predicted by using the existing estimations $\hat{\boldsymbol{\theta}}(1),\hat{\boldsymbol{\theta}}(2),\ldots,\hat{\boldsymbol{\theta}}(k)$ to predict the system output while fully taking into consideration the time variance in parameters [24,250–252].

In summary, the design processes of the control input law and the parameter estimation are very similar in some adaptive control systems. This inspires us to design and analyze a new adaptive control system by utilizing the "symmetric similarity."

10.2.1 Concepts and Design Principle of Symmetric Similarity

To study the symmetric similarity for the adaptive control system design in a rigorous and intensive way, the concept of symmetric similarity structure of the cost functions (a cost function of parameter estimation and a cost function of control input) of an adaptive control system is given as follows.

Concept 10.1

The cost function of parameter estimation and the cost function of control input for an adaptive control system are said to have *symmetric similarity structure*, if the two cost functions are in the same form when interchanging the desired output and the current control input in the cost function of the control input with the actual output and the estimated parameter in the cost function of parameter estimation, respectively.

It is worth pointing out that Concept 10.1 and some other contents of this chapter are not as rigorous as previous chapters in the sense of mathematics. However, for the convenience of delivering the main idea in the following, here we still try to present it in a mathematical way like Concept 10.1.

Example 10.1

The following cost function of the control input:

$$J(u(k)) = [y^*(k+1) - f(y(k),\ldots,y(k-n_y),u(k),\ldots,u(k-n_u),\hat{\boldsymbol{\theta}}(k))]^2$$

and the cost function of parameter estimation

$$J(\hat{\boldsymbol{\theta}}(k+1)) = [y(k+1) - f(y(k),\ldots,y(k-n_y),u(k),\ldots,u(k-n_u),\hat{\boldsymbol{\theta}}(k+1))]^2$$

have symmetric similarity structure.

Example 10.2

The following cost function of the control input:

$$J(u(k)) = [y^*(k+1) - f(y(k),\ldots,y(k-n_y),u(k),\ldots,u(k-n_u),\hat{\boldsymbol{\theta}}(k))]^2 + \lambda \| u(k) - u(k-1) \|^2$$

and the cost function of parameter estimation

$$J(\hat{\boldsymbol{\theta}}(k+1)) = [y(k+1) - f(y(k),\ldots,y(k-n_y),u(k),\ldots,u(k-n_u),\hat{\boldsymbol{\theta}}(k+1))]^2 + \mu \| \hat{\boldsymbol{\theta}}(k+1) - \hat{\boldsymbol{\theta}}(k) \|^2$$

have symmetric similarity structure.

Example 10.3

The cost function of control input (4.3) and the cost function of PPD estimation (4.5) in Chapter 4 have symmetric similarity structure.

Concept 10.2

An adaptive control system is said to have *symmetric similarity structure* if the cost function of control input and the cost function of parameter estimation have symmetric similarity structure, and the controller algorithm and the parameter estimation algorithm are derived by using the same optimization procedure.

For example, the adaptive control systems proposed in Refs. [124,242,243,245] have symmetric similarity structures.

On the basis of the above concepts of symmetric similarity structure, the design principle of symmetric similarity structure is presented as follows:

> *Design principle of symmetric similarity*: For a time-varying parameter estimation algorithm, obtained by minimizing a cost function of time-varying parameter, there must exist a corresponding "symmetric similarity" adaptive control algorithm, obtained by minimizing a cost function of the control input, and vice versa. The two algorithms together constitute an adaptive control system with a "symmetric similarity structure."

The above *design principle of symmetric similarity* provides a new way to design adaptive control systems. It puts the design methods of the parameter estimator and the controller into a unified framework. Thus, the dual design and analysis tasks for the estimator and the controller within a adaptive control system are simplified as a single design and analysis problem of parameter estimation algorithm or control input algorithm.

10.2.2 Adaptive Control with Symmetric Similarity Structure

On the basis of the concepts and principle of symmetric similarity structure proposed in Section 10.2.1, we can verify whether an existing adaptive control system has symmetric similarity structure or not, and design some new adaptive control systems.

10.2.2.1 Adaptive Control with Symmetric Similarity Structure for Linear Systems

Projection-Based Adaptive Control — Consider SISO LTI system

$$A(q^{-1})y(k) = B(q^{-1})u(k), \tag{10.2}$$

where $A(q^{-1}) = 1 + a_1q^{-1} + \cdots + a_nq^{-n}$, $B(q^{-1}) = b_1q^{-1} + \cdots + b_nq^{-n}$, and q^{-1} is a shifting operator with $q^{-1}y(k) = y(k-1)$.

Equation (10.2) can be rewritten in the following one-step-ahead predictor form:

$$y(k+1) = \boldsymbol{H}^T(k)\boldsymbol{\theta}, \tag{10.3}$$

where

$$\boldsymbol{\theta} = [-a_1,\ldots,-a_n,b_1,\ldots,b_n]^T,$$

$$\boldsymbol{H}(k) = [y(k),\ldots,y(k-n+1),u(k),\ldots,u(k-n+1)]^T.$$

Consider the following two cost functions of control input and parameter estimation with symmetric similarity structure, respectively:

$$J(u(k)) = \min \frac{1}{2} \left| u(k) - u(k-1) \right|^2, \qquad (10.4)$$
$$\text{s.t. } y^*(k+1) = \boldsymbol{H}^T(k)\hat{\boldsymbol{\theta}}(k).$$

$$J(\hat{\theta}(k)) = \min \frac{1}{2} \left\| \hat{\boldsymbol{\theta}}(k) - \hat{\boldsymbol{\theta}}(k-1) \right\|^2, \qquad (10.5)$$
$$\text{s.t. } y(k) = \boldsymbol{H}^T(k-1)\hat{\boldsymbol{\theta}}(k).$$

Minimizing cost functions (10.4) and (10.5), respectively, one can obtain the following control law and estimation law with the symmetric similarity structure:

$$u(k) = u(k-1) + \frac{\hat{b}_1(k)(y^*(k+1) - \boldsymbol{H}'^T(k)\hat{\boldsymbol{\theta}}'(k) - \hat{b}_1(k)u(k-1))}{\hat{b}_1(k)^2}, \qquad (10.6)$$

$$\hat{\boldsymbol{\theta}}(k) = \hat{\boldsymbol{\theta}}(k-1) + \frac{\boldsymbol{H}(k-1)[y(k) - \boldsymbol{H}^T(k-1)\hat{\boldsymbol{\theta}}(k-1)]}{\boldsymbol{H}^T(k-1)\boldsymbol{H}(k-1)}, \qquad (10.7)$$

where

$$\boldsymbol{H}'(k) = [y(k), \ldots, y(k-n+1), u(k-1), \ldots, u(k-n+1)]^T,$$

$$\hat{\boldsymbol{\theta}}'(k) = [-\hat{a}_1(k), \ldots, -\hat{a}_n(k), \hat{b}_2(k), \ldots, \hat{b}_n(k)]^T.$$

To avoid zero denominator in (10.6) and (10.7), two constants, λ and μ, are added respectively; then we have

$$u(k) = u(k-1) + \frac{\hat{b}_1(k)(y^*(k+1) - \boldsymbol{H}'^T(k)\hat{\boldsymbol{\theta}}(k) - \hat{b}_1(k)u(k-1))}{\lambda + \hat{b}_1(k)^2}, \qquad (10.8)$$

$$\hat{\boldsymbol{\theta}}(k) = \hat{\boldsymbol{\theta}}(k-1) + \frac{\boldsymbol{H}(k-1)[y(k) - \boldsymbol{H}^T(k-1)\hat{\boldsymbol{\theta}}(k-1)]}{\mu + \boldsymbol{H}^T(k-1)\boldsymbol{H}(k-1)}. \qquad (10.9)$$

The adaptive control system using above control law and estimation law is with symmetric similarity structure.

For system (10.2), the other two extended cost functions of control input and parameter estimation can also be considered as follows to design the adaptive control system with symmetric similarity structure:

$$J(u(k)) = |y^*(k+1) - \boldsymbol{H}^T(k)\hat{\boldsymbol{\theta}}(k)|^2 + \lambda\ |u(k) - u(k-1)|^2, \tag{10.10}$$

$$J(\hat{\theta}(k)) = |y(k) - \boldsymbol{H}^T(k-1)\hat{\theta}(k)|^2 + \mu\|\hat{\theta}(k) - \hat{\theta}(k-1)\|^2, \tag{10.11}$$

where $\lambda > 0$ and $\mu > 0$ are two weighting factors added to punish the variations of control input and parameter estimation, respectively. The large variation of control input between two consecutive time instants may cause saturation or abrasion of the actuator. Similarly, the large variation of parameter estimation between two consecutive time instants due to noise or sensor faults would lead to instability of the adaptive control system.

Cost functions (10.10) and (10.11) have symmetric similarity structure. Minimizing them, one can obtain the following controller algorithm and estimation law, respectively, which have symmetric similarity structure:

$$u(k) = u(k-1) + \frac{\hat{b}_1(k)}{\lambda + \hat{b}_1(k)^2}(y^*(k-1) - \boldsymbol{H}'^T(k)\hat{\boldsymbol{\theta}}(k) - \hat{b}_1(k)u(k-1)), \tag{10.12}$$

$$\hat{\boldsymbol{\theta}}(k) = \hat{\boldsymbol{\theta}}(k-1) + \frac{\boldsymbol{H}(k-1)[y(k) - \boldsymbol{H}^T(k-1)\hat{\boldsymbol{\theta}}(k-1)]}{\mu + \boldsymbol{H}^T(k-1)\boldsymbol{H}(k-1)}. \tag{10.13}$$

Although Equations (10.12) and (10.13) have the same forms with Equations (10.8) and (10.9), respectively, they are derived from different theories essentially.

According to Concept 10.1, the cost function of control input (10.4) and the cost function of parameter estimation (10.11) are neither symmetric nor similar. Thus, the control law (10.6) and the parameter estimation law (10.13), derived from these two cost functions, respectively, do not have symmetric similarity structure as a result. For the same reason, the cost function of control input (10.10) and the cost function of parameter estimation (10.5), as well as the control law (10.12) and the parameter estimation law (10.7), do not have symmetric similarity structure.

RLS-Based Adaptive Control — Considering system (10.3) again, we design a cost function of control input which is symmetrically similar to the RLS-based cost function of parameter estimation, and develop an RLS-based adaptive control system with symmetric similarity structure according to Concept 10.2.

The common RLS-based cost function of parameter estimation is

$$J(\theta) = \frac{1}{2}\sum_{k=1}^{N}((y(k) - \boldsymbol{H}^T(k-1)\boldsymbol{\theta})^2 + (\boldsymbol{\theta} - \hat{\boldsymbol{\theta}}(0))^T P^{-1}(0)(\boldsymbol{\theta} - \hat{\boldsymbol{\theta}}(0))), \tag{10.14}$$

where $\hat{\boldsymbol{\theta}}(0)$ and $P(-1)$ are given, and $P(-1)$ is positive definite.

By minimizing the cost function (10.14), one can obtain the following RLS-based parameter estimation algorithm:

$$\hat{\boldsymbol{\theta}}(k) = \hat{\boldsymbol{\theta}}(k-1) + \frac{P(k-2)\boldsymbol{H}(k-1)}{1+\boldsymbol{H}^T(k-1)P(k-2)\boldsymbol{H}(k-1)}(y(k) - \boldsymbol{H}^T(k-1)\hat{\boldsymbol{\theta}}(k-1)),$$
$$P(k-1) = P(k-2) - \frac{P(k-2)\boldsymbol{H}(k-1)\boldsymbol{H}^T(k-1)P(k-2)}{1+\boldsymbol{H}^T(k-1)P(k-2)\boldsymbol{H}(k-1)}, \tag{10.15}$$

Although RLS algorithm (10.15) has a fast convergence rate, the gain $(P(k-2)H(k-1))/(1+H^T(k-1)P(k-2)H(k-1))$ may become very small only after several iterations (generally, 10 or 20 steps). As a direct result, the RLS algorithm cannot deal with time-varying parameters. The corresponding improved algorithm is given as Equations (2.53)–(2.55). From the design principle of symmetric similarity, the control law, which is symmetrically similar to the RLS-based parameter estimation algorithm (10.15), cannot address the tracking problem in general, but is valid in dealing with the case of $y^*(k+1) = \text{const}$. The details are as follows.

The cost function of control input with the symmetric similarity structure to the cost function of parameter estimation (10.14) is

$$J(u(k)) = \frac{1}{2}\sum_{t=1}^{k}((y^*(t+1) - y(t+1))^2 + (u(k) - u(0))^T P^{-1}(0)(u(k) - u(0))). \tag{10.16}$$

By the same optimization procedure, one can obtain the following controller algorithm:

$$u(k) = u(k-1) + \frac{P(k-2)\hat{b}_1(k)}{1+\hat{b}_1(k)^2 P(k-2)}(y^*(k+1) - \boldsymbol{H}'^T(k)\hat{\boldsymbol{\theta}}(k) - \hat{b}_1(k)u(k-1)),$$
$$P(k-1) = P(k-2) - \frac{P(k-2)^2\hat{b}_1(k)^2}{1+\hat{b}_1(k)^2 P(k-2)}, \tag{10.17}$$

where

$$\boldsymbol{H}'(k) = [y(k),\ldots,y(k-n+1),u(k-1),\ldots,u(k-n+1)]^T,$$
$$\hat{\boldsymbol{\theta}}'(k) = [-\hat{a}_1(k),\ldots,-\hat{a}_n(k),\hat{b}_2(k),\ldots,\hat{b}_n(k)]^T.$$

But how to design an RLS-based adaptive control system with symmetric similarity structure to deal with the tracking problem? A typical method is to revise the control law by adding nonzero time-varying gain, such as least squares with covariance modification, least-squares with covariance resetting, least-squares with time-varying forgetting factors, improved least-squares, and so on. For example, the least-squares algorithm with covariance resetting is

$$
\begin{aligned}
u(k) &= u(k-1) + \frac{P(k-2)\hat{b}_1(k)}{1+\hat{b}_1(k)^2 P(k-2)}(y^*(k+1) - \boldsymbol{H}'^T(k)\hat{\boldsymbol{\theta}}(k) - \hat{b}_1(k)u(k-1)), \\
P(k-1) &= P(k-2) - \frac{P^2(k-2)\hat{b}_1(k)^2}{1+\hat{b}_1(k)^2 P(k-2)}, \\
P(k-1) &= \sigma_k I, \quad \text{if } P(k-1) \le \varepsilon \text{ or } k = k_i, i = 1,2,\ldots
\end{aligned}
\tag{10.18}
$$

10.2.2.2 Adaptive Control with Symmetric Similarity Structure for Nonlinear Systems

For simplicity, rewrite nonlinear system (10.1) as

$$
y(k+1) = f(\boldsymbol{Y}(k), u(k), \boldsymbol{U}(k-1), \boldsymbol{\theta}(k)), \tag{10.19}
$$

where $\boldsymbol{Y}(k) = [y(k),\ldots,y(k-n_y)]$, $\boldsymbol{U}(k) = [u(k),\ldots,u(k-n_u)]$, and $\boldsymbol{\theta}(k)$ is an unknown slowly time-varying parameter.

The cost function of control input is given as

$$
J(u(k)) = [y^*(k+1) - f(\boldsymbol{Y}(k), u(k), \boldsymbol{U}(k-1), \hat{\boldsymbol{\theta}}(k))]^2 + \lambda \| u(k) - u(k-1) \|^2, \tag{10.20}
$$

and the cost function of parameter estimation is

$$
\begin{aligned}
J(\hat{\boldsymbol{\theta}}(k+1)) &= [y(k+1) - f(\boldsymbol{Y}(k), u(k), \boldsymbol{U}(k-1), \hat{\boldsymbol{\theta}}(k+1))]^2 \\
&\quad + \mu \| \hat{\boldsymbol{\theta}}(k+1) - \hat{\boldsymbol{\theta}}(k) \|^2.
\end{aligned}
\tag{10.21}
$$

Clearly, (10.20) and (10.21) have symmetric similarity structure.

Taking first-order Taylor expansion on $f(\boldsymbol{Y}(k), u(k), \boldsymbol{U}(k-1), \hat{\boldsymbol{\theta}}(k+1))$ at the point $[\boldsymbol{Y}(k), u(k), \boldsymbol{U}(k-1), \hat{\boldsymbol{\theta}}(k)]$ with respect to the variable $\hat{\boldsymbol{\theta}}(k+1)$, one has

$$
\begin{aligned}
f(\boldsymbol{Y}(k), u(k), \boldsymbol{U}(k-1), \hat{\boldsymbol{\theta}}(k+1)) &\cong f(\boldsymbol{Y}(k), u(k), \boldsymbol{U}(k-1), \hat{\boldsymbol{\theta}}(k)) \\
&\quad + f_\theta^T(k)(\hat{\boldsymbol{\theta}}(k+1) - \hat{\boldsymbol{\theta}}(k)),
\end{aligned}
\tag{10.22}
$$

where

$$f_{\theta}(k) = \left[\frac{\partial f(\boldsymbol{Y}(k), u(k), \boldsymbol{U}(k-1), \hat{\boldsymbol{\theta}}(k+1))}{\partial \hat{\boldsymbol{\theta}}}\right]\Bigg|_{\boldsymbol{\theta}=\hat{\boldsymbol{\theta}}(k)}.$$

Substitute (10.22) into (10.21) and let $(\partial J(\hat{\boldsymbol{\theta}}(k+1)))/(\partial \hat{\boldsymbol{\theta}}(k+1)) = 0$. According to the matrix inverse lemma, the projection type parameter estimation algorithm is given as

$$\hat{\boldsymbol{\theta}}(k+1) = \hat{\boldsymbol{\theta}}(k) + \frac{\eta f_{\theta}(k)(y(k+1) - f(\boldsymbol{Y}(k), u(k), \boldsymbol{U}(k-1), \hat{\boldsymbol{\theta}}(k)))}{\mu + \|f_{\theta}(k)\|^2}, \tag{10.23}$$

where η is the step factor added to make the parameter estimation algorithm more flexible and μ is a positive constant added to avoid a singular denominator.

Taking first-order Taylor expansion on $f[\boldsymbol{Y}(k), u(k-1), \boldsymbol{U}(k-1), \hat{\boldsymbol{\theta}}(k)]$ at the point of $[\boldsymbol{Y}(k), u(k-1), \boldsymbol{U}(k-1), \hat{\boldsymbol{\theta}}(k)]$ with respect to variable $u(k)$ leads to

$$\begin{aligned} f(\boldsymbol{Y}(k), u(k), \boldsymbol{U}(k-1), \hat{\boldsymbol{\theta}}(k)) = {} & f(\boldsymbol{Y}(k), u(k-1), \boldsymbol{U}(k-1), \hat{\boldsymbol{\theta}}(k)) \\ & + f_u(k)(u(k) - u(k-1)), \end{aligned} \tag{10.24}$$

where

$$f_u(k) = \frac{\partial f(\boldsymbol{Y}(k), u(k), \boldsymbol{U}(k-1), \hat{\boldsymbol{\theta}}(k))}{\partial u(k)}\Bigg|_{u(k)=u(k-1)}$$

Substitute (10.24) into (10.20) and let $(\partial J(u(k)))/(\partial u(k)) = 0$. By using the matrix inverse lemma, the projection type control law is proposed as

$$u(k) = u(k-1) + \frac{\rho f_u(k-1)(y^*(k+1) - f(\boldsymbol{Y}(k), u(k-1), \boldsymbol{U}(k-1), \hat{\boldsymbol{\theta}}(k)))}{\lambda + \|f_u(k-1)\|^2}, \tag{10.25}$$

where ρ is a step factor added to make the control law more flexible and λ is a positive constant added to avoid a singular denominator.

The adaptive control system, composed of parameter estimate algorithm (10.23) and controller algorithm (10.25), satisfies Concept 10.1 and thus is having symmetric similarity structure.

Remark 10.1

It can be seen from (10.20) and (10.21) that: (1) the physical meaning of $\lambda > 0$ is to punish the change of control input to avoid large control input variation between two consecutive time instants, and thus to avoid the saturation and abrasion of the actuator in practice. Similarly, the physical meaning of $\mu > 0$ is to punish the change of parameter estimation to avoid the large parameter estimation variation between two consecutive time instants, and thus to avoid the instability of adaptive control systems caused by the inaccurate data due to the noises and sensor faults in practice. (2) By properly selecting λ and μ, one can restrain the execution of Taylor linearization within an appropriate range. Thus, it is very important to select proper weighting factors to guarantee good performance of the above adaptive control system.

It is easy to derive the RLS-based adaptive control system with symmetric similarity structure, similar to the projection-based adaptive control system.

10.2.3 MFAC with Symmetric Similarity Structure

Consider a general discrete-time nonlinear SISO system

$$y(k+1) = f(y(k),\dots,y(k-n_y),u(k),\dots,u(k-n_u)), \tag{10.26}$$

where $u(k) \in R$ and $y(k) \in R$ are the system output and the control input at time instant k, n_y and n_u are two unknown integers, and $f(\cdots) : R^{n_u+n_y+2} \mapsto R$ is an unknown nonlinear function.

In Chapter 4, three kinds of MFAC schemes are introduced: CFDL–MFAC, PFDL–MFAC, and FFDL–MFAC schemes. Both controllers and estimators in these three control schemes are based on the projection algorithms, and thus they have symmetric similarity. This section will further explore the design of MFAC with symmetric similarity structure based on the RLS algorithm. Since the fundamental idea and design procedure are the same as in Section 10.2.2, here we just summarize the three classes of MFAC schemes with symmetric similarity structure for the purpose of comparison in Tables 10.1–10.3, based on the projection and RLS algorithms, respectively, and omit rigorous derivations for simplicity.

Any parameter estimation algorithm and controller algorithm in Table 10.1 can be integrated as a CFDL data model based MFAC scheme. But only the MFAC scheme consisting of the projection-based parameter estimation algorithm and the projection-based controller algorithm has symmetric similarity structure, as well as the MFAC scheme consisting of the RLS-based parameter estimation algorithm and the RLS-based controller algorithm.

Table 10.1 CFDL–MFAC Schemes

PPD Estimation Algorithm	
Projection algorithm	$\hat{\phi}_c(k) = \hat{\phi}_c(k-1) + \frac{\eta \Delta u(k-1)}{\mu + \Delta u(k-1)^2}(\Delta y(k) - \hat{\phi}_c(k-1)\Delta u(k-1))$
RLS algorithm	$\hat{\phi}_c(k) = \hat{\phi}_c(k-1) + \frac{P_\phi(k-2)\Delta u(k-1)(\Delta y(k) - \hat{\phi}_c(k-1)\Delta u(k-1))}{1 + \Delta u(k-1)^2 P_\phi(k-2)}$ $P_\phi(k-1) = P_\phi(k-2) - \frac{\Delta u(k-1)^2 P_\phi(k-2)^2}{1 + \Delta u(k-1)^2 P_\phi(k-2)}$ or other modified RLS algorithms for time-varying parameters shown in Chapter 2
Controller Algorithm	
Projection algorithm	$u(k) = u(k-1) + \frac{\rho \hat{\phi}_c(k)(y^*(k+1) - y(k))}{\lambda + \lvert\hat{\phi}_c(k)\rvert^2}$
RLS algorithm	$u(k) = u(k-1) + \frac{P_u(k-2)\hat{\phi}_c(k)(y^*(k+1) - y(k))}{1 + \hat{\phi}_c(k)^2 P_u(k-2)}$ $P_u(k-1) = P_u(k-2) - \frac{P_u(k-2)^2 \hat{\phi}_c(k)^2}{1 + \hat{\phi}_c(k)^2 P_u(k-2)}$ or other modified RLS algorithm shown in Chapter 2.

10.2.4 Simulations

From the aforementioned discussion and design procedure, it is easy to see that: (1) the parameter estimation algorithm and the control algorithm of an adaptive control system with symmetric similarity structure are computed iteratively; thus, they are suitable for both linear and nonlinear systems. Further, the design of the parameter estimation and controller algorithm can be realized by a uniform procedure with the same structure, which is the distinctive "superior property" of the adaptive control system with symmetric similarity structure, while most of the other existing adaptive control systems do not have such a property. (2) For the traditional adaptive control systems, the constant in the denominator of the projection algorithm is added to avoid zero denominator, without any physical meaning. Instead, the constant in the denominator of the parameter estimation algorithm proposed in this chapter is to punish the variation of the parameter estimation. Similarly, the constant in the denominator of the controller algorithm proposed in this chapter is to punish the variation of the control input. Such constants added in the denominator make

Table 10.2 PFDL–MFAC Schemes

PG Estimation Algorithm	
Projection algorithm	$\hat{\boldsymbol{\phi}}_{p,L}(k)=\hat{\boldsymbol{\phi}}_{p,L}(k-1)+\dfrac{\eta\Delta\boldsymbol{U}_L(k-1)(y(k)-y(k-1)-\hat{\boldsymbol{\phi}}_{p,L}^T(k-1)\Delta\boldsymbol{U}_L(k-1))}{\mu+\lVert\Delta\boldsymbol{U}_L(k-1)\rVert^2}$
RLS algorithm	$\hat{\boldsymbol{\phi}}_{p,L}(k)=\hat{\boldsymbol{\phi}}_{p,L}(k-1)+\dfrac{P_\phi(k-2)\Delta\boldsymbol{U}_L(k-1)(\Delta y(k)-\hat{\boldsymbol{\phi}}_{p,L}^T(k-1)\Delta\boldsymbol{U}_L(k-1))}{1+\Delta\boldsymbol{U}_L^T(k-1)P_\phi(k-2)\Delta\boldsymbol{U}_L(k-1)}$ $P_\phi(k-1)=P_\phi(k-2)-\dfrac{P_\phi(k-2)\Delta\boldsymbol{U}_L(k-1)\Delta\boldsymbol{U}_L^T(k-1)P_\phi(k-2)}{1+\Delta\boldsymbol{U}_L^T(k-1)P_\phi(k-2)\Delta\boldsymbol{U}_L(k-1)}$ or other modified RLS algorithms shown in Chapter 2.
Controller Algorithm	
Projection algorithm	$u(k)=u(k-1)+\dfrac{\rho_1\hat{\phi}_1(k)\left(y^*(k+1)-y(k)-\hat{\phi}_1(k)\sum_{i=2}^{L}\rho_i\hat{\phi}_i(k)\Delta u(k-i+1)\right)}{\lambda+\lvert\hat{\phi}_1(k)\rvert^2}$
RLS algorithm	$u(k)=u(k-1)+\dfrac{P_u(k-2)\hat{\phi}_1(k)((y^*(k+1)-y(k)-\hat{\boldsymbol{\phi}}'^T_{p,L}(k)\Delta\boldsymbol{U}_L'^T(k))}{1+\hat{\phi}_1(k)^2P_u(k-2)}$ $P_u(k-1)=P_u(k-2)-\dfrac{P_u(k-2)^2\hat{\phi}_1(k)^2}{1+\hat{\phi}_1(k)^2P_u(k-2)}$ where $\hat{\boldsymbol{\phi}}'_{p,L}(k)=\left[\hat{\phi}_2(k),\ldots,\hat{\phi}_L(k)\right]$ and $\Delta\boldsymbol{U}_L'(k)=\left[\Delta u(k-1),\ldots,\Delta u(k-L+1)\right]$. or other modified RLS algorithms shown in Chapter 2.

the designed adaptive control systems capable of dealing with nonminimum phase systems, which is difficult to achieve by the traditional adaptive control systems. Further, it can be proved that the extensions of weighted one-step-ahead controller algorithm can stabilize any linear system [76]. (3) The control system design procedure can be simplified when one designs an adaptive control system according to the design principle of symmetric similarity, which makes the controller design and estimator design be a uniform "adaptor" design with low cost.

Up to now, it is very difficult to analyze the "superior" properties of adaptive control systems with symmetric similar structures in a rigorous way due to the lack of effective mathematical tools. Some new advanced mathematical methods

Table 10.3 FFDL–MFAC Schemes

PG Estimation Algorithm	
Projection algorithm	$\hat{\boldsymbol{\phi}}_{f,L_y,L_u}(k) = \hat{\boldsymbol{\phi}}_{f,L_y,L_u}(k-1) + \frac{\eta \Delta \boldsymbol{H}_{L_y,L_u}(k-1)(y(k)-y(k-1)-\hat{\boldsymbol{\phi}}^T_{f,L_y,L_u}(k-1)\Delta \boldsymbol{H}_{L_y,L_u}(k-1)}{\mu + \Vert \Delta \boldsymbol{H}_{L_y,L_u}(k-1) \Vert^2}$
RLS algorithm	$\hat{\boldsymbol{\phi}}_{f,L_y,L_u}(k) = \hat{\boldsymbol{\phi}}_{f,L_y,L_u}(k-1) + \frac{P_\phi(k-2)\Delta \boldsymbol{H}_{L_y,L_u}(k-1)(\Delta y(k)-\hat{\boldsymbol{\phi}}^T_{f,L_y,L_u}(k-1)\Delta \boldsymbol{H}_{L_y,L_u}(k-1))}{1+\Delta \boldsymbol{H}^T_{L_y,L_u}(k-1)P_\phi(k-2)\Delta \boldsymbol{H}_{L_y,L_u}(k-1)}$ $P_\phi(k-1) = P_\phi(k-2) - \frac{P_\phi(k-2)\Delta \boldsymbol{H}_{L_y,L_u}(k-1)\Delta \boldsymbol{H}^T_{L_y,L_u}(k-1)P_\phi(k-2)}{1+\Delta \boldsymbol{H}^T_{L_y,L_u}(k-1)P_\phi(k-2)\Delta \boldsymbol{H}_{L_y,L_u}(k-1)}$ or other modified RLS algorithms shown in Chapter 2.
Controller Algorithm	
Projection algorithm	$u(k) = u(k-1) + \frac{\rho_{L_y+1}\hat{\phi}_{L_y+1}(k)(y^*(k+1)-y(k))}{\lambda + \lvert\hat{\phi}_{L_y+1}(k)\rvert^2}$ $- \frac{\hat{\phi}_{L_y+1}(k)\sum_{i=1}^{L_y}\rho_i\hat{\phi}_i(k)\Delta y(k-i+1)}{\lambda + \lvert\hat{\phi}_{L_y+1}(k)\rvert^2}$ $- \frac{\hat{\phi}_{L_y+1}(k)\sum_{i=L_y+2}^{L_y+L_u}\rho_i\hat{\phi}_i(k)\Delta u(k+L_y-i+1)}{\lambda + \lvert\hat{\phi}_{L_y+1}(k)\rvert^2}$
RLS algorithm	$u(k) = u(k-1) + \frac{P_u(k-2)\hat{\phi}_{L_y+1}(k)}{1+\hat{\phi}_{L_y+1}(k)^2P_u(k-2)}$ $\times (y^*(k+1) - y(k) - \hat{\boldsymbol{\phi}}'^{T}_{f,L_y,L_u}(k)\Delta \boldsymbol{H}'_{L_y,L_u}(k))$ $P_u(k-1) = P_u(k-2) - \frac{P_u(k-2)^2\hat{\phi}_{L_y+1}(k)^2}{1+\hat{\phi}_{L_y+1}(k)^2P_u(k-2)}$ where $\hat{\boldsymbol{\phi}}'_{f,L_y,L_u}(k) = \left[\hat{\phi}_1(k),\ldots,\hat{\phi}_{L_y}(k),\hat{\phi}_{L_y+2}(k),\ldots,\hat{\phi}_{L_y+L_u}(k)\right]$, and $\Delta \boldsymbol{H}'_L(k) = \left[\Delta y(k),\ldots,\Delta y(k-L_y+1),\Delta u(k-1),\ldots,\Delta u(k-L_u+1)\right]$ or other modified RLS algorithms shown in Chapter 2.

are expected to be introduced. However, with the rapid development of computer technology in recent years, one can find another way to verify the "superiorities" of an adaptive control systems with symmetric similarity structure by simulations, as we will show in the following section.

10.2.4.1 Adaptive Control with Symmetric Similarity Structure

The following simulation examples are used to verify the effectiveness of the projection-based adaptive control scheme with symmetric similarity structure.

Example 10.4

Consider the following nonlinear system:

$$y(k+1) = y(k)\sin(\theta_1(k)u(k)) - \theta_2(k)u^2(k) + \theta_3(k)u(k-1). \tag{10.27}$$

The initial values are $y(1) = y(2) = 0.1$ and $u(1) = 0.3$. The desired output signal is $y^*(k) = 1 + 0.5 \times (-1)^{\text{round}(k/200)}$.

When a constant parameter $\theta(k) = [0.3,0.6,2]^T$ is considered, the following simulation gives the comparison results between the adaptive control scheme with symmetric similarity structure and the adaptive control scheme without symmetric similarity structure.

For the projection-based adaptive control scheme with symmetric similarity structure, (10.23) and (10.25) are applied with $\rho = 0.2$, $\lambda = 1$, $\eta = 1$, and $\mu = 0.1$. For the adaptive control scheme without symmetric similarity structure, we adopt the RLS-based parameter estimation algorithm with forgetting factors (2.53)–(2.55) and the projection-based controller algorithm (10.25), with $\rho = 0.2$, $\lambda = 0.1$, $P(0) = 1000I$, and $\alpha = 0.95$.

From the simulation results shown in Figure 10.1, it can be seen that the control performances of the two schemes are almost the same except for the different control input signals. However, the adaptive control system with symmetric similarity structure has a smaller computational burden, is simple to implement, and has a very low input energy cost.

When time-varying parameters $\theta_1(k)=0.3$, $\theta_2(k)=0.6$, and $\theta_3(k)=0.5\sin(k/500)$ are considered, we still apply the above two control schemes. The superiority of adaptive control system with symmetric similarity structure can be seen from the simulation results shown in Figure 10.2.

The following example is to verify the effectiveness of the RLS-based adaptive control scheme with symmetric similarity structure.

Example 10.5

Consider the following nonlinear system:

$$y(k+1) = \frac{\theta_1(k)y(k)}{1+\theta_2(k)y^2(k)} + \theta_3(k)u^3(k). \tag{10.28}$$

The initial values are $y(1) = 1$, $y(2)= 0.5$, and $u(1)= 0.1$. The desired output signal is $y^*(k) = 2 \times (-1)^{\text{round}(k/400)}$.

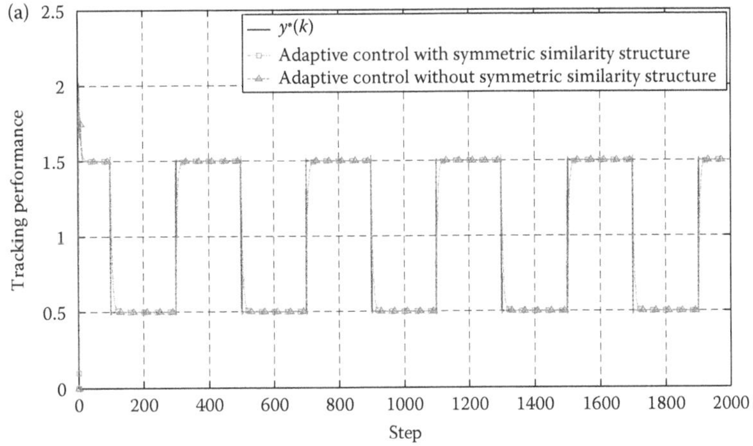

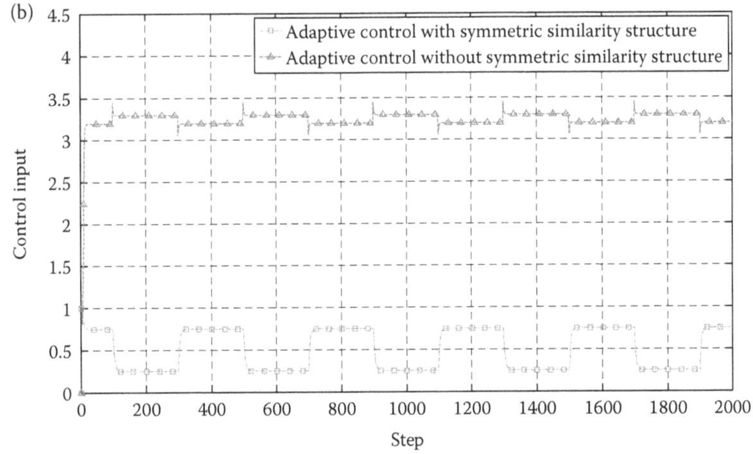

Figure 10.1 Simulation comparison between the projection-based adaptive control scheme with symmetric similarity structure and the adaptive control scheme without symmetric similarity structure for nonlinear system (10.27) with time-invariant parameters. (a) Tracking performance. (b) Control input.

The simulation comparison between the adaptive control scheme with symmetric similarity structure and the adaptive control scheme without symmetric similarity structure is carried out for $\theta(k) = [1,1,1]^T$.

The RLS-based parameter estimation algorithm with forgetting factor (2.53)–(2.55) and the corresponding RLS-based controller algorithm are used as the adaptive control system with symmetric similarity structure. In the parameter estimation algorithm, the covariance matrix and the forgetting factor are set to $P_\theta(0) = 10000I$, $\alpha_\theta = 0.99$. In the controller algorithm, the covariance matrix and the forgetting factor are set to $P_u(0) = 0.5$, $\alpha_u = 0.99$. If $P_u(k) < 0.1$, we reset $P_u(k) = P_u(0)$. For the adaptive control scheme without symmetric similarity

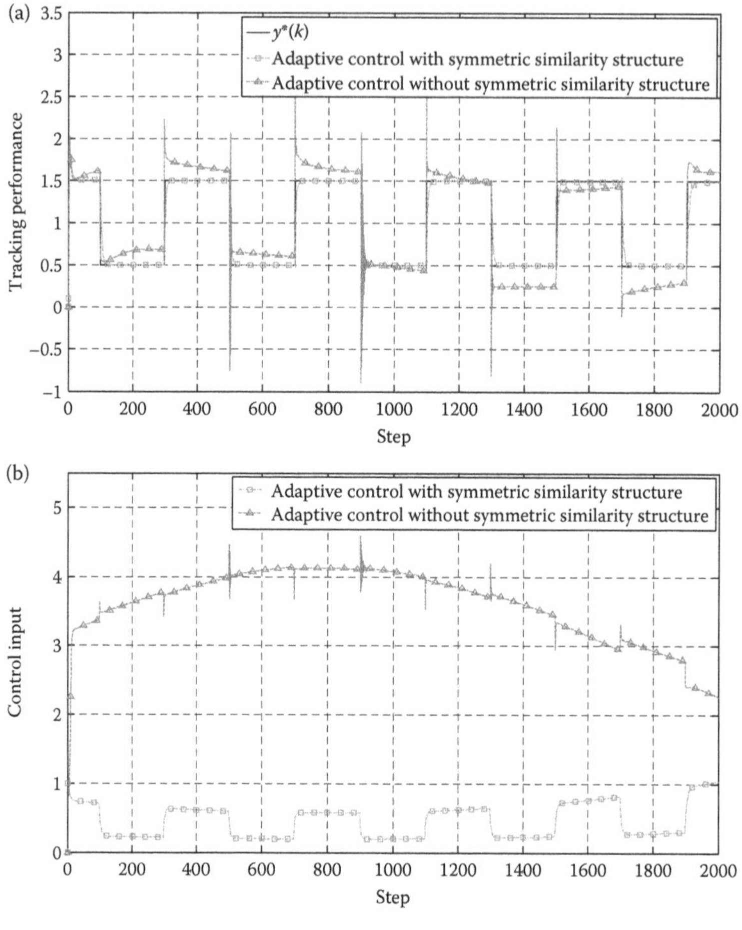

Figure 10.2 Simulation comparison between the projection-based adaptive control scheme with symmetric similarity structure and the adaptive control scheme without symmetric similarity structure for nonlinear system (10.27) with time-varying parameters. (a) Tracking performance. (b) Control input.

structure, we adopt the RLS-based parameter estimation algorithm with forgetting factor (2.53)–(2.55) and the projection-based controller algorithm (10.25). The controller parameters are $\rho = 0.5$, $\lambda = 0.5$, $P_\theta(0) = 10000I$, and $\alpha_\theta = 0.99$.

The simulation results are shown in Figure 10.3. It is obvious that the RLS-based adaptive control scheme with symmetric similarity structures gives better control performance and needs a very low input energy cost.

10.2.4.2 MFAC with Symmetric Similarity Structure

Example 10.6

Consider the following nonlinear system:

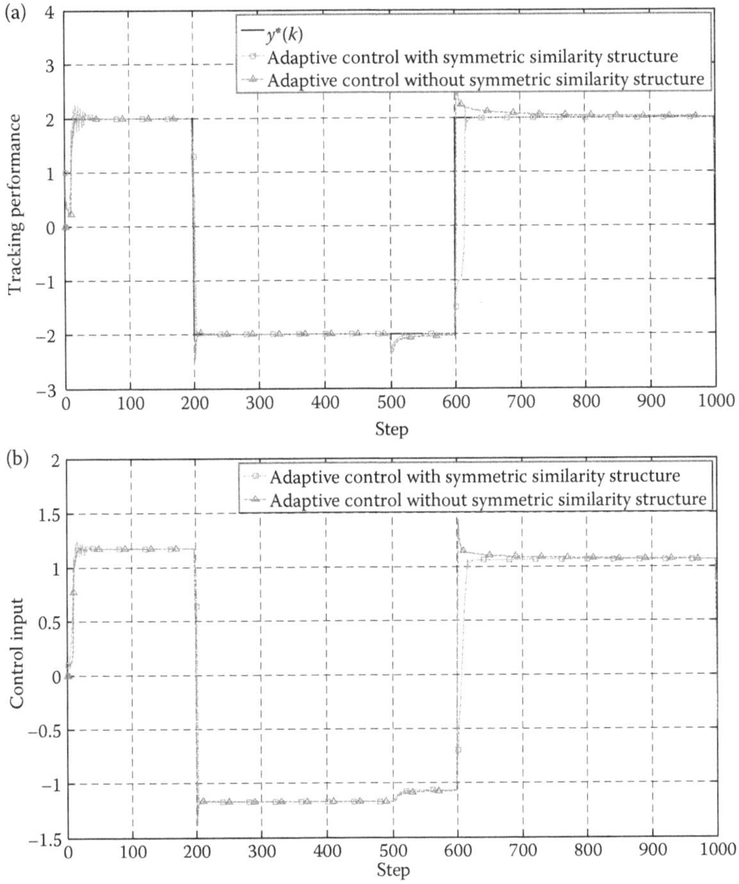

Figure 10.3 Simulation comparison between the RLS-based adaptive control scheme with symmetric similarity structure and the adaptive control scheme without symmetric similarity structure for nonlinear system (10.28). (a) Tracking performance. (b) Control input.

$$y(k+1) = \begin{cases} \dfrac{\left(a(k)+2.5\right)y(k)y(k-1)}{1+y^2(k)+y^2(k-1)} + 0.7\sin\left(0.5\left(y(k)+y(k-1)\right)\right) & \\ \quad \times\cos\left(0.5\left(y(k)+y(k-1)\right)\right) + 1.2u(k) + 1.4u(k-1), & k \le 200 \\ 0.55y(k) + 0.46y(k-1) + 0.07y(k-2) + 0.1u(k) & \\ \quad + 0.02u(k-1) + 0.03u(k-2), & 200 < k \le 400 \\ -0.1y(k) - 0.2y(k-1) - 0.3y(k-2) + 0.1u(k) & \\ \quad + 0.02u(k-1) + 0.03u(k-2), & 400 < k \le 600 \end{cases} \tag{10.29}$$

where $a(k) = 2\sin(k/50)$.

System (10.29) consists of three subsystems: the first one is a nonminimum phase nonlinear system, which is obtained by adding the term $1.4u(k-1)$ into the system of Ref. [163], and the other two are taken from Ref. [168]. It is worth pointing out that the second subsystem is an open-loop unstable system.

The initial values are $y(1)=0.1$, $y(2)=0.1$, $y(3)=0.5$, $u(1)=0.3$, $u(2)=0.2$, and $\hat{\phi}_{f,L_y,L_u}(3)=[0,0,0]^T$. The pseudo-orders are set to $L_u=2$ and $L_y=1$. The desired trajectory is given as $y^*(k)=5\times(-1)^{\mathrm{round}(k/100)}$.

The simulation comparison between the MFAC scheme with symmetric similarity structure and the MFAC scheme without symmetric similarity structure for nonlinear system (10.29) is carried out. For the MFAC scheme with symmetric similarity structure, we apply the projection-based estimation algorithm and the projection-based controller algorithm, shown in Table 10.3. The controller parameters are set to $\eta=\rho=0.6$, $\lambda=0.003$, and $\mu=0.01$. For the MFAC scheme without symmetric similarity structure, we apply the RLS-based estimation algorithm and the projection-based controller algorithm, shown in Table 10.3. The controller parameters are set as $P(0)=1000I$, $\rho=0.6$, and $\lambda=0.003$.

The simulation results are shown in Figure 10.4. It is obvious that the control performance of MFAC with symmetric similarity structure is better.

Example 10.7

Consider the following nonlinear system [162]:

$$y(k+1)=\frac{y(k)}{1+y(k)^2}+u(k)^3. \tag{10.30}$$

The initial values are set as $y(1)=0$, $y(2)=0$, $y(3)=0$, $y(4)=0.1$, $y(5)=0.1$, $y(6)=0.5$, $u(1)=0$, $u(2)=0$, $u(3)=0$, $u(4)=0.3$, $u(5)=0.2$ and PG is $\hat{\phi}_{f,L_y,L_u}(6)=[0,0,0,0,0,0]^T$. The pseudo-orders are set as $L_u=3$ and $L_y=3$. The desired trajectory is given as $y^*(k)=5+2.5\times(-1)^{\mathrm{round}(k/100)}$.

For the MFAC scheme with symmetric similarity structure, we apply the projection-based estimation algorithm and the projection-based controller algorithm, shown in Table 10.3. The controller parameters are set to $\eta=\rho=0.3$, $\lambda=1.8$, and $\mu=1$. For the MFAC scheme without symmetric similarity structure, we apply the RLS-based estimation algorithm and the projection-based controller algorithm, shown in Table 10.3. The controller parameters are set to $P(0)=1000I$, $\rho=0.3$, and $\lambda=1.8$.

The simulation results are shown in Figure 10.5. It is obvious that the control performance of the MFAC with symmetric similarity structures is better.

10.3 Similarity between MFAC and MFAILC

Taking the model-based adaptive control and the MFAC as examples, the concepts and design principle of symmetric similarity are discussed for adaptive control systems in previous sections. Some numerical examples are provided to verify the "superiority" of the control systems with symmetric similarity structure. The design principle of symmetric similarity makes it more convenient to design, analyze, and

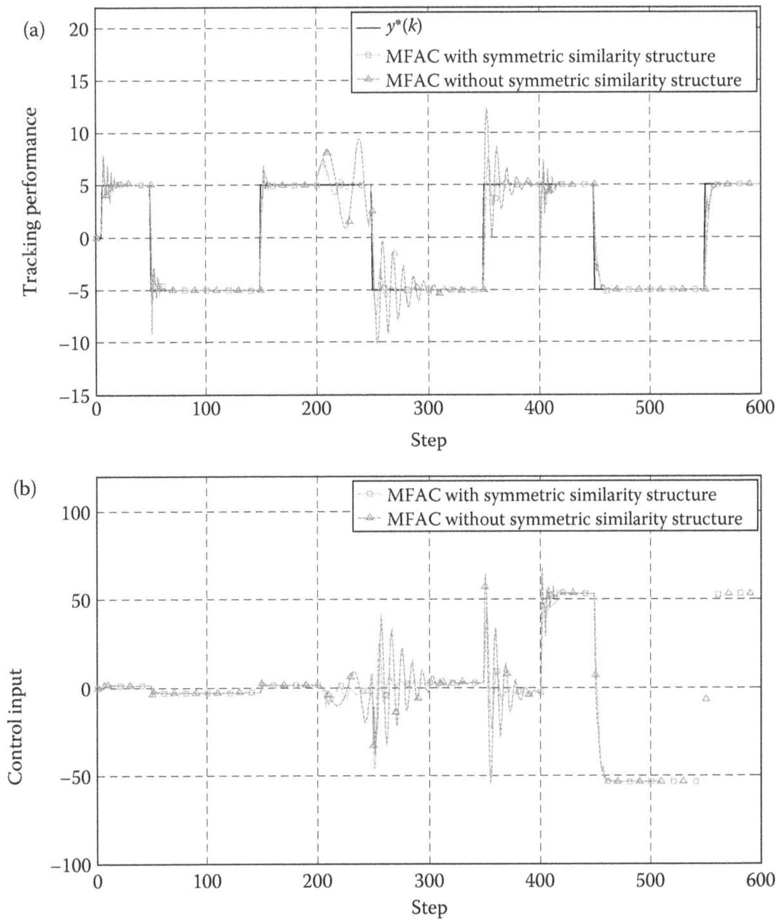

Figure 10.4 Simulation comparison between the MFAC scheme with symmetric similarity structure and the MFAC scheme without symmetric similarity structure for nonlinear system (10.29). (a) Tracking performance. (b) Control input.

implement an adaptive control system. In addition, such concepts and design principles provide a new perspective for control theory research, and they can be used to develop some new adaptive control systems.

In this section, the similarities between MFAC and MFAILC are discussed from a broader viewpoint by comparing their design and analysis methods, which have been given in Chapters 4 and 7, respectively.

From a systematic viewpoint, the derivation and analysis tools of MFAC and MFAILC are identical in nature. The only difference is that the design and analysis

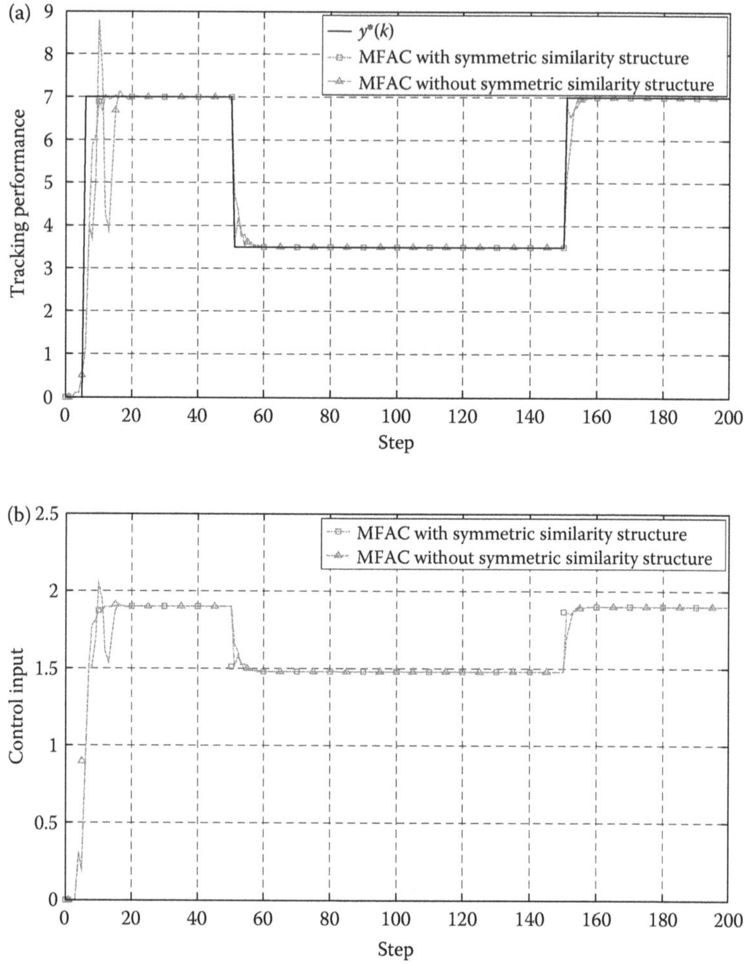

Figure 10.5 Simulation comparison between the MFAC scheme with symmetric similarity structure and the MFAC scheme without symmetric similarity structure for nonlinear system (10.30). (a) Tracking performance. (b) Control input.

of MFAC are conducted along the time axis, while the design and analysis of MFAILC are along the iteration axis. In other words, the MFAILC approach can be obtained by "erecting up" the MFAC approach from the horizontal time axis to the vertical iteration axis. On the other hand, we can also say that the MFAC approach can be obtained by "laying down" the MFAILC approach from the vertical iteration axis to the horizontal time axis. Since the detailed design and analysis of these two control schemes have been provided in previous chapters correspondingly, this section mainly summarizes and discusses their symmetric similarity features.

Taking CFDL–MFAC and CFDL–MFAILC as an example, the comparison table is provided in Table 10.4 where the similarities and differences between these two control approaches are clearly illustrated. There also exists a similar comparative relationship between PFDL/FFDL–MFAC and PFDL/FFDL–MFAILC. The details are omitted due to page limitation.

From Table 10.4, it is obvious that CFDL–MFAC and CFDL–MFAILC are similar in nature, which can be further summarized as follows:

1. The controlled plants are similar, that is, both of them are presented for a class of completely unknown nonlinear systems.
2. The dynamic linearization methods for nonlinear systems are similar, that is, both of them are developed by introducing a parameter called the "pseudo partial derivative" based on Cauchy's mean-value theorem.
3. The design process and structures of controllers and estimators are similar and both of them are derived by the same optimization procedure.
4. The convergence analysis methods are similar, that is, both of them are analyzed by using the contraction mapping tool.

While there also exist some differences between the two control approaches

1. The controlled plant of MFAILC is repeatable on a finite time interval. The feature of MFAILC is to update the control input signal and PPD parameters along the iteration axis. Instead, MFAC is conducted in the infinite time interval without any requirement of repeatability or periodicity, and it updates the control input signal and the PPD parameter along the time axis.
2. What MFAILC achieves is pointwise convergent over the entire finite time interval asymptotically along the iteration axis, while MFAC achieves asymptotic convergence along the time axis.
3. MFAC is a feedback control scheme, while MFAILC is a feedforward control scheme.

Essentially, the MFAC approach belongs to the category of nonlinear adaptive control theory, while MFAILC belongs to the category of iterative learning control theory. From a much broader viewpoint, the similarity between MFAC and MFAILC can be extended to similarity between the well-established adaptive control theory and iterative learning control theory.

10.4 Similarity between Adaptive Control and Iterative Learning Control

Adaptive control, especially for linear systems, has been well established in theory and extensively applied in practice. After adaptive control was proposed in the last

Table 10.4 Comparisons between CFDL–MFAC and CFDL–MFAILC

	MFAC	*MFAILC*
Plant	$y(k+1) = f(y(k),\ldots,y(k-n_y), u(k),\ldots,u(k-n_u))$, $k \in \{0,1,\ldots,\infty\}$ (Equation (4.1))	$y(k+1,i) = f\big(y(k,i),\ldots,y(k-n_y,i), u(k,i),\ldots,u(k-n_u,i)\big)$, $k \in \{0,1,\ldots,T\}$, $i \in \{0,1,\ldots,\infty\}$ (Equation (7.1))
Key Assumption	The control system satisfies the generalized *Lipschitz* condition along the time axis (Assumption 3.2)	The control system satisfies the generalized *Lipschitz* condition along the iteration axis (Assumption 7.2)
Dynamic Linearization	$\Delta y(k+1) = \phi_c(k)\Delta u(k)$, $k \in \{0,1,\ldots,\infty\}$ Dynamical linearization along the time axis (Theorem 3.1)	$\Delta y(k+1,i) = \phi_c(k,i)\Delta u(k,i)$, $k \in \{0,1,\ldots,T\}$, $i \in \{0,1,\ldots,\infty\}$ Dynamical linearization along the iteration axis (Theorem 7.1)
Controller Algorithm	$u(k) = u(k-1) + \dfrac{\rho\hat{\phi}_c(k)(y^*(k+1) - y(k))}{\lambda + \lvert\hat{\phi}_c(k)\rvert^2}$ Control input is updated along the time axis (Equation (4.9))	$u(k,i) = u(k,i-1) + \dfrac{\rho\hat{\phi}_c(k,i)}{\lambda + \lvert\hat{\phi}_c(k,i)\rvert^2} e(k+1,i-1)$ Control input is updated along the iteration axis (Equation (7.15))

Estimation Algorithm	$\hat{\phi}_c(k) = \hat{\phi}_c(k-1) + \frac{\eta \Delta u(k-1)}{\mu + \Delta u(k-1)^2} \times (\Delta y(k) - \hat{\phi}_c(k-1)\Delta u(k-1))$ Parameter estimation algorithm is updated along the time axis (Equation (4.7))	$\hat{\phi}_c(k,i) = \hat{\phi}_c(k,i-1) + \frac{\eta \Delta u(k,i-1)}{\mu + \lvert \Delta u(k,i-1) \rvert^2} \times (\Delta y(k+1,i-1) - \hat{\phi}_c(k,i-1)\Delta u(k,i-1))$ Parameter estimation algorithm is updated along the iteration axis (Equation (7.14))
Convergence Property	$\lim_{k\to\infty} \lvert y^* - y(k+1) \rvert = 0$ $\{y(k)\}$ and $\{u(k)\}$ are bounded, $\forall k \in \{0,1,\ldots,\infty\}$. Asymptotic tracking tasks along the time axis	$\lim_{i\to\infty} \lvert y^*(k+1) - y(k+1,i) \rvert = 0,$ $\forall k = 0,1,\ldots,T$ $\{y(k,i)\}$ and $\{u(k,i)\}$ are bounded, $\forall k \in \{0,1,\ldots,T\}, \forall i \in \{0,1,\ldots,\infty\}$. Asymptotic convergence over the entire finite time interval along the iteration axis
Analysis Tool	Contraction mapping based analysis methods along time axis k	Contraction mapping based analysis methods along iteration axis i

century, its typical design methods including analysis tools have been emerging gradually. On the other hand, iterative learning control (ILC), proposed in 1984 originally for repeatable systems, is a branch of control theory and has been widely applied in practice.

As pointed out in the last section, adaptive control and iterative learning control are two different control methods applied to different control backgrounds for different control tasks.

For the discrete-time adaptive control, (1) it is designed to deal with slowly time-varying or time-invariant parametric systems with known structures. (2) Its control objective is to find a control input sequence such that the system output tracks the desired signals asymptotically, and the tracking error converges along the time axis. In other words, adaptive control systems are conducted along the time axis. (3) The stability and convergence analysis is mainly based on the *Lyapunov* stability theory and the key technical lemma (KTL). For iterative learning control: (1) The controlled plant must be repeatable over a finite time interval. The repeatability means that not only the dynamics of the controlled system is repeated but also all of the initial states and the desired tracking trajectory are repeated. (2) The control objective of ILC is to find a control input profile in the finite interval such that the system output can perfectly track the desired trajectory over the entire finite time interval, and the tracking error converges along the iteration axis. In other words, iterative learning control is conducted along the iteration axis. (3) The stability and convergence analysis is mainly based on contraction mapping.

Adaptive control and iterative learning control are two different control approaches; both of them, however, can deal with systems with uncertainties. In fact, "adaptation" and "learning" have some similar properties and strong inherent relationships. It is of interest to study the relationships between the two control approaches and to build a bridge linking them. Once we construct such a bridge, some more "superior" control schemes might be developed according to the laws of crossover and mutation between these two approaches. On the basis of the discussion in the previous sections, we will further show the similarities between discrete-time adaptive control and discrete-time iterative learning control, as well as the principle of similarity for controller design. By extending the well-known adaptive control design and analysis methods, a new "superior" control approach, discrete-time adaptive iterative learning control, is proposed to solve a new control problem (perfect tracking over a finite time interval for arbitrarily fast time-varying parametric systems), while relaxing the assumptions on repetitiveness in traditional ILC.

The discrete-time adaptive ILC proposed in this section is similar to the discrete-time adaptive control from the angle of the controller structure, estimator structure, analysis tool, and so on. For repeatable discrete-time systems with arbitrarily fast time-varying parametric uncertainties, the proposed approach can guarantee perfect tracking performance over a finite time interval, although both the initial states and the desired trajectory are iteration-varying. Therefore, it gives a more "superior" control performance than adaptive control and iterative learning control alone [258,261]. This section will provide a detailed comparison between discrete-time adaptive control and adaptive iterative learning control. A table about the similarity and contrast, including all aspects of the control plant, assumptions, control objective, analysis tool, convergence result, and so on, is provided for readers to carefully explore the similarities between these two methods.

10.4.1 Adaptive Control for Discrete-Time Nonlinear Systems

10.4.1.1 Problem Formulation

For simplicity and without loss of generality, a simple discrete-time system is considered:

$$x(k+1) = \theta\xi(x(k)) + u(k), \tag{10.31}$$

where $x(k) \in R$ denotes the measurable system state, $u(k) \in R$ denotes the control input, θ is an unknown constant parameter, $\xi(x(k))$ is a known nonlinear function and is bounded for any bounded $x(k)$, and $k = 0,1,\ldots,\infty$ denotes the sampling time instant.

The desired reference trajectory, $r(k+1)$, which is bounded for all time $k = 0,1, \ldots,\infty$, and is known.

Define the tracking error as $e(k) = x(k) - r(k)$. The control objective is to find a proper control input sequence $u(k)$ such that the system state $x(k)$ can track the desired trajectory $r(k)$, that is, $\lim_{k\to\infty} e(k) = 0$.

10.4.1.2 Adaptive Control Design

Two discrete-time adaptive control schemes are presented with an RLS-based estimator and a projection-based estimator, respectively.

RLS-Based Discrete-Time Adaptive Control Scheme is designed as follows.

$$u(k) = r(k+1) - \hat{\theta}(k)\xi(x(k)), \tag{10.32}$$

$$\hat{\theta}(k) = \hat{\theta}(k-1) + P(k-1)\xi(x(k-1))e(k), \tag{10.33}$$

$$P(k-1) = P(k-2) - \frac{P^2(k-2)\xi^2(x(k-1))}{1 + P(k-2)\xi^2(x(k-1))}, \tag{10.34}$$

where $\hat{\theta}(k)$ is the estimation of the unknown system parameter θ at time k, the initial value $\hat{\theta}(0)$ can be chosen arbitrarily, and $P(0)$ is a positive constant.

Projection-Based Discrete–Time Adaptive Control Scheme is designed as follows.

$$u(k) = r(k+1) - \hat{\theta}(k)\xi(x(k)), \tag{10.35}$$

$$\hat{\theta}(k) = \hat{\theta}(k-1) + \frac{a\xi(x(k-1))}{c + \xi(x(k-1))^2} e(k), \tag{10.36}$$

where $\hat{\theta}(k)$ is the estimation of the unknown system parameter θ at time k, the initial value $\hat{\theta}(0)$ can be chosen arbitrarily, $0 < a < 2$, and $c > 0$.

10.4.1.3 Stability and Convergence Analysis

For the rigorous analysis of stability and convergence properties for the designed adaptive control systems, two assumptions and the key technical lemma should be introduced.

Assumption 10.1

The nonlinear function $\xi(x(k))$ satisfies the linear growth condition, that is,

$$\left|\xi(x(k))\right| \leq c_1^0 + c_2^0\left|x(k)\right|, \quad \forall k$$

where $0 < c_1^0 < \infty$ and $0 < c_2^0 < \infty$.

Assumption 10.2

All of the unknown system parameter θ, the desired trajectory $r(k)$, and the initial state $x(0)$ are bounded for all $k \in \{0,1,\ldots\}$.

Remark 10.2

Assumption 10.1, the linear growth condition, is a common requirement for the analysis of the discrete-time adaptive control system using the key technical lemma. And it should be noted that in Assumption 10.2, we only need the existence of such bounds, without requiring their exact values.

Key Technical Lemma [76]

If the following conditions are satisfied for some given sequences $s(k)$, $\boldsymbol{\sigma}(k)$, $\{b_1(k)\}$, and $\{b_2(k)\}$

a. $$\lim_{k\to\infty}\frac{s(k)^2}{b_1(k)+b_2(k)\boldsymbol{\sigma}(k)^T\boldsymbol{\sigma}(k)}=0, \quad (10.37)$$

   where $\{b_1(k)\}$, $\{b_2(k)\}$, and $\{s(k)\}$ are real-valued scalar sequences and $\boldsymbol{\sigma}(k)$ is a real-valued p-dimensional vector.

b. Uniform boundedness condition

$$0 < b_1(k) < K < \infty \quad \text{and} \quad 0 \le b_2(k) < K < \infty \quad (10.38)$$

   for all $k \ge 1$.

c. Linear boundedness condition

$$\|\boldsymbol{\sigma}(k)\| \le C_1 + C_2 \max_{0\le\tau\le k}|s(\tau)|, \quad (10.39)$$

   where $0 < C_1 < \infty$ and $0 < C_2 < \infty$.

It follows that

1. $\lim_{k\to\infty} s(k) = 0$.
2. $\{\|\boldsymbol{\sigma}(k)\|\}$ is bounded.

The stability and convergence analysis is provided as follows for the RLS-based discrete-time adaptive control. By a similar procedure, the readers themselves can analyze the stability and convergence of the projection-based discrete-time adaptive control system.

Theorem 10.1

For discrete-time system (10.31) satisfying Assumptions 10.1 and 10.2, the RLS-based adaptive control scheme (10.32)–(10.34) guarantees the following properties:

a. The parameter estimation $\hat{\theta}(k)$ is bounded for all k, and

$$|\hat{\theta}(k)-\theta| \le |\hat{\theta}(k-1)-\hat{\theta}| \le \cdots \le |\hat{\theta}(0)-\theta|.$$

b. The tracking error converges to zero as time k goes to infinity, that is,

$$\lim_{k\to\infty} e(k) = 0.$$

c. Both the control input $u(k)$ and the system state $x(k)$ are bounded for all k.

Proof

There are three parts for this proof. The first part proves the boundedness of $\hat{\theta}(k)$. Then, the convergence of tracking error is shown in the second part. BIBO stability for the system is given in the last part.

Boundedness of $\hat{\theta}(k)$ Define the parameter error as $\varphi(k) = \theta - \hat{\theta}(k)$. According to controller algorithm (10.32), the tracking error dynamics is

$$\begin{aligned} e(k+1) = x(k+1) - r(k+1) &= \theta\xi(x(k)) + u(k) - r(k+1) \\ &= \theta\xi(x(k)) + (r(k+1) - \hat{\theta}(k)\xi(x(k))) - r(k+1) \\ &= \varphi(k)\xi(x(k)). \end{aligned} \tag{10.40}$$

Subtracting θ from both sides of parameter estimation algorithm (10.33), and using (10.40), we have

$$\begin{aligned} \varphi(k) &= \varphi(k-1) - P(k-1)\xi(x(k-1))e(k) \\ &= (1 - P(k-1)\xi^2(x(k-1)))\varphi(k-1). \end{aligned} \tag{10.41}$$

To evaluate the relationship between $\varphi(k)$ and $\varphi(k-1)$, further inspection of the term $1 - P(k-1)\xi^2(x(k-1))$ in (10.41) is needed. From (10.34), one can derive

$$P(k-1) = \frac{P(k-2)}{1 + P(k-2)\xi^2(x(k-1))}. \tag{10.42}$$

Therefore

$$1 - P(k-1)\xi^2(x(k-1)) = 1 - \frac{P(k-2)\xi^2(x(k-1))}{1 + P(k-2)\xi^2(x(k-1))} = \frac{1}{1 + P(k-2)\xi^2(x(k-1))}. \tag{10.43}$$

According to (10.42)

$$P(k-1)^{-1} = P(k-2)^{-1} + \xi^2(x(k-1)). \tag{10.44}$$

Thus, $P(k) > 0$ holds for any k.

Taking the absolute value on both sides of (10.41) and according to (10.43) and (10.44), we obtain

$$|\varphi(k)| = \frac{1}{1 + P(k-2)\xi^2(x(k-1))}|\varphi(k-1)| \le |\varphi(k-1)|. \tag{10.45}$$

Clearly, inequality (10.45) shows the nonincreasing property of the parameter estimation error $\varphi(k)$. In terms of Assumption 10.2, θ is bounded. Hence, the parameter estimation $\hat{\theta}(k)$ is bounded for all k. Then, conclusion (a) of Theorem 10.1 is obtained.

Asymptotic Convergence of Tracking Error Define a nonnegative function $V(k) = P(k-1)^{-1}\varphi(k)^2$; then

$$\Delta V(k) = V(k) - V(k-1) = P(k-1)^{-1}\varphi(k)^2 - P(k-2)^{-1}\varphi(k-1)^2. \tag{10.46}$$

According to (10.40) and (10.41), Equation (10.46) is rewritten as

$$\begin{aligned}\Delta V(k) &= P(k-1)^{-1}\left(\varphi(k-1) - P(k-1)\xi(x(k-1))e(k)\right)^2 - P(k-2)^{-1}\varphi(k-1)^2\\ &= \left[P(k-1)^{-1} - P(k-2)^{-1}\right]\varphi(k-1)^2\\ &\quad - 2\varphi(k-1)\xi(x(k-1))e(k) + P(k-1)\xi(x(k-1))^2 e(k)^2. \end{aligned}\tag{10.47}$$

From (10.40) and (10.44), we have

$$\left[P(k-1)^{-1} - P(k-2)^{-1}\right]\varphi(k-1)^2 = \xi(x(k-1))^2\varphi(k-1)^2 = e(k)^2 \tag{10.48}$$

and

$$-2\varphi(k-1)\xi(x(k-1))e(k) = -2e(k)^2. \tag{10.49}$$

Substituting (10.48) and (10.49) into (10.47) yields

$$\Delta V(k) = -\left[1 - P(k-1)\xi(x(k-1))^2\right]e(k)^2. \tag{10.50}$$

Again, using (10.43), Equation (10.50) becomes

$$\Delta V(k) = -\frac{e(k)^2}{1 + P(k-2)\xi^2(x(k-1))} \le 0. \tag{10.51}$$

Thus, $V(k)$ is a nonnegative and nonincreasing function. Summing up (10.51) from 0 to k leads to

$$V(k) = V(0) - \sum_{j=1}^{k} \frac{e(j)^2}{1 + P(j-2)\xi^2(x(j-1))}. \tag{10.52}$$

Since $V(0) = P(-1)^{-1}\varphi(0)^2$ is bounded, and $V(k)$ is nonnegative and nonincreasing, one can immediately conclude from (10.52) that

$$\lim_{k\to\infty} \sum_{j=1}^{k} \frac{e(j)^2}{1 + P(j-2)\xi^2(x(j-1))} < \infty. \tag{10.53}$$

that is,

$$\lim_{k\to\infty} \frac{e(k)^2}{1 + P(k-2)\xi^2(x(k-1))} = 0. \tag{10.54}$$

In terms of the linear growth condition in Assumption 10.1 and the definition of the tracking error, one obtains

$$\left|\xi(x(k-1))\right| \le c_1^0 + c_2^0\left|x(k-1)\right| \le c_1^0 + c_2^0\left(\left|e(k-1)\right| + \left|r(k-1)\right|\right). \tag{10.55}$$

With Assumption 10.2, $|r(k)|$ is bounded for all k. Then, there must exist constants c_1 and c_2 such that

$$\left|\xi(x(k-1))\right| \le c_1 + c_2\left|e(k-1)\right|, \tag{10.56}$$

where $c_1 = c_1^0 + c_2^0 \sup_k \left|r(k)\right|$ and $c_2 = c_2^0$.

From (10.56), we have

$$\left|\xi(x(k-1))\right| \le c_1 + c_2 \max_{0\le\tau\le k}\left|e(k)\right|.$$

Then, according to the KTL, conclusion (b) of Theorem 10.1 holds.

BIBO Stability We have shown the asymptotic convergence of tracking error, and the reference trajectory $r(k)$ is bounded, so the system output $x(k)$ is bounded as a direct result.

According to controller algorithm (10.32), it is easy to obtain that

$$|u(k)| \le |r(k+1)| + |\hat{\theta}(k)| \; |\xi(x(k))| \le |r(k+1)| + |\hat{\theta}(k)| \left(c_1^0 + c_2^0 |x(k)|\right). \tag{10.57}$$

Since $\hat{\theta}(k)$ and $x(k)$ are bounded for all k, in terms of Assumption 10.2, $r(k)$ being bounded, it is clear that $u(k)$ is bounded for all k. Thus, conclusion (c) is obtained. ■

10.4.2 Adaptive ILC for Discrete-Time Nonlinear Systems

10.4.2.1 Problem Formulation

Different from (10.31) in the previous section, the control plant considered here is a repetitive system over a finite time interval $\{0,1,\ldots,T\}$ with unknown time-varying parametric uncertainty

$$x(k+1,i) = \theta(k)\xi(x(k,i)) + u(k,i), \quad k \in \{0,1,\ldots,T\}, \tag{10.58}$$

where $x(k,i) \in R$, $u(k,i) \in R$, $\theta(k) \in R$, and $\xi(x(k,i))$ are the measurable system state, the control input, an unknown time-varying parameter, and a known nonlinear function, at time k of iteration i, respectively. $k \in \{0,1,\ldots,T\}$ denotes the time instant, and $i \in \{0,1,\ldots,\infty\}$ denotes the iteration number.

The target trajectory is given as $r(k,i)$, $k \in \{0,1,\ldots,T\}$, which is a function of time k and iteration i. Thus, the target trajectory considered here can vary with iteration i arbitrarily. It is a fundamental relaxation of the assumption in traditional ILC that the tracking trajectory must be the same at all iterations.

Define the tracking error as $e(k,i) = x(k,i) - r(k,i)$, $k \in \{0,1,\ldots,T\}$, $i \in \{0,1,\ldots,\infty\}$. The control objective is to find an appropriate control input sequence $u(k,i)$ such that the system state $x(k,i)$ follows the target trajectory $r(k,i)$, that is,

$$\lim_{i\to\infty} e(k,i) = 0. \quad \forall k \in \{0,1,\ldots,T\}, \; i \in \{0,1,\ldots,\infty\}.$$

10.4.2.2 Adaptive Iterative Learning Control Design

Two discrete-time adaptive iterative learning control schemes, based on an RLS estimator and a projection estimator, respectively, are shown as follows.

RLS-Based Adaptive Iterative Learning Control is designed as follows.

$$u(k,i) = r(k+1,i) - \hat{\theta}(k,i)\xi(x((k,i)), \tag{10.59}$$

$$\hat{\theta}(k,i) = \hat{\theta}(k,i-1) + P(k,i-1)\xi(x(k,i-1))e(k+1,i-1), \tag{10.60}$$

$$P(k,i-1) = P(k,i-2) - \frac{P^2(k,i-2)\xi^2\left(x(k,i-1)\right)}{1 + P(k,i-2)\xi^2(x(k,i-1))}. \tag{10.61}$$

$\hat{\theta}(k,i)$ is the estimation of the unknown time-varying parameter $\theta(k)$ at iteration i, the initial value $\hat{\theta}(k,0)$, $k \in \{0,1,\ldots,T\}$, can be chosen arbitrarily. Similarly, the initial value $P(k,0) = P_0$ can also be chosen arbitrarily.

Projection-based Adaptive Iterative Learning Control is designed as follows.

$$u(k,i) = r(k+1,i) - \hat{\theta}(k,i)\xi(x(k,i)), \tag{10.62}$$

$$\hat{\theta}(k,i) = \hat{\theta}(k,i-1) + \frac{a\xi(x(k,i-1))}{c + \xi(x(k,i-1))^2} e(k+1,i-1), \tag{10.63}$$

where $\hat{\theta}(k,i)$ is the estimation of the unknown time-varying parameter $\theta(k)$ at iteration i, the initial value $\hat{\theta}(k,0)$ can be chosen arbitrarily, $0 < a < 2$, and $c > 0$.

10.4.2.3 Stability and Convergence Analysis

The stability and convergence analysis is based on the following two assumptions and the KTL-like lemma.

Assumption 10.3

The nonlinear function $\xi(x(k,i))$ satisfies the linear growth condition, that is

$$\left|\xi(x(k,i))\right| \le c_1^0 + c_2^0\left|x(k,i)\right|, \quad \forall k \in \{0,\ldots,T\} \ \text{ and } \ \forall i \in \{0,1,\ldots,\infty\}$$

where $0 < c_1^0 < \infty$ and $0 < c_2^0 < \infty$.

Assumption 10.4

All of the unknown system parameter $\theta(k)$, the target trajectory $r(k,i)$, and the initial state $x(0,i)$ are bounded for all $k \in \{0,1,\ldots,T\}$ and $i \in \{0,1,\ldots,\infty\}$.

Remark 10.3

Assumption 10.3 is also called the linear growth condition. The only difference with Assumption 10.1 lies in the fact that Assumption 10.3 holds for all time instants $k \in \{0,1,...,T\}$ and iteration $i = 0,1,...$. Similarly, we only need the existence of the bounds in Assumption 10.4 without requiring their exact values.

KTL-Like Lemma

For any $k \in \{0,1,...,T\}$ and $i \in \{0,1,...,\infty\}$, if the following conditions are satisfied for some given sequences $\{s(k,i)\}$, $\{\boldsymbol{\sigma}(k,i)\}$, $\{b_1(k,i)\}$, and $\{b_2(k,i)\}$:

a. $$\lim_{i\to\infty} \frac{s(k,i)^2}{b_1(k,i) + b_1(k,i)\boldsymbol{\sigma}(k,i)^T \boldsymbol{\sigma}(k,i)} = 0, \tag{10.64}$$

   where $\{b_1(k,i)\}$,$\{b_2(k,i)\}$, and $\{s(k,i)\}$ are real-valued scalar sequences and $\boldsymbol{\sigma}(k,i)$ is a real-valued p-dimensional vector.

b. Uniform boundedness condition

   $$0 < b_1(k,i) < K < \infty \quad \text{and} \quad 0 < b_2(k,i) < K < \infty \tag{10.65}$$

   for all $k \in \{0,1,...,T\}$ and $i \in \{0,1,...,\infty\}$.

c. The linear bounded condition

   $$\|\boldsymbol{\sigma}(k,i)\| \le C_1 + C_2 \max_{0\le\tau\le i} \max_{k\in\{0,\cdots,T\}} |s(k,\tau)|, \tag{10.66}$$

   where $0 < C_1 < \infty$ and $0 < C_2 < \infty$.

It follows that

1. $\lim_{i\to\infty} s(k,i) = 0\cdot$
2. $\{\|\boldsymbol{\sigma}(k,i)\|\}$ is a bounded sequence.

Proof

If $\{s(k,i)\}$ is a bounded sequence for all $k \in \{0,1,...,T\}$ and $i \in \{0,1,...,\infty\}$, then by (10.66), $\{\boldsymbol{\sigma}(k,i)\}$ is a bounded sequence. In terms of (10.64) and (10.65), it follows that

$$\lim_{i\to\infty} s(k,i) = 0.$$

Now, assume that $\{s(k,i)\}$ is unbounded. It follows that there exists a subsequence $\{i_n\}$ such that

$$\lim_{i_n \to \infty} \left| s(k,i_n) \right| = \infty,$$

and

$$\left| s(k,i) \right| \le \left| s(k,i_n) \right| \quad \forall i \le i_n.$$

Now, for subsequence $\{i_n\}$, we have

$$\begin{aligned}
\left| \frac{s(k,i_n)}{\left[b_1(k,i_n) + b_1(k,i_n)\boldsymbol{\sigma}(k,i_n)^T \boldsymbol{\sigma}(k,i_n) \right]^{1/2}} \right| &\ge \frac{\left| s(k,i_n) \right|}{\left[K + K \left\| \boldsymbol{\sigma}(k,i_n) \right\|^2 \right]^{1/2}} \\
&\ge \frac{\left| s(k,i_n) \right|}{K^{1/2} + K^{1/2} \left\| \boldsymbol{\sigma}(k,i_n) \right\|} \\
&\ge \frac{\left| s(k,i_n) \right|}{K^{1/2} + K^{1/2} \left[C_1 + C_2 \left| s(k,i_n) \right| \right]}.
\end{aligned}$$

Hence

$$\lim_{k_n \to \infty} \left| \frac{s(k,i_n)}{\left[b_1(k,i_n) + b_1(k,i_n)\boldsymbol{\sigma}(k,i_n)^T \boldsymbol{\sigma}(k,i_n) \right]^{1/2}} \right| \ge \frac{1}{K^{1/2} C_2} > 0.$$

This contradicts (10.64) and hence the assumption that $\{s(k,i)\}$ is unbounded is false and the results follow. ■

Remark 10.4

Different from the key technical lemma, the KTL-like lemma is formulated and proved along the iteration axis i. All of the properties to be proven in this lemma are satisfied over the entire finite interval $k \in \{0,1,\ldots,T\}$ in a pointwise manner and asymptotically in the iteration domain.

The stability and convergence analysis is provided as follows for the RLS-based discrete-time adaptive ILC. By a similar procedure, the readers themselves can give the analysis of projection-based discrete-time adaptive ILC.

Theorem 10.2

For discrete-time system (10.58), which is repeatable over a finite time interval, under Assumptions 10.3 and 10.4, the RLS-based adaptive ILC scheme (10.59)–(10.61) guarantees the following properties:

a. $\forall k \in \{0,1,\ldots,T\}$ and $\forall i \in \{0,1,\ldots\}$, the parameter estimation $\hat{\theta}(k,i)$ is bounded, and

$$\left|\hat{\theta}(k,i)-\theta(k)\right| \le \left|\hat{\theta}(k,i-1)-\theta(k)\right| \le \cdots \le \left|\hat{\theta}(k,0)-\theta(k)\right|.$$

b. The tracking error converges to zero as the iteration number i goes to infinity

$$\lim_{i\to\infty} e(k,i) = 0, \quad \forall k \in \{0,1,\ldots,T\}.$$

c. Both the control input $u(k,i)$ and the system output $y(k,i)$ are bounded for all k and i.

Proof

There are three parts for this proof. The first part proves the boundedness of $\hat{\theta}(k,i)$. Then, the convergence of the tracking error is shown in the second part. BIBO stability for the system is given in the last part.

Boundedness of $\hat{\theta}(k,i)$ Define the parameter error as $\varphi(k,i) = \theta(k) - \hat{\theta}(k,i)$. According to controller algorithm (10.59), the tracking error dynamics is

$$\begin{aligned} e(k+1,i) &= x(k+1,i) - r(k+1,i) = \theta(k)\xi(x(k,i)) + u(k,i) - r(k+1,i) \\ &= \theta(k)\xi(x(k,i)) + \left(r(k+1,i) - \hat{\theta}(k,i)\xi(x(k,i))\right) - r(k+1,i) \\ &= \varphi(k,i)\xi(x(k,i)). \end{aligned} \tag{10.67}$$

Subtracting $\theta(k)$ from both sides of the parameter estimation algorithm (10.60), and using (10.67), we have

$$\begin{aligned} \varphi(k,i) &= \varphi(k,i-1) - P(k,i-1)\xi(x(k,i-1))e(k+1,i-1) \\ &= \left(1 - P(k,i-1)\xi^2(x(k,i-1))\right)\varphi(k,i-1). \end{aligned} \tag{10.68}$$

To evaluate the relationship between $\varphi(k,i)$ and $\varphi(k,i-1)$ described by (10.68), further inspection of the term $1 - P(k,i-1)\xi^2(x(k,i-1))$ in (10.68) is needed. From (10.61), one can derive

$$P(k,i-1) = \frac{P(k,i-2)}{1 + P(k,i-2)\xi^2(x(k,i-1))}. \tag{10.69}$$

Thus

$$\begin{aligned} 1 - P(k,i-1)\xi^2(x(k,i-1)) &= 1 - \frac{P(k,i-2)\xi^2(x(k,i-1))}{1 + P(k,i-2)\xi^2(x(k,i-1))} \\ &= \frac{1}{1 + P(k,i-2)\xi^2(x(k,i-1))}. \end{aligned} \tag{10.70}$$

From (10.69), we have

$$P(k,i-1)^{-1} = P(k,i-2)^{-1} + \xi^2(x(k,i-1)). \tag{10.71}$$

Hence $P(k,i) > 0$ holds for all k and i.

Taking the absolute value on both sides of (10.68) and according to (10.70) and (10.71), we obtain

$$|\varphi(k,i)| = \frac{1}{1 + P(k,i-2)\xi^2(x(k,i-1))}|\varphi(k,i-1)| \le |\varphi(k,i-1)|. \tag{10.72}$$

Clearly, (10.72) shows a nonincreasing property of $|\varphi(k,i)|$. In terms of Assumption 10.4, $\theta(k)$ is bounded. Hence, the parameter estimation $\hat{\theta}(k,i)$ is bounded $\forall k \in \{0,1,\ldots,T\}$ and $\forall i \in \{0,1,\ldots\}$. Then, conclusion (a) of Theorem 10.2 is obtained.

Asymptotic Convergence of Tracking Error Define a nonnegative function $V(k,i) = P(k,i-1)^{-1}\,\phi(k,i)^2$, $k \in \{0,1,\ldots,T-1\}$ then

$$\begin{aligned} \Delta V(k,i) &= V(k,i) - V(k,i-1) \\ &= P(k,i-1)^{-1}\varphi(k,i)^2 - P(k,i-2)^{-1}\varphi(k,i-1)^2. \end{aligned} \tag{10.73}$$

According to (10.67) and (10.68), Equation (10.73) is rewritten as

$$\begin{aligned} \Delta V(k,i) &= P(k,i-1)^{-1}\big(\varphi(k,i-1) - P(k,i-1)\xi(x(k,i-1))e(k+1,i-1)\big)^2 \\ &\quad - P(k,i-2)^{-1}\varphi(k,i-1)^2 \\ &= (P(k,i-1)^{-1} - P(k,i-2)^{-1})\varphi(k,i-1)^2 \\ &\quad - 2\varphi(k,i-1)\xi(x(k,i-1))e(k+1,i-1) \\ &\quad + P(k,i-1)\xi(x(k,i-1))^2 e(k+1,i-1)^2. \end{aligned} \tag{10.75}$$

From (10.67) and (10.71), we have

$$\left[P(k,i-1)^{-1} - P(k,i-2)^{-1}\right]\varphi(k,i-1)^2 = \xi(x(k,i-1))^2\varphi(k,i-1)^2 = e(k+1,i-1)^2, \tag{10.75}$$

and

$$-2\varphi(k,i-1)\xi(x(k,i-1))e(k+1,i-1) = -2e(k+1,i-1)^2. \tag{10.76}$$

Substituting (10.75) and (10.76) into (10.74) yields

$$\Delta V(k,i) = -\left[1 - P(k,i-1)\xi(x(k,i-1))^2\right]e(k+1,i-1)^2. \tag{10.77}$$

Using (10.70) again, Equation (10.77) becomes

$$\Delta V(k,i) = -\frac{e(k+1,i-1)^2}{1 + P(k,i-2)\xi^2(x(k,i-1))} \le 0. \tag{10.78}$$

Thus, $V(k,i)$ is a nonnegative and nonincreasing function.
Summing up (10.78) from 0 to i leads to

$$V(k,i) = V(k,0) - \sum_{j=1}^{i} \frac{e(k+1,j-1)^2}{1 + P(k,j-2)\xi^2(x(k,j-1))}. \tag{10.79}$$

Since $V(k,0) = P(k,-1)^{-1}\varphi(k,0)^2$ is bounded, and $V(k,i)$ is nonnegative and nonincreasing, one can immediately conclude from (10.79) that

$$\lim_{i\to\infty}\sum_{j=1}^{i} \frac{e(k+1,j-1)^2}{1 + P(k,j-2)\xi^2(x(k,j-1))} < \infty, \tag{10.80}$$

and

$$\lim_{i\to\infty} \frac{e(k+1,i)^2}{1 + P(k,i-1)\xi^2(x(k,i))} = 0. \tag{10.81}$$

In terms of the linear growth condition of Assumption 10.3 and the definition of the tracking error, one obtains

$$\left|\xi(x(k,i))\right| \le c_1^0 + c_2^0\left|x(k,i)\right| \le c_1^0 + c_2^0\left(\left|e(k,i)\right| + \left|r(k,i)\right|\right). \tag{10.82}$$

With Assumption 10.4, $|r(k,i)|$ is bounded. Then, there must exist constants c_1^* and c_2^* such that

$$\left|\xi(x(k,i))\right| \le c_1^* + c_2^*\left|e(k,i)\right|, \tag{10.83}$$

where $c_1^* = c_1^0 + c_2^0 \sup_i \max_{k\in\{0,\dots,T\}}\left|r(k,i)\right|$ and $c_2^* = c_2^0$.

According to Assumptions 10.3 and 10.4, $\left|e(0,i)\right| \le \left|x(0,i)\right| + \left|r(0,i)\right|$ is bounded. Then, from (10.83), we have

$$\begin{aligned}\left|\xi(x(k,i))\right| &\le c_1^* + c_2^*\left|e(k,i)\right| \\ &\le c_1^* + c_2^*\left(\max_{\tau\le i}\left|e(0,\tau)\right| + \max_{\tau\le i}\max_{k\in\{0,\dots,T-1\}}\left|e(k+1,\tau)\right|\right) \\ &\le c_1 + c_2\max_{\tau\le i}\max_{k\in\{0,\dots,T-1\}}\left|e(k+1,\tau)\right|,\end{aligned} \tag{10.84}$$

where $c_1 = c_1^* + c_2^*\sup_i\left|e(0,i)\right|$ and $c_1 = c_2^*$.

In terms of the KTL-like lemma, conclusion (b) of Theorem 10.2 holds, that is,

$$\lim_{i\to\infty} e(k+1,i) = 0,$$

for all $k \in \{0,1,\dots,T\}$.

BIBO Stability The asymptotic convergence of the tracking error has been proven and the reference trajectory $r(k,i)$ is bounded, so the system output $x(k,i)$ is bounded as a direct result.

According to controller algorithm (10.59), it is easy to obtain that

$$\begin{aligned}\left|u(k,i)\right| &\le \left|r(k+1,i)\right| + \left|\hat{\theta}(k,i)\right|\ \left|\xi(x(k,i))\right| \\ &\le \left|r(k+1,i)\right| + \left|\hat{\theta}(k,i)\right|\left(c_1^0 + c_2^0\left|x(k,i)\right|\right).\end{aligned} \tag{10.85}$$

Since $\left|\hat{\theta}(k,i)\right|$ and $x(k,i)$ are bounded $\forall k \in \{0,1,\dots,T\}$ and $\forall i \in \{0,1,\dots\}$, in terms of Assumption 10.4, it is clear that $u(k,i)$ is bounded for all k and i. Thus, conclusion (c) is obtained. ■

Remark 10.5

It should be pointed out that traditional discrete-time adaptive control approaches (10.32)–(10.36) are not suitable for system (10.58) due to the time-varying parameters and finite time tracking task. The proposed discrete-time adaptive ILC (10.59)–(10.63) are conducted along the iterative learning axis i, and not along the time axis k; therefore, it can fulfill the perfect tracking task of the controlled systems with any time-varying parametric uncertainty. Furthermore, parameter updating law (10.60) and (10.63) is noncausal with respect to time, that is, $e(k+1,i-1)$ is used for computing $\hat{\theta}(k,i)$. It is common in ILC but rare in other online control updating algorithms. The noncausal term plays a key role in convergence analysis.

10.4.3 Comparison between Adaptive Control and Adaptive ILC

From Table 10.5, it is easy to find similarities between the proposed discrete-time adaptive ILC and the discrete-time adaptive control:

1. The control plants are similar and the assumptions of adaptive control are similar to that of adaptive ILC.
2. The controller structures and estimator structures are similar.
3. The key technical lemma is similar to the KTL-like lemma.
4. Stability and convergence analyses are similar.

The only difference is that the controller design, the stable analysis, and the actuator implementation of the discrete-time adaptive control system are all conducted along the time axis. Instead, the discrete-time adaptive ILC system is conducted along the iteration axis. The discrete-time adaptive ILC can be obtained by "erecting up" the discrete-time adaptive from the horizontal time axis to the vertical iteration axis. Or the discrete-time adaptive control can be obtained by "laying down" the discrete-time adaptive ILC from the vertical iteration axis to the horizontal time axis. Thereafter, the discrete-time adaptive control and the discrete-time adaptive ILC are similar in nature despite some trivial differences existing in, for example, the application background, control plants, control tasks, and so on. As an interesting result, these two control system design and analysis methods have complemented each other. More works on this topic can be found in Refs. [208,211,253–261].

Table 10.5 Comparison between Discrete-Time Adaptive Control and Discrete-Time Adaptive ILC

	Discrete-Time Adaptive Control	*Discrete-Time Adaptive ILC*
Plant	$x(k+1) = \theta\xi(x(k)) + u(k)$, $k \in \{0,1,\ldots,\infty\}$ 1. The system runs over an infinite time interval 2. The system is not required to be repetitive 3. The unknown parameter should be constant	$x(k+1,i) = \theta(k)\xi(x(k,i)) + u(k,i)$, $k \in \{0,1,\ldots,T\}, i \in \{0,1,\ldots,\infty\}$ 1. The system runs over a finite time interval 2. The system is repetitive 3. The unknown parameter can vary with time arbitrarily
	Linearly parametric systems in time domain	Linearly parametric systems in time-and-iteration domain
Key Assumption	The nonlinear function $\xi(x(k))$ satisfies the linear growth condition for all $k \in \{0,1,\ldots,\infty\}$	The nonlinear function $\xi(x(k,i))$ satisfies the linear growth condition for all $k \in \{0,1,\ldots,T\}$ and $i \in \{0,1,\ldots,\infty\}$
	Linear growth condition in time domain	Linear growth condition in iteration domain
Desired Trajectory	$r(k)$ is bounded for all $k \in \{0,1,\ldots,\infty\}$. Tracking tasks over infinite time interval	$r(k,i)$ is bounded for all $k \in \{0,1,\ldots,T\}$ and $i \in \{0,1,\ldots,\infty\}$. Tracking tasks over finite time interval
	Asymptotically tracking along the time axis	Asymptotically tracking along the iteration axis
Controller Algorithm	$u(k) = r(k+1) - \hat{\theta}(k)\xi(x(k))$, $k \in \{0,1,\ldots,\infty\}$	$u(k,i) = r(k+1,i) - \hat{\theta}(k,i)\xi(x(k,i))$, $k \in \{0,1,\ldots,T-1\}$ and $i \in \{0,1,\ldots,\infty\}$
	The learning gain is updated along the time axis	The learning gain is updated along the iteration axis

Parameter Estimation Algorithm	RLS	$\hat{\theta}(k) = \hat{\theta}(k-1) + P(k-1)\xi(x(k-1))e(k)$ $P(k-1) = P(k-2) - \frac{P^2(k-2)\xi^2(x(k-1))}{1+P(k-2)\xi^2(x(k-1))}$	$\hat{\theta}(k,i) = \hat{\theta}(k,i-1) + P(k,i-1)\xi(x(k,i-1))e(k+1,i-1)$ $P_{k-1}(k,i-1) = P_{k-2}(k,i-2) - \frac{P^2(k,i-2)\xi^2(x(k,i-1))}{1+P(k,i-2)\xi^2(x(k,i-1))}$
	Projection Algorithm	$\hat{\theta}(k) = \hat{\theta}(k-1) - \frac{a\xi(x(k-1))}{c+\xi(x(k-1))^2}e(k),$	$\hat{\theta}(k,i) = \hat{\theta}(k,i-1) - \frac{a\xi(x(k,i-1))}{c+\xi(x(k,i-1))^2}e(k+1,i-1)$
		Iteratively updating estimator along the time axis	Iteratively updating estimator along the iteration axis
Convergence Property		$\forall k$, $\hat{\theta}(k)$ is bounded, $\lvert\hat{\theta}(k)-\theta\rvert \le \cdots \le \lvert\hat{\theta}(0)-\theta\rvert$, and $\lim_{k\to\infty} e(k) = 0$ The system is BIBO stable	$\forall k \in \{0,\ldots,T-1\}$ and $\forall i \in \{0,1,\ldots,\infty\}$ $\hat{\theta}(k,i)$ is bounded, $\lvert\hat{\theta}(k,i)-\theta(k)\rvert \le \cdots \le \lvert\hat{\theta}(k,0)-\theta(k)\rvert$, and $\lim_{i\to\infty} e(k,i) = 0$ The system is BIBO stable
		Analysis along the time axis	Analysis along the iteration axis
Analysis Tool		KTL	KTL-like lemma
		KTL w.r.t. time axis	KTL-like lemma w.r.t. iteration axis

10.5 Conclusions

In this chapter, the concepts, design principles, and properties for adaptive control systems with symmetric similarity structure are presented. The adaptive control and MFAC with symmetric similarity structure are explored, and their "superior" performance is verified by illustrative simulations. Meanwhile, the similarities between MFAC and MFAILC are discussed in detail. Furthermore, the similarities between adaptive control and iterative learning control are studied from a broader viewpoint. As a result, by applying crossover and mutation operation to these two approaches, a new discrete-time adaptive iterative learning control approach with more "superior" control performance is presented.

Utilizing symmetric similarity in control systems can simplify controller design and analysis greatly. The control system with symmetric similarity structure can be implemented easily. For example, the control law and the estimation law of the MFAC scheme with symmetric similarity structure can be implemented by the same computing procedure. On the other hand, some new results of controller design and control system analysis can also be easily obtained with the control system design principle of the symmetric similarity structure. For example, it is possible to extend any new result in adaptive control to adaptive ILC, which is symmetrically similar to adaptive control. Furthermore, MFAC and MFAILC, periodic adaptive control and iterative learning control, and so on also have symmetric similarity structure. Hence, it is possible to learn from each other for controller design and analysis as well as to generate new results.

Chapter 11

Applications

11.1 Introduction

Model-based control theory has achieved great success in theory and applications, especially in the fields of industrial processes and aerospace. However, it is a challenging task to apply the model-based control methods in practical processes and plants since modeling a practical industrial process is very difficult and sometimes impossible. Even if the model of the controlled plant is established, unmodeled dynamics is also inevitable, which may lead to unpredictable problems when the model-based methods are applied in practical applications. Modeling and unmodeled dynamics are major obstacles to the use of elegant model-based controllers in practice, and a huge gap between theoretical results and practical application effectiveness always exists. Thus, it is of great significance to study the data-driven model-free control methods and their applications, which only use the input/output measurement data of the controlled plant to design the control system without involving the plant mathematical model.

MFAC was first proposed for a class of nonlinear systems in 1994 [23]. Since then, the MFAC theory has been developed and improved. The MFAC approach has been successfully implemented in many practical applications, for example, the linear motor [29,30], the pH neutralization process [262,263], freeway traffic control [209,210,264,265], the antirolling system for large warships [266,267], liquid-level control [25,268,269], the glass furnace control system [270], the magnetic suspension switched reluctance motor [271], MR dampers [272], the power plant main-steam pressure control system [273], ball mill load control [274,275], large radio telescope control [276], cable-mesh reflector control [277], feed cabin supporting structure control [278], electrical furnace control [279], the multipoint forming process [280], the underwater towed heave compensation system [281], injection molding [30], the welding process [282–285], vertical drilling control [286], macroeconomic dynamic control [287], the multiwire saw system [288], the electrode regulator system of arc furnace [289], load control of wind turbine [290,291], the blood pump [292,293], the voltage source PWM rectifier [294], and so on.

In this chapter, MFAC schemes are applied in several typical practical controlled plants, such as the three-tank water system, the permanent magnet linear motor, the freeway traffic control system, the welding process, and the wind turbine machine. Readers can find many other practical applications of the MFAC methods in relevant references.

11.2 Three-Tank Water System

Liquid level is one of the most important parameters in the industrial process. Three-tank water-level control system is a typical nonlinear system with large time delay and time-varying parameters. Many controlled plants in industrial processes can be fully or partly described as a three-tank system.

The experimental equipment of three-tank water system is manufactured by Zhejiang Tianhuang Science and Technology Industrial Company. The experiment purpose of applying the MFAC approach in the three-tank system is to verify the effectiveness of this approach. Furthermore, two other typical data-driven control approaches, VRFT and IFT, are also applied to the three-tank water system to compare the drawbacks and advantages of these approaches.

11.2.1 Experimental Setup

The experimental setup is shown in Figure 11.1a. The three-tank water system consists of a host computer, a set of control instruments, actuator, three water tanks, and three liquid level detectors. The actuator includes an electric regulating valve and a three-phase magnetic pump. The pump is used to deliver water flow to the water tanks, and its maximum flow is 32 L/min. The electric regulating valve is used to control the pump flow rate, and its repetitive precision is ±1%. The diffusing silicon pressure sensor with an accuracy of 0.25 cm is used to detect the liquid level. The diameter and height of each tank are 35 and 20 cm, respectively. Each tank has a manual inlet valve and a manual outlet valve.

A schematic diagram of the three-tank system is shown in Figure 11.1b. The system output signal is the liquid level of the lower tank, and the control input signal is the pump flow rate of the upper tank. The objective is to control the pump flow rate of the upper tank such that the liquid level of the lower tank remains at the desired one. The long water supply pipeline results in a large time delay of the control input signal. Moreover, three tanks and the actuator have nonlinear characteristics. Thus, the controlled plant is a large time-delay nonlinear system.

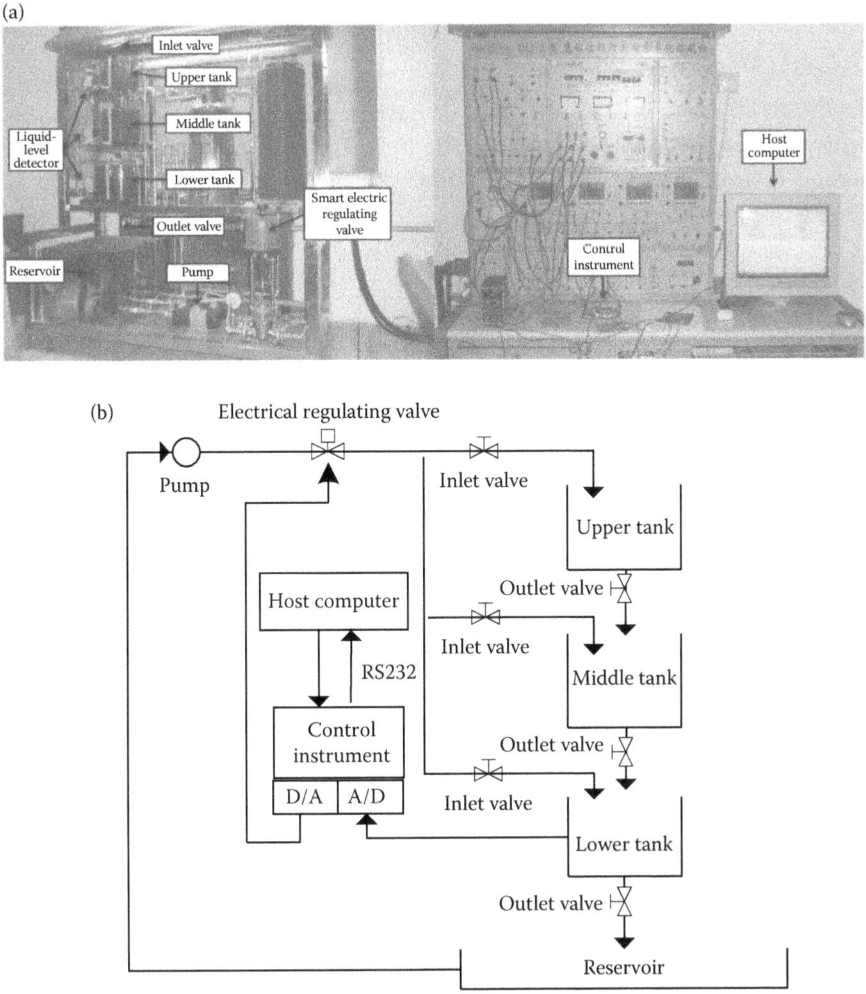

Figure 11.1 Three-tank water system. (a) Experimental equipment. (b) Schematic diagram.

The original control instrument can only realize the PID control method. Thus, some necessary hardware and software modifications and control instruments have been developed to use the MFAC and the other two typical data-driven control methods in this equipment. The hardware includes the ARM7 processor, the 10-bit AD/DA interface, and the communication interface. The software program of the host computer, which is written in C++ Builder, includes the control algorithm, controller parameter setting, monitoring and displaying control input, system output signals, and so on.

11.2.2 Three Data-Driven Control Schemes

11.2.2.1 MFAC Scheme

The design of the MFAC scheme depends only on the input and output measurements without any *a priori* knowledge about the system. According to the design method in Section 4.3, the PFDL–MFAC scheme for the liquid level $h_d(k)$ is designed as follows:

$$\hat{\boldsymbol{\phi}}_{p,L}(k) = \hat{\boldsymbol{\phi}}_{p,L}(k-1) + \frac{\eta \Delta \boldsymbol{U}_L(k-1)(h(k) - h(k-1) - \hat{\boldsymbol{\phi}}^T_{p,L}(k-1)\Delta \boldsymbol{U}_L(k-1))}{\mu + \|\Delta \boldsymbol{U}_L(k-1)\|^2}, \tag{11.1}$$

$$\hat{\boldsymbol{\phi}}_{p,L}(k) = \hat{\boldsymbol{\phi}}_{p,L}(1), \text{ if } \|\hat{\boldsymbol{\phi}}_{p,L}(k)\| \le \varepsilon \text{ or } \|\Delta \boldsymbol{U}_L(k-1)\| \le \varepsilon \text{ or } \operatorname{sign}(\hat{\phi}_1(k)) \ne \operatorname{sign}(\hat{\phi}_1(1)), \tag{11.2}$$

$$u(k) = u(k-1) + \frac{\rho_1 \hat{\phi}_1(k)(h_d(k+1) - h(k))}{\lambda + |\hat{\phi}_1(k)|^2} - \frac{\hat{\phi}_1(k)\sum_{i=2}^{L} \rho_i \hat{\phi}_i(k)\Delta u(k-i+1)}{\lambda + |\hat{\phi}_1(k)|^2}, \tag{11.3}$$

where $h(k)$ is the liquid level of the lower tank (cm), $u(k)$ is the pump flow (L/min), $\hat{\boldsymbol{\phi}}_{p,L}(1)$ is the initial value of $\hat{\boldsymbol{\phi}}_{p,L}(k)$, $\lambda > 0$, $\mu > 0$, $\eta \in (0,2]$, $\rho_i \in (0,1]$, $i = 1,2,\ldots,L$, and ε is a small positive constant.

11.2.2.2 VRFT Method

For an unknown LTI SISO plant $P(z)$, a fixed structure controller $C(z,\Theta)$ with an unknown Θ and an invertible reference model $M(z)$ are prespecified; then the control objective of VRFT [51,52,117,295] is to find the optimal controller parameters using a batch of the measured input/output data of the open-loop or closed-loop controlled system such that the closed-loop control system approximates the reference model $M(z)$. The optimal controller parameter is obtained by minimizing the following model reference cost function:

$$J(\Theta) = \left\| \left(\frac{P(z)C(z,\Theta)}{1 + P(z)C(z,\Theta)} - M(z) \right) W(z) \right\|_2^2, \tag{11.4}$$

where $W(z)$ is a weighting function and z is the one-step-ahead shifting operator.

In practice, the following controller structure is widely used for the VRFT method:

$$u(k) = u(k-1) + K_1 e(k) + K_2 e(k-1) + \cdots + K_{n_c} e(k - n_c + 1), \tag{11.5}$$

where $\Theta = [K_1, K_2, \ldots, K_{n_c}]^T$ denotes the controller parameter and n_c is the controller order.

11.2.2.3 IFT Method

For an unknown LTI SISO plant $P(z)$, a fixed structure controller $C(z,\Theta)$ with an unknown Θ and a desired output signal $y^*(k)$ are prespecified; then the control objective of IFT [40,43,45,296,297] is to find the optimal controller parameter using the measured input/output data obtained from two designed experiments, including a normal experiment and a gradient experiment [40,45,296,297]. The controller parameter is obtained by minimizing the following cost function:

$$J(\Theta) = \frac{1}{2N} \sum_{k=1}^{N} \left(y(k,\Theta) - y^*(k) \right)^2, \tag{11.6}$$

where N is the number of input/output data pairs. Theoretically, if the closed-loop system is stable at each iteration, then the local optimal controller parameters can be found as the iterative number approaches infinity [45]. Here, the fixed controller structure like (11.5) is adopted in IFT.

The main differences between IFT and VRFT are: (1) IFT method is based upon a gradient-descent approach and is therefore an iterative technique. Consequently, it requires a large number of experiments on the true plant to iteratively optimize the controller parameters. (2) VRFT method obtains the optimal controller parameters merely using a batch of measured I/O data, but does not require any iterative experiment; thus it is a "one-shot" batch method.

11.2.3 Experimental Investigation

The experimental investigation is carried out in two experiments. One is to verify the effectiveness of three data-driven control methods and investigate the influence on control performance by choosing different controller parameters, and the other is to investigate the robustness of three data-driven control methods as the system structure is changed.

11.2.3.1 Experiment I

The sampling period adopted is 1 s and the experiment time is 400 s. The inlet valve of the upper water tank and the outlet valves of all three tanks are open, while the inlet valves of the middle and lower water tanks are closed. Thus, the objective is to control the pump flow rate of the upper water tank such that the liquid level of the lower water tank remains at $h_d = 5$(cm). The initial values of the system are $u(0) = 0$(L/min) and $h = 0$(cm).

PFDL–MFAC Scheme—Generally, larger control input linearization constant L will lead to a better control performance. However, larger L will bring about more parameters, which should be adjusted online, resulting in heavier online computation. To obtain a proper L, two experiments are performed to compare the control performances of the PFDL–MFAC schemes with $L = 3$ and $L = 5$.

The initial values of PG are set to be $\hat{\boldsymbol{\phi}}_{p,L}(0) = [0.5,0,0]^T$ for $L = 3$ and $\hat{\boldsymbol{\phi}}_{p,L}(0) = [0.5,0,0,0,0]^T$ for $L = 5$, respectively. The other controller parameters in two PFDL–MFAC schemes are set to be $\lambda = 1$, $\eta = 1$, $\mu = 1$, $\rho_i = 1$, $i = 1,\ldots,L$. From the simulation results shown in Figure 11.2, we can see that the selection of $L = 5$ leads to a better control performance.

VRFT Method—The VRFT method requires that the controller structure is known. Thus, the order n_c of controller (11.5) must be given before the controller parameters are tuned by using the VRFT method. To obtain good designed parameters n_c and Θ for the VRFT, four cases are considered:

$$\{n_c, \text{the number of I/O pairs}\} = \{(3, 200) \quad (5, 200) \quad (3, 400) \quad (5, 400)\}$$

Since the system's delay is about 24 s, the reference model is set to $M = z^{-24}$. From the experimental results shown in Figure 11.3, we can see that case 4 gives the best performance with the corresponding VRFT control law as

$$\begin{aligned} u(k) &= u(k-1) + 12.04e(k) - 11.78e(k-1) - 2.045e(k-2) \\ &\quad + 0.747e(k-3) + 1.365e(k-4). \end{aligned}$$

IFT Method—Similar to VRFT, IFT also requires that the controller structure is known. Thus, the order n_c of controller (11.5) must also be given before the controller parameters are tuned. However, the criterion functions of tuning parameter in IFT and VRFT are different. To obtain good designed parameters n_c and Θ for the IFT, the same four cases are considered:

$$\{n_c, \text{the number of I/O pairs}\} = \{(3, 200) \quad (5, 200) \quad (3, 400) \quad (5, 400)\}$$

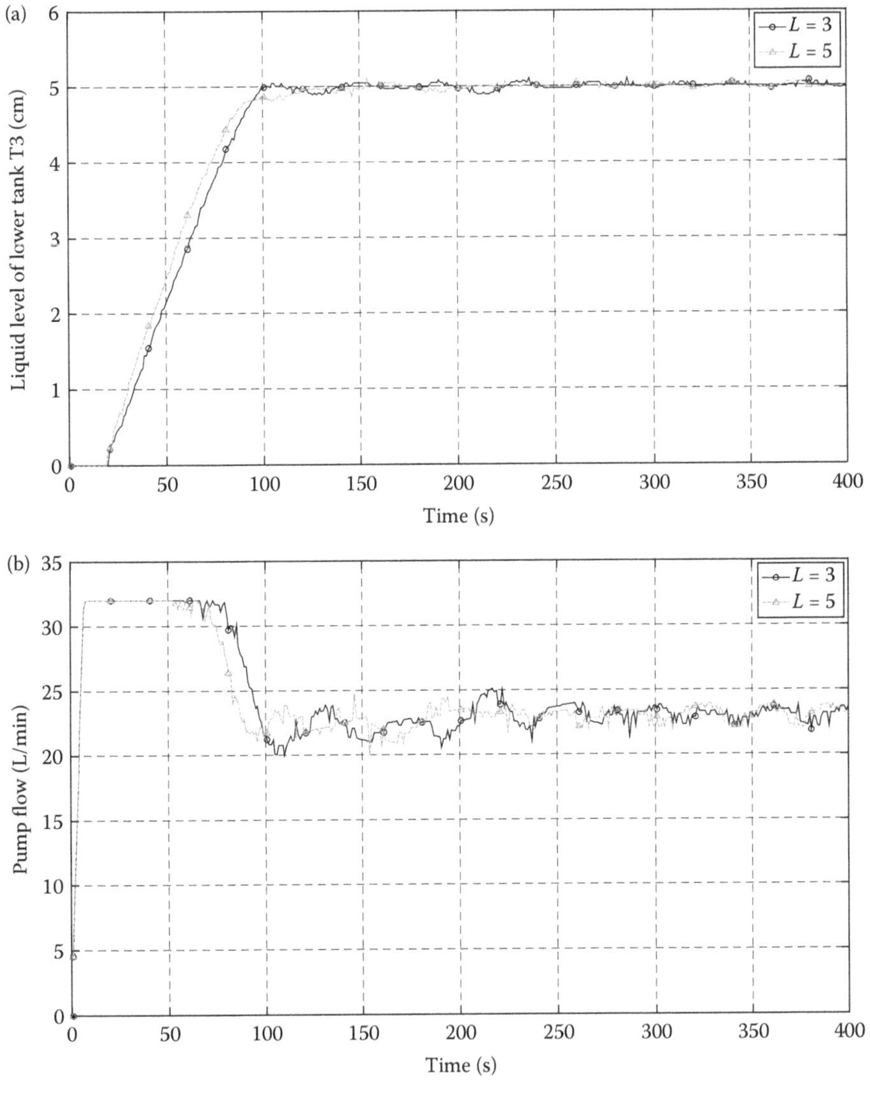

Figure 11.2 Control performances of PFDL–MFAC schemes with $L = 3$ and $L = 5$. (a) Liquid level of lower water tank. (b) Pump flow of upper water tank.

Since the iterative experiments are very time-consuming, only five iterations are carried out for each case. The measured I/O data obtained from the experiments are used to tune the controller parameters. From the experimental results shown in Figure 11.4, we can see that case 4 gives the best performance with the corresponding IFT control law as

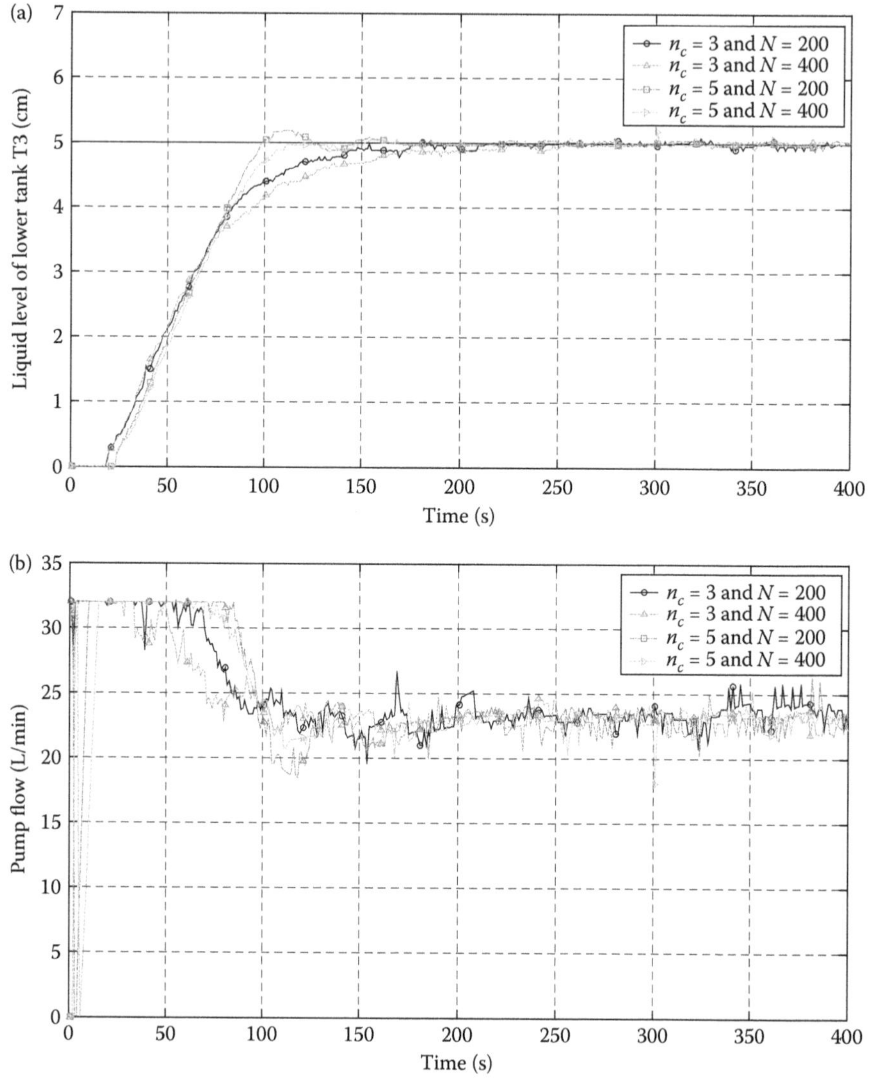

Figure 11.3 Control performances for VRFT methods with different controller orders and data pairs. (a) Liquid level of lower water tank. (b) Pump flow of upper water tank.

$$u(k) = u(k-1) + 8.62e(k) - 8.922e(k-1) + 0.555e(k-2)$$
$$-0.585e(k-3) + 0.851e(k-4).$$

From the above experimental results, we can see that all the three data-driven control methods can work well. But it is worth pointing out that: (1) From the aspect

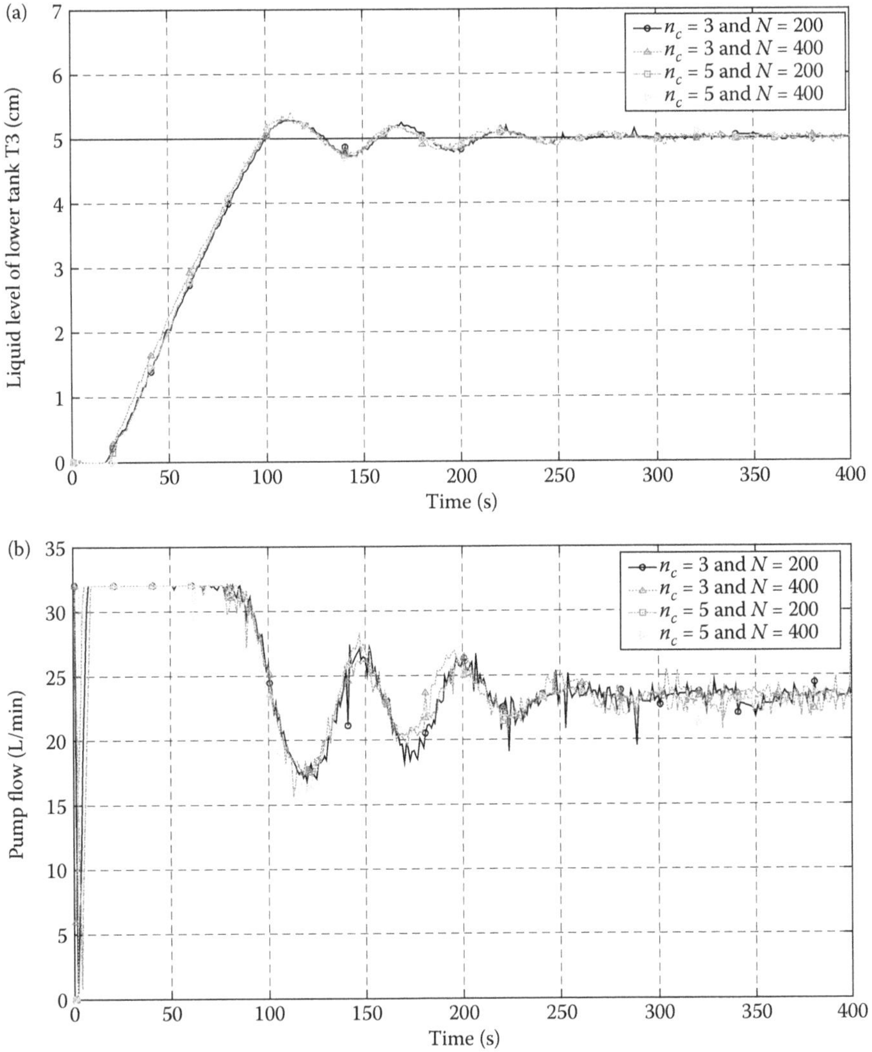

Figure 11.4 Control performances for IFT methods with different controller orders and data pairs. (a) Liquid level of lower water tank. (b) Pump flow of upper water tank.

of control performance, MFAC is better than the other two data-driven control methods. For IFT, the control performance is not satisfactory. The reasons may come from the limited number of I/O data pairs collected. (2) From the aspect of the number of I/O data pairs, MFAC is the online adaptive control approach, and only utilizes the real-time measured I/O data to update controller parameters without the need for an offline data collection procedure, while VRFT and IFT are the offline methods, the data collection procedure is inevitable, and the higher the number of I/O data pairs for controller parameters tuning, the better the control performance. (3) The parameter tuning procedure for IFT is time-consuming compared with VRFT, since many iterative experiments are needed to find the optimal controller parameters by using the IFT method. It took 5 h even for only five iterations.

11.2.3.2 Experiment II

Using the best controller parameters for three data-driven control methods obtained in the first experiment, the following experiment is carried out to investigate the robustness of three data-driven control methods with changing system structure.

The sampling period, the experiment time, the desired liquid level, and the initial values of the system are the same as above. We redo Experiment I with the best controller parameters obtained until the controlled process has reached the steady state, and then we close the inlet valve of the upper water tank and open the inlet valve of the lower water tank at $k = 250$ s. Consequently, the controlled plant becomes a one-tank water system, and the control input becomes the pump flow of the lower water tank. This implies that the controlled system's structure, delay, and order have changed. From the aspect of system dynamics, the controlled plant in this experiment is a nonlinear system with time-varying structure. The experimental results of three data-driven control methods are shown in Figure 11.5.

To compare the control performance of the three control methods more clearly, three numerical indexes, including root mean square (RMS) of tracking error as $e_{\text{RMS}} = \sqrt{(1/N)(\sum_{k=1}^{N})e(k)^2}$, integral time absolute error (ITAE) as $e_{\text{ITAE}} = \sum_{k=1}^{N} k\left|e(k)\right|$, and total sum of squares (TSS) of the control input change as $\Delta u_{\text{TSS}} = \sum_{k=1}^{N} \Delta u(k)^2$, are shown in Table 11.1.

The experimental results show that all three data-driven methods can work well for the three-tank water system, even if the system structure is time-varying. However, the following are noted: (1) The MFAC approach is easy to implement. The control performance of MFAC is better than that of the other two approaches. Smaller vibration of the control input signal of MFAC means less abrasion to the actuator valve. (2) The MFAC is an adaptive control approach without needing any offline data collection procedure, whereas VRFT and IFT are offline approaches, and the data collection procedure is inevitable. (3) The IFT method is very time-consuming since it requires a large number of iterative experiments on the true plant to optimize the controller parameters.

11.3 Permanent Magnet Linear Motor

The permanent magnet linear motor is a kind of experiment that can convert electric energy directly into mechanical energy and drive the motor to move in a straight line without any intermediate mechanism. Compared with the traditional motor

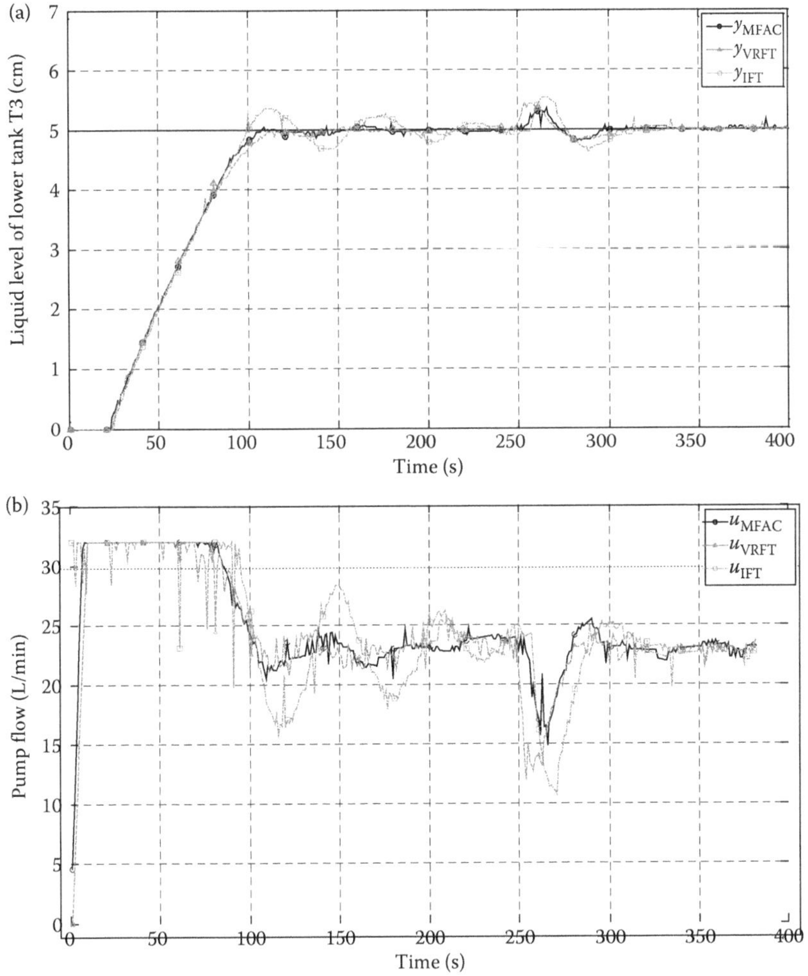

Figure 11.5 Performance comparison among three data-driven control approaches. (a) Liquid level of lower water tank. (b) Pump flow of upper and lower water tank.

Table 11.1 RMS and ITAE of the Tracking Errors, TSS of Control Input Change

	MFAC	*VRFT*	*IFT*
e_{RMS}	0.01675	0.01679	0.01702
e_{ITAE}	132	144	179
Δu_{TSS}	333	481	1950

using mechanical transmission, the linear motor has its great advantages, such as simple structure, noncontact, nonabrasion, low noise, high speed, high precision, and so on. Recently, the linear motor has found many successful applications in high-speed, high-precision equipments.

Owing to the variation of motor parameters, friction, slot effect and end effect of the motor, and other factors during the running of the linear motor, it is very difficult to build an accurate model for the linear motor [298]. Furthermore, the control performance of the linear motor by applying model-based control methods is not good enough because of external disturbances and various uncertainties such as unknown load. For a dual-axis linear motor gantry system, the accurate control problem becomes more difficult compared with the single-axis linear motor since there exists a coupling between the two driving axes. Therefore, to improve control performance, a possible way, besides adopting high-performance hardware, is to develop advanced control methods [299–302].

For improving the dynamic stiffness of the motor, adaptive robust control methods have been proposed to suppress external disturbances and cope with parameter perturbation. However, it is difficult to obtain good performance due to the complex nonlinear behavior in the motor model dynamics. Moreover, this kind of control methods usually has a high computational complexity that significantly influences real-time performance for linear motor servomechanisms; therefore, its application to the linear motor is very restricted [299]. Since the PID control is not capable of dealing with the various time-varying disturbances in motor operation, it is difficult to meet the high-speed, high-precision requirement of the linear motor [302]. Although the control method with the disturbance observer can compensate for the observed disturbance to some extent, it needs an accurate model of the motor [301]. The neural-network-based controller is designed by using merely I/O data of the motor, and has been applied to linear motor control, but its disadvantages are also inevitable, such as difficulties in training the networks, heavy computational burden, and incapability of dealing with various uncertainties in the linear motor [300]. Based on this observation, it is of significant importance to develop a controller independent of the model for precise linear motor servomechanisms.

In this section, the MFAC scheme is applied to a linear motor and a dual-axis linear motor gantry system. The experimental results show that, by applying this control scheme, the accuracy of position tracking as well as the ability to reject disturbance increases.

11.3.1 Permanent Magnet Linear Motor System

11.3.1.1 Experimental Setup

The device in this experiment is a linear motor servomechanism as shown in Figure 11.6, which was developed by Zhengzhou Winner Motor Technology Co. Ltd. The system consists of a U-shaped permanent magnet synchronous linear servo

Figure 11.6 Hardware configuration of the linear motor servomechanism.

motor, a mechanical moving platform, precise linear guideways, a grating encoder as the position detector, and a digital servo driver. The technical parameters of the linear motor are as follows: maximum speed is 1 m/s, maximum acceleration is 60 m/s^2, maximum stroke is 380 mm, rated thrust is 30 N, and peak thrust is 80 N. Resolution of the grating is 5 μm.

The linear motor servomechanism is a double-loop system. The inner loop for speed control is implemented by a linear motor driver developed by Elmo Co., and the outer loop for position control is based on the MFAC scheme. The schematic diagram of the system is depicted in Figure 11.7. The experimental steps are as follows:

Step 1 Construct the designed MFAC scheme on PC via MATLAB®/Simulink®, and replace the input interface and the output interface with cSPACE modules in Simulink. Compile the MFAC algorithm and then generate the DSP code automatically, which will be downloaded into DSP via the USB interface.

Step 2 Using the cSPACE card, compare the position signal of the linear motor collected from the grating encoder with the desired sinusoidal position signal to generate the position error signal. Call the MFAC algorithm in the DSP to calculate the control signal. Then send this control signal to the linear motor servo driver via the D/A decoder.

Step 3 Amplify the control signal from the cSPACE card and then drive the linear motor.

11.3.1.2 MFAC Scheme

The CFDL data model of the linear motor servomechanism is described as

$$\Delta y(k+1) = \phi_c(k)\Delta u(k) \tag{11.7}$$

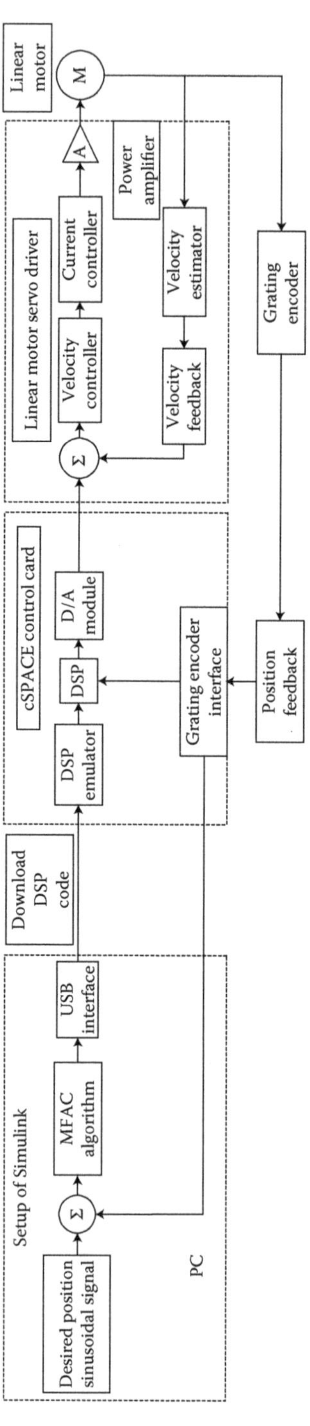

Figure 11.7 Schematic diagram of the MFAC based linear motor servo mechanism.

where the control input $u(k)$ represents the input voltage (V) of the linear motor at time k, the system output $y(k)$ represents the position (mm) of the linear motor at time k, $\Delta y(k+1) \triangleq y(k+1) - y(k)$, and $\Delta u(k) \triangleq u(k) - u(k-1)$.

Remark 11.1

Linear motor servomechanisms satisfy the two assumptions for the CFDL model transformation as given in Chapter 3. First, the linear motor is a continuous moving system whose dynamics satisfies the smoothness condition. Second, a bounded change of input voltage must cause a bounded change of motor position when the control voltage is not out of the permitted range, namely, that the generalized *Lipschitz* condition is satisfied.

For a desired position trajectory $y_d(k)$, the CFDL–MFAC scheme for a linear motor is as follows:

$$\hat{\phi}_c(k) = \hat{\phi}_c(k-1) + \frac{\eta \Delta u(k-1)(y(k) - y(k-1) - \hat{\phi}_c(k-1)\Delta u(k-1))}{\mu + |\Delta u(k-1)|^2}, \tag{11.8}$$

$$\hat{\phi}_c(k) = \hat{\phi}_c(1), \text{ if } |\hat{\phi}_c(k)| \le \varepsilon \text{ or } |\Delta u(k-1)| \le \varepsilon \text{ or } \operatorname{sign}(\hat{\phi}_c(k)) \ne \operatorname{sign}(\hat{\phi}_c(1)), \tag{11.9}$$

$$u(k) = u(k-1) + \frac{\rho \hat{\phi}_c(k)\left(y_d(k+1) - y(k)\right)}{\lambda + |\hat{\phi}_c(k)|^2}, \tag{11.10}$$

where $\hat{\phi}_c(1)$ is the initial value of $\hat{\phi}_c(k)$, $\lambda > 0$, $\mu > 0$, $\eta \in (0,2]$, and $\rho \in (0,1]$ are adjustable parameters, and ε is a small positive constant.

11.3.1.3 Experiments

Experiment I: Performance Test and Analysis of the Linear Motor Servo Driver. The performance of the outer loop control strategy is largely dependent on the dynamic response of the linear motor servo driver. Therefore, it is necessary to debug the driver and analyze its fundamental performance, such that all the performance indexes required by subsequent experiments are sufficiently guaranteed by the servo driver.

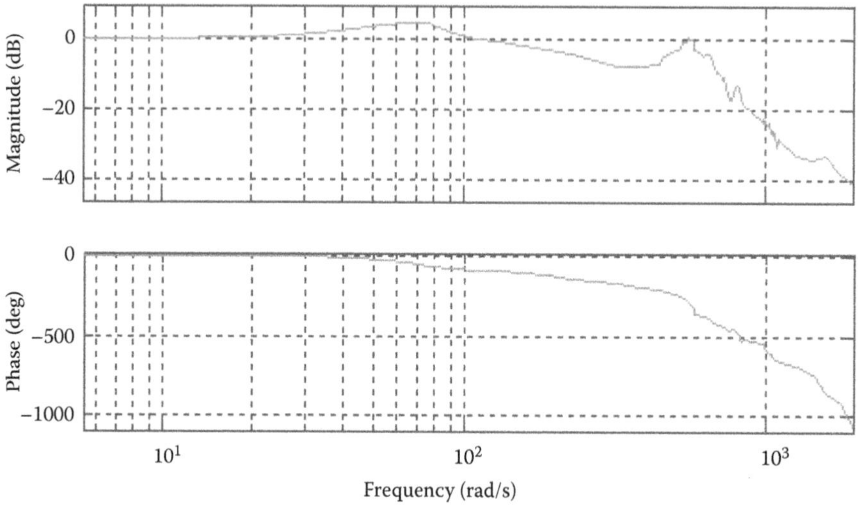

Figure 11.8 Frequency characteristics of the linear motor servo driver.

After calibrating and setting the parameters of the servo driver via Elmo Composer, the frequency characteristics are plotted in Figure 11.8, from which we find that the speed closed-loop system has a fast response to low frequency (below 5 Hz).

The speed trajectory of the speed closed-loop system to track a step signal with 20,000 counts per second is illustrated in Figure 11.9. In such circumstances, the corresponding speed is about 0.1 m/s, by noting that the resolution of the grating is 5 μm. Figure 11.9 shows that this linear motor servo driver has a fast response, a short settling time, and a small steady error to track this speed step

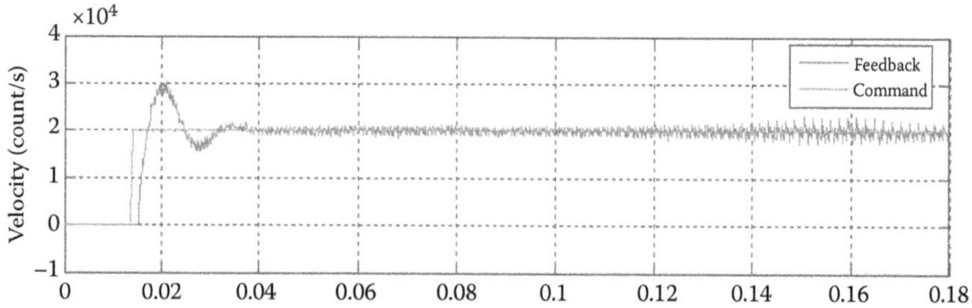

Figure 11.9 Velocity tracking performance of the linear motor servo driver.

signal. For position tracking control considered in this section, the accuracy of the studied linear motor is high enough to meet the requirement of the following experiments.

Experiment II: Control of the Linear Motor Servomechanism. The objective of Experiment II is to study the performance and robustness of several data-driven control methods in position control of the linear motor, in case that the speed of the inner loop is stable.

For the above linear motor servomechanism, two kinds of speed and one kind of load disturbance are studied in the following experiments. The three typical cases are listed in Table 11.2. To realize the linear motor's position tracking, the PID controller, the neural-network controller, and the MFAC algorithm are used for comparison. The sampling time is 0.005 s.

Case I.

PID control: The three parameters are set to $k_P = 0.9$, $k_I = 18$, and $k_D = 0$, which have been carefully tuned. The simulation result is shown in Figure 11.10a, and the position of the tracking error is less than 0.6 mm.

The BP neural-network-based controller: The three-layer feedback network structure NN_{1-20-1} is used. The threshold of the tracking error is set to 0.0001 mm. The simulation result is illustrated in Figure 11.10b, which shows that the position tracking error is less than 0.4 mm.

Table 11.2 Experiment Cases

Case	*Desired Position Sinusoidal Signal (mm)*	*Load Disturbance*	*Experiment Objective*
I	$90\sin(2\pi kT/1000)$ (average speed is about 72 mm/s)	No load	To illustrate the position tracking performance when the motor runs at low speed
II	$90\sin(2\pi kT/200)$ (average speed is about 360 mm/s)	No load	To illustrate the position tracking performance when the motor runs at high speed
III	$90\sin(2\pi kT/200)$ (average speed is about 360 mm/s)	2 kg load is added after the motor is stable	To illustrate the robustness of control methods when there is load disturbance

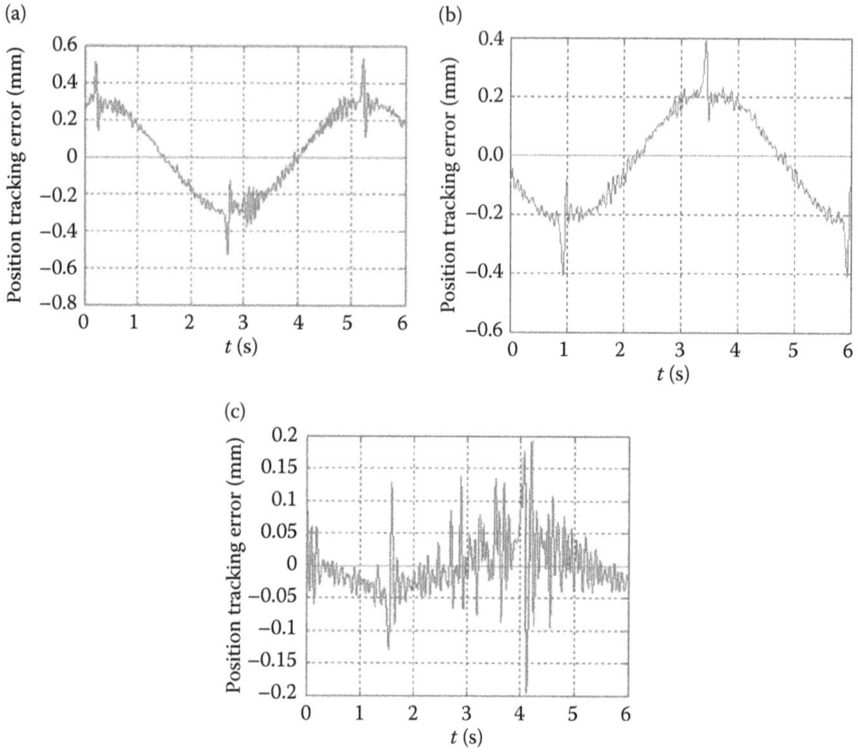

Figure 11.10 Position tracking error by using three control methods under Case 1. (a) PID control. (b) NN-based control. (c) CFDL-MFAC method.

CFDL–MFAC scheme (11.8)–(11.10): The parameters are $\eta = 1.5$, $\mu = 1.5$, $\rho = 0.01$, $\lambda = 4$, and $\varepsilon = 0.001$. $\hat{\phi}(1)$ as the initial value of $\hat{\phi}(k)$ is 2. The simulation result is shown in Figure 11.10c, which indicates that the tracking error of position is less than 0.2 mm. The control performance is better than that of the first two methods.

If the magnitude of the desired sinusoidal position signal is reduced to 30 mm, namely, the stroke of the linear motor is shortened, we may find a very small change in the position error, less than 0.01 mm, than previous experiments for all the three control methods. If the magnitude of the desired sinusoidal position signal is increased to 150 mm, namely, a long stroke is considered, a similar control performance is obtained.

Case II.

PID control: The three parameters are set to $k_P = 1$, $k_I = 24$, and $k_D = 0$. The simulation is shown in Figure 11.11a, and the maximum position tracking error is about 3 mm. Compared with the result of PID in Case 1, the tracking error becomes larger and the control accuracy is lowered.

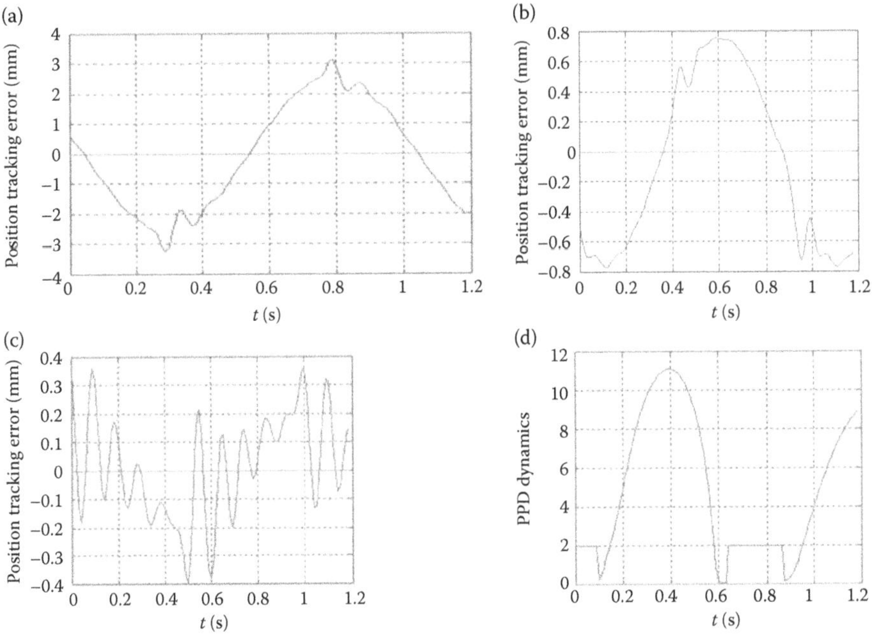

Figure 11.11 Performance comparison among PID, NN, and MFAC under Case 2. (a) PID control. (b) NN-based control. (c) MFAC method. (d) PPD dynamics of MFAC.

The position tracking error when using the same neural-network-based controller with the same parameter setting as Case 1 is illustrated in Figure 11.11b, which shows that the maximum error is less than 0.8 mm and the control performance is much better than the PID control result.

Compared with PID control and neural-network-based control, the same CFDL–MFAC scheme (11.8)–(11.10) with the same parameter setting except for $\lambda = 1.3$ in this case still gives the best control performance. The simulation result in Figure 11.11c indicates that thc position tracking error is less than 0.4 mm.

For the permanent magnet linear motor, if the motor's speed increases, the current of the motor driver will increase. Hence, those disturbances related to current change significantly influence the dynamic performance of the linear motor. From the above experimental results, we find that the MFAC scheme guarantees good tracking performance even when the motor runs at high speed. It reflects an intrinsic characteristic of MFAC that this method is independent of the system model. The PPD estimation of the MFAC algorithm, as shown in Figure 11.11d, leads to the adaptive adjustment of the relationship between the change of position and the change of voltage.

Further investigation and analysis of the CFDL–MFAC scheme reveals that when we choose $\lambda > 8$ and do not modify any other parameters, there will be a very large phase displacement whose maximal value could reach 180°; if we choose a larger parameter as $\lambda > 10$, the system would become lazy and even inactive. The reason is that the change of input voltage may tend to zero and the PPD may not change as time evolves due to overlarge λ; therefore, the ability to adjust the changes of voltage and position of the motor is lost. On the other hand, if we decrease λ, the position signal will become a triangle wave, even with a flattened

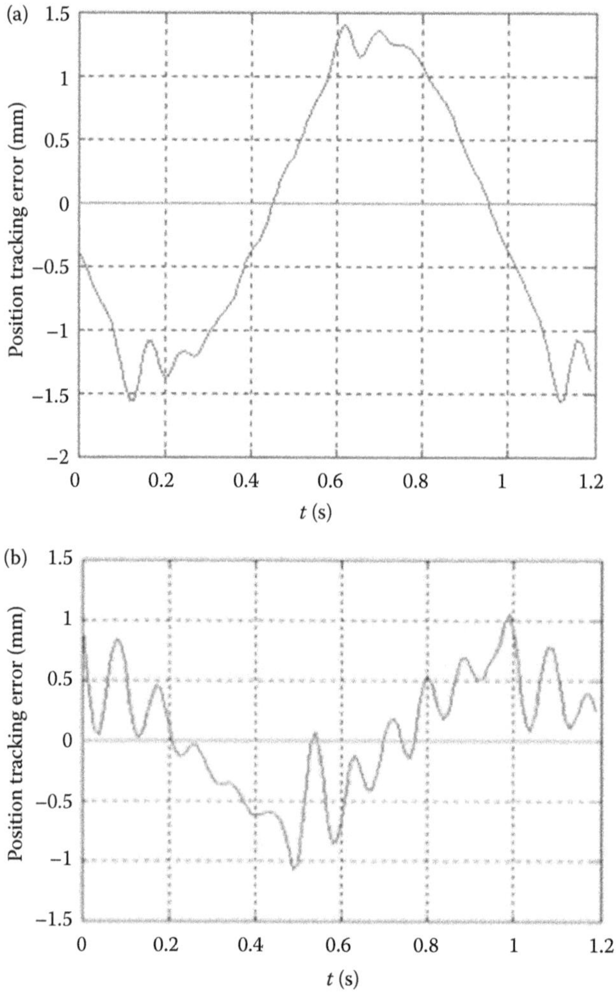

Figure 11.12 Position tracking error by using two control methods under Case 3. (a) NN-based control. (b) MFAC method.

top, and there will be a very large tracking error. When $\lambda < 0.5$, the system will become oscillatory. The reason is that a small λ may lead to an overlarge change of input voltage, such that the system cannot run steadily. From the above analysis, we know that the choice of λ greatly influences the system's dynamic performance. It is consistent with the conclusions drawn by theoretical analysis.

Case III.

PID controller: The system cannot be stabilized no matter how we tune the PID parameters. The linear motor oscillates, and then stops when it reaches the maximum stroke.

The best control performance by using neural-network-based control is demonstrated in Figure 11.12a, which shows that the position tracking error is less than 1.5 mm. Compared with the result in Case 2, the position tracking error is much larger.

CFDL–MFAC scheme (11.8)–(11.10): The best tracking performance is still obtained as shown in Figure 11.12b, which indicates that the position tracking error is less than 1 mm.

The voltage, position, speed, and acceleration of the motor are related to the load. From the experiments in Case 3, we find that MFAC has strong robustness in dealing with the load variation. Furthermore, the performance of the PID controller is very sensitive to parameter regulation. A small change in the PID parameters may cause great variation of the control performance. The NN-based control is quite time-consuming. Comparatively speaking, the CFDL–MFAC scheme is easy to design, use, and implement. It is worth pointing out that a better tracking performance can be obtained by using the PFDL–MFAC scheme.

11.3.2 Dual-Axis Linear Motor Gantry System

11.3.2.1 Experimental Setup

The dual-axis linear motor gantry system, LMG2A-CB6-CC8, was manufactured by Hiwin Technologies Corporation, Taiwan. The two D1 drivers are from Mega-Fabs Motion Systems Ltd, Israel. The DTC-8B four-channel interface panel, a programmable multi-axis controller (PMAC), was manufactured by Delta Tau Data Systems Inc. Company, USA. The hardware configuration is shown in Figure 11.13. The dual-axis linear motor gantry system includes two axes, that is, the *X*-axis and *Y*-axis, to realize positioning and contour manufacturing in a plane. It uses linear guideways to achieve linear motions driven by a single-sided coreless motor. For each linear motor in each axis of the gantry system, the maximum speed is 3 m/s and the maximum acceleration is 50 m/s^2. The maximum force is 330 N in the *X*-axis and the maximum force is 585 N in the *Y*-axis. The encoder resolution is 10^{-1} μm.

The MFAC control system diagram for the dual-axis linear motors is shown in Figure 11.14. The inner loop, that is, the speed loop, is closed by a D1 driver, and

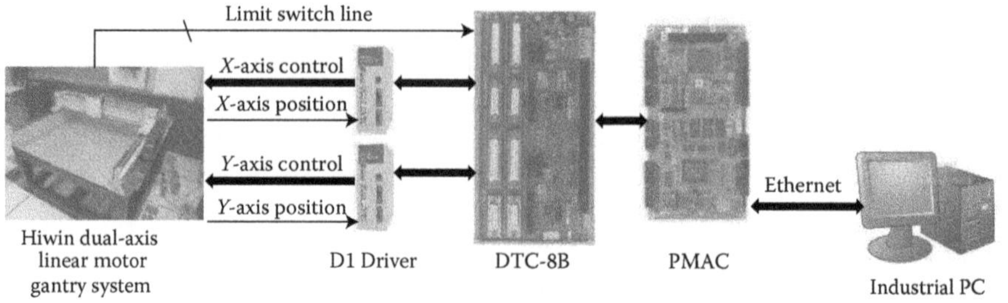

Figure 11.13 Hardware configuration of the dual-axis linear motor gantry system.

the outer loop, that is, the position loop, is realized by MFAC. The designed MFAC algorithms were realized by Visual C++ on an industrial personal computer, which generates the control inputs according to the desired output position signal, the real motor position signal from PMAC, and the limit switch signal, and then sends the control input signal to the PMAC. The PMAC acquires the motor position signal from the encoder and the limit switch signal, and then sends these signals to the host computer; and PMAC will simultaneously convert the digital control input to −10 V ~ +10 V analog voltage signal through the D/A converter and send it to the D1 driver. The analog control input is applied to the D1 drivers to drive the linear motor to move.

11.3.2.2 MFAC Scheme for the Dual-Axis Linear Motor Gantry System

The two linear motors of the dual-axis linear motor gantry system can be described by the FFDL data model as follows:

$$\Delta y_i(k+1) = \phi_{i,1}(k)\Delta y_i(k) + \phi_{i,2}(k)\Delta u_i(k), \quad i = x, y,$$

where $u_i(k)$ is the i axis control input (V) of the gantry system at time k, $y_i(k)$ is the i axis motor rotor output position (μm) at time k. $\Delta y_i(k+1) = y_i(k+1) - y_i(k)$ and $\Delta u_i(k) = u_i(k) - u_i(k-1)$.

Based on the aforementioned data model, the FFDL–MFAC scheme (4.65)–(4.67) is used to realize motor control along the *X*-axis and *Y*-axis of the dual-axis linear motor gantry system coordinately.

11.3.2.3 Experiments

The aim of the following experiment is to verify the effectiveness of the FFDL–MFAC scheme in coordinated control of the dual-axis linear motor gantry system.

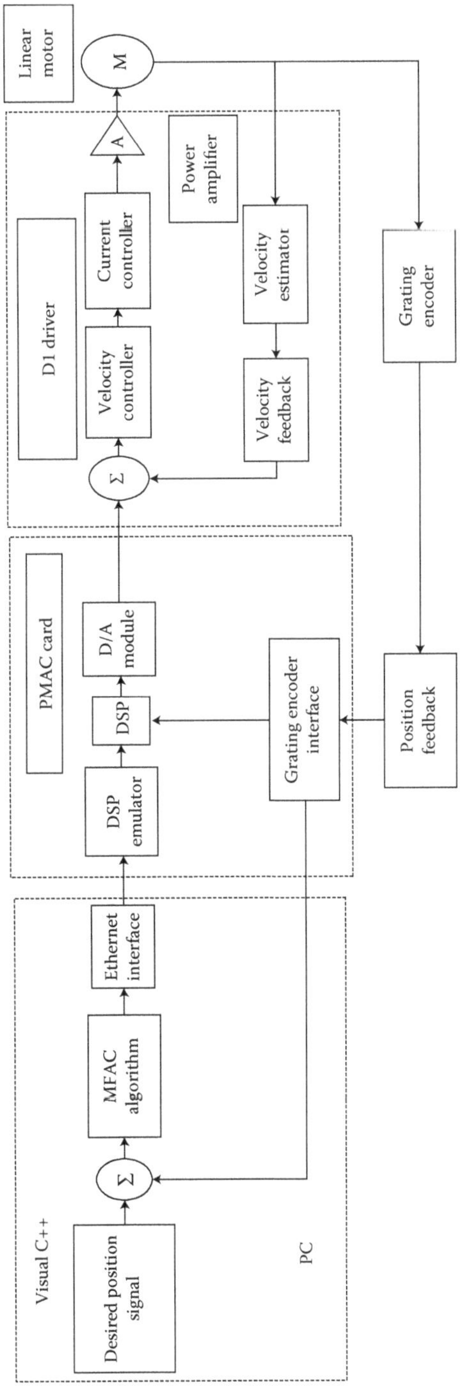

Figure 11.14 Diagram of the dual-axis linear motor gantry system using MFAC.

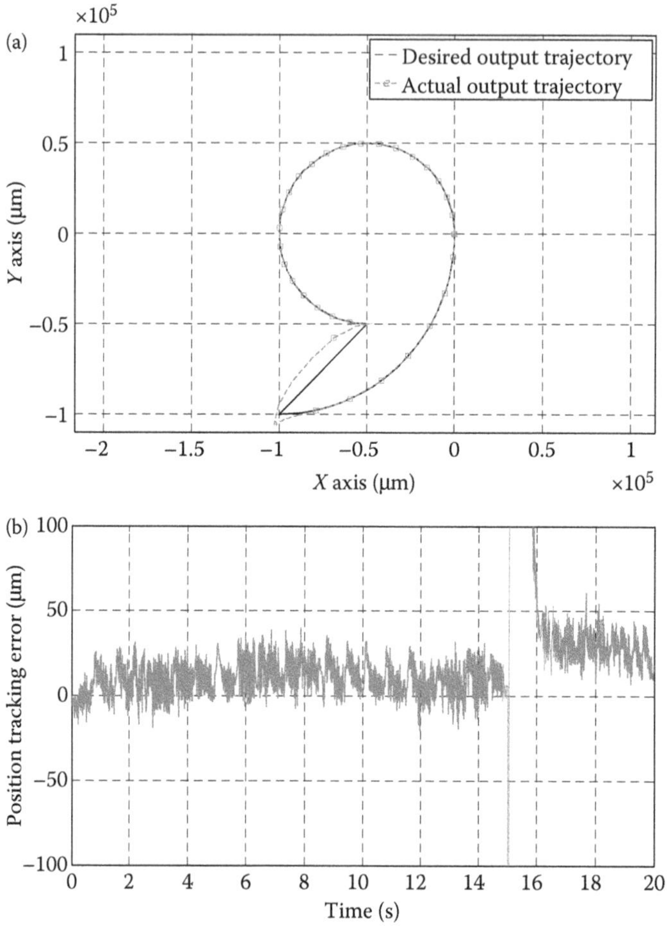

Figure 11.15 (a) Tracking performance of MFAC applied to the dual-axis linear motor gantry system. (b) Position tracking error.

The desired output trajectory is given as a circle with radius R_d and angular frequency ω_d (rad/s):

$$r_d(\omega_d, t) = R_d(e^{j\omega_d t} - 1) = R_d(\cos\omega_d t - 1) + iR_d \sin\omega_d t.$$

With this description, it is natural that the desired outputs in the X-axis and Y-axis are $x_d(t) = R_d(\cos\omega_d t - 1)$ and $y_d(t) = R_d \sin\omega_d t$, respectively.

To test the robustness of the control method with the desired output signal changes, we set the radius R_d and angular frequency ω_d as follows:

$$\begin{cases} R_d = 0.05, \omega_d = 0.1\pi, & 0\text{ s} \le t < 15\text{ s}, \\ R_d = 0.1, \omega_d = 0.1\pi, & 15\text{ s} \le t < 20\text{ s}. \end{cases}$$

The experiment lasts for 20 s and the sampling time of the control system is 0.005 s.

Applying the FFDL–MFAC scheme with suitable controller parameters to the linear motors of the dual-axis linear motor gantry system, the experimental results are shown in Figure 11.15. In Figure 11.15a, we can see that the control scheme can guarantee the system to track the desired trajectory quickly even if there exists a desired trajectory jump at the 15th second; Figure 11.15b shows that the steady-state error is less than 50 μm. The experiment shows that MFAC for the dual-axis linear motor gantry system has satisfying transient and steady-state performance.

11.4 Freeway Traffic System

Freeway traffic control becomes increasingly important in the field of traffic engineering because of the rapid development of freeway infrastructure and the fast-growing traffic demand in metropolitan areas in both developed and developing countries. However, it is difficult to satisfy the tremendous traffic demand, caused by the increasing population and economic development, relying only on building or reconstructing the urban traffic infrastructure. Freeway traffic congestion is becoming one of the most important problems for urban transportation systems. Freeway traffic congestion leads to travel time delay, inefficient utilization of the freeway infrastructure, and decreasing traffic safety. Moreover, traffic congestion will also increase fuel consumption and air pollution in practice. Hence, to prevent traffic jam and to utilize the freeway infrastructure efficiently, the last choice for the traffic administration is to control the freeway flow.

In general, there are three main freeway traffic control methods: ramp metering, mainline speed control, and corridor control. Among these methods, ramp metering is the most popular one [303]. Ramp metering can be implemented by means of traffic lights installed at the entrance of an on-ramp, which is used to meter the number of entering vehicles and prevent traffic volume from exceeding the freeway capacity. Ramp metering, when properly applied, is an effective way to ease freeway congestion and improve the efficiency of freeway infrastructure. Mainline speed control utilizes changeable message signs indicating the desired speed information to guide the driver's behavior so as to make the traffic flow density be homogeneous and avoid traffic congestion. However, this is a suggestive strategy without any rules to punish the violation. Similarly, corridor control is also a suggestive traffic flow control method. It utilizes changeable message signs displaying the real-time traffic flow information to guide the vehicles in a congested segment to an alternative recommended route for avoiding further intensive congestion.

From the viewpoint of system control, ramp metering is a typical regulating problem and numerous control methods can be applied, for example, mathematics programming [304], linear quadratic regulator [305], PID-like controller [306,307], function approximation based on neural network [308], optimization and optimal control theory [309], iterative learning control [67,204,265], and so on. These methods, according to Ref. [303], can be further classified into three strategies: fixed-time control strategy, locally actuated control strategy, and system control strategy. The fixed-time strategy is based on a simple and static traffic model together with historical data; the locally actuated control strategy acts in real time on the basis of traffic conditions in the immediate vicinity of the on-ramp; and the system control strategy calculates a new metering action according to real-time freeway traffic conditions of the entire traffic system. Among these strategies, the control performance of the fixed-time control strategy is the poorest one, while it was reported that the local control strategy is far easy to design and implement, and has proven to be noninferior to the more sophisticated system control method in most cases [303,310].

Among numerous ramp metering strategies, ALINEA, a typical PI controller, is the most extensively applied local control strategy at present. Furthermore, the design of the ALINEA controller does not require any model information of the traffic system *a priori* and thus it is easy to implement. However, it is well known that the PI controller is not good at dealing with highly nonlinear time-varying systems with structure and parametric uncertainties, and there is no theoretical analysis result that can guarantee its stability. Hence, it is hard to obtain good control performance by applying the ALINEA control method [311] when the structure and parameters of traffic systems have a certain sudden variation, which often occurs due to weather change and daily travel peak in the morning and afternoon.

It should be noted that the traffic control problem has its own unique characteristics. First, ramp metering is not needed if the traffic is too light since there is no traffic jam under such conditions. Instead, ramp metering is also unnecessary when the traffic is too dense and exceeds some critical value because a breakdown will happen anyway [67,204]. Second, although all the states of traffic flow systems are measurable due to the increasing growth of information and computer technology, as we all know, it is too expensive and requires massive capital investment for traffic flow data measuring. Furthermore, it is extremely difficult to construct an exact mathematical model of the traffic flow system by using the measured traffic flow data due to the complexity of the traffic system itself. Needless to say, the real traffic system is affected by various external disturbances and uncertainties, such as weather, geometry of the road, drivers' behavior, and so on. Finally, the traffic flow system is with strong nonlinearity and all the system parameters are time-varying due to exogenous uncertainties. Moreover, the off-ramp traffic volume is uncontrollable. Therefore, the control problems of traffic flow are very challenging. Even though the traffic model is known *a priori*, the model-based traffic flow control system is still difficult to design and implement due to the strong nonlinearity of the traffic system itself.

In this section, the MFAC scheme is applied to freeway traffic via ramp metering. The numerical simulations are provided using the existing traffic flow model to verify the effectiveness of the MFAC scheme.

11.4.1 Macroscopic Traffic Model

The space and time discretized traffic flow model [312] used in this section was proposed by Papageorgiou in 1989. It divides a freeway into several sections, and each section contains one on-ramp and one off-ramp only, as shown in Figure 11.16.

The mathematical formulation of the discretized traffic flow model is given as follows:

$$\rho_i(k+1) = \rho_i(k) + \frac{T}{L_i}[q_{i-1}(k) - q_i(k) + r_i(k) - s_i(k)], \tag{11.11}$$

$$q_i(k) = \rho_i(k)v_i(k), \tag{11.12}$$

$$\begin{aligned} v_i(k+1) = v_i(k) &+ \frac{T}{\tau}[V(\rho_i(k)) - v_i(k)] \\ &+ \frac{T}{L_i}v_i(k)[v_{i-1}(k) - v_i(k)] - \frac{\nu T}{\tau L_i}\frac{[\rho_{i+1}(k) - \rho_i(k)]}{[\rho_i(k) + \kappa]}, \end{aligned} \tag{11.13}$$

$$V(\rho_i(k)) = v_{\text{free}}\left(1 - \left[\frac{\rho_i(k)}{\rho_{\text{jam}}}\right]^l\right)^m, \tag{11.14}$$

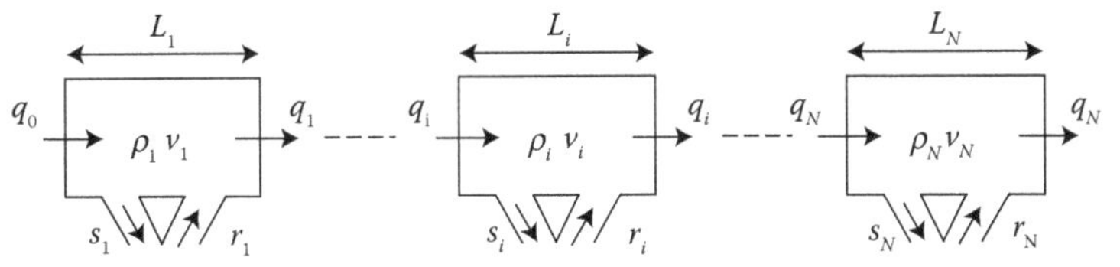

Figure 11.16 A freeway segment subdivided into sections.

where T is the sample time interval in hours. $k \in \{0,1,...K\}$ denotes the kth time interval, $i \in \{1,...,N\}$ is the ith section of a freeway, and N is the total section number. The model variables are listed below:

$\rho_i(k)$: density in section i at time kT (veh/lane/km).
$v_i(k)$: space mean speed in section i at time kT (km/h).
$q_i(k)$: traffic flow leaving section i and entering section $i+1$ at time kT (veh/h).
$r_i(k)$: on-ramp traffic volume for section i at time kT (veh/h).
$s_i(k)$: off-ramp traffic volume for section i at time kT (veh/h), which is regarded as an unknown disturbance.
L_i: length of freeway in section i (km).
v_{free} and ρ_{jam}: the free speed and the maximum possible density per lane, respectively. They are two important parameters in the traffic flow model, since their accuracy affects the accuracy of the traffic flow model greatly.
τ, γ, κ, l, m: constant parameters that reflect particular characteristics of a given traffic system and depend upon the freeway geometry, vehicle characteristics, drivers' behaviors, and so on.

Equations (11.11)–(11.14) constitute the macroscopic traffic model. Equation (11.11) is the well-known conservation equation, (11.12) is the flow equation, (11.13) is the empirical dynamic speed equation, and (11.14) represents the density-dependent equilibrium speed.

The macroscopic traffic model should satisfy the condition $T < L_{\min}/v_{\text{free}}$, that is, the sampling time T should be chosen such that the vehicle with v_{free} can be detected in the shortest section during the time period $[kT, (k+1)T]$. In other words, if the selected sampling time T is too large, some useful information would be lost since vehicles running in the shortest section may not be detected, which prevents the dynamic control algorithm from working efficiently.

We assume that the traffic flow rate entering section 1 during the time period between kT and $(k+1)T$ is $q_0(k)$ and the mean speed of the traffic entering section 1 is equal to the mean speed of the traffic flow in section 1, that is, $v_0(k) = v_1(k)$. We also assume that the mean speed and traffic density of the traffic exiting section $N+1$ are equal to those of section N, that is, $v_{N+1}(k) = v_N(k)$ and $\rho_{N+1}(k) = \rho_N(k)$. The boundary conditions are summarized as follows:

$$\rho_0(k) = q_0(k)/v_1(k), \tag{11.15}$$

$$v_0(k) = v_1(k), \tag{11.16}$$

$$\rho_{N+1}(k) = \rho_N(k), \tag{11.17}$$

$$v_{N+1}(k) = v_N(k),\ \forall k \in \{0,1,\ldots,K\}. \tag{11.18}$$

From the viewpoint of system control, one can control the traffic flow density of a section by metering its corresponding on-ramp entering volume to eliminate the traffic jams, and utilize the freeway infrastructure effectively. Motivated by such a consideration, on-ramp traffic volume $r_i(k)$ can be regarded as the control input signal of the traffic flow system in this section, and the traffic flow density $\rho_i(k)$ is regarded as the output signal of the traffic system. Then, based on this observation, some existing control methods can be applied to such freeway traffic systems.

11.4.2 Control Scheme

For the traffic system, the control objective is to seek an appropriate on-ramp traffic flow $r_i(k)$ such that the traffic density $\rho_i(k)$ tracks the desired traffic density $\rho_{i,d}(k)$. It is worth pointing out that the off-ramp traffic volume $s_i(k)$ is an uncontrollable variable, and is regarded as exogenous disturbance here.

Obviously, even though the freeway model is known, it is difficult to design a proper control input using the traditional model-based control approaches such as optimal control and adaptive control because of the strong nonlinearity and uncertainties in the freeway traffic flow model.

The macroscopic traffic flow model (11.11) is reformulated in the following form:

$$\begin{aligned}\rho_i(k+1) &= \rho_i(k) + \frac{T}{L_i}[v_{i-1}(k)\rho_{i-1}(k) - v_i(k)\rho_i(k) + r_i(k) - s_i(k)] \\ &= \left(1 - \frac{T}{L_i}v_i(k)\right)\rho_i(k) + \frac{T}{L_i}v_{i-1}(k)\rho_{i-1}(k) + \frac{T}{L_i}r_i(k) - \frac{T}{L_i}s_i(k). \qquad (11.19)\end{aligned}$$

Let $a_i(k) = 1 - (T/L_i)v_i(k)$, $b_i(k) = (T / L_i)v_{i-1}(k)$, $c_i(k) = (T/L_i)$. Equation (11.19) is rewritten as

$$\rho_i(k+1) = a_i(k)\rho_i(k) + b_i(k)\rho_{i-1}(k) + c_i(k)r_i(k) - c_i(k)s_i(k). \qquad (11.20)$$

Since Equation (11.20) is continuously differentiable with respect to all variables in a compact set, and the finite variation of traffic flow would not cause an infinite increase of traffic density, the macroscopic traffic flow model (11.20) can be transformed into the following CFDL data model:

$$\Delta\rho_i(k+1) = \phi_{i,c}(k)\Delta r_i(k), \qquad (11.21)$$

where $\Delta\rho_i(k+1) = \rho_i(k+1) - \rho_i(k)$, $\Delta r_i(k) = r_i(k) - r_i(k-1)$.

For a given desired traffic density $\rho_{i,d}(t)$, the CFDL–MFAC scheme is designed as follows:

$$\hat{\phi}_{i,c}(k)=\hat{\phi}_{i,c}(k-1)+\frac{\eta\Delta r_i(k-1)(\rho_i(k)-\rho_i(k-1)-\hat{\phi}_{i,c}(k-1)\Delta r_i(k-1))}{\mu+\left|\Delta r_i(k-1)\right|^2}, \quad (11.22)$$

$$\hat{\phi}_{i,c}(k)=\hat{\phi}_{i,c}(1), \text{ if } |\hat{\phi}_{i,c}(k)|\le \varepsilon \text{ or } \left|\Delta r_i(k-1)\right|\le\varepsilon \text{ or } \operatorname{sign}(\hat{\phi}_{i,c}(k))\ne\operatorname{sign}(\hat{\phi}_{i,c}(1)), \quad (11.23)$$

$$r_i(k)=r_i(k-1)+\frac{\rho\hat{\phi}_{i,c}(k)(\rho_{i,d}(k+1)-\rho_i(k))}{\lambda+|\hat{\phi}_{i,c}(k)|^2}, \quad (11.24)$$

where $\hat{\phi}_{i,c}(1)$ is the initial value of $\hat{\phi}_{i,c}(k)$, $\mu>0$, $\lambda>0$, $\eta\in(0,2]$, $\rho\in(0,1]$, and ε is a small positive constant. Using the on-ramp traffic volume and the traffic density of the mainstream on the freeway, the control input and the estimation of PPD are computed by (11.24) and (11.22), respectively.

Note that the queuing demands actually impose certain constraints on the control inputs of ramp metering, that is, the on-ramp volumes cannot exceed the current demands plus the existing waiting queues at on-ramps $i\in I_{on}$ at time k; thus

$$r_i(k)\le d_i(k)+\frac{l_i(k)}{T}, \quad i\in I_{on}, \quad (11.25)$$

where $d_i(k)$ denotes the traffic demand flow (veh/h) at time k at the ith on-ramp and $l_i(k)$ denotes the length (veh) of a possibly existing waiting queue at time k at the ith on-ramp. I_{on} denotes the set of indexes of the sections where an on-ramp exists. On the other hand, the waiting queue is the accumulation of the difference between the demands and the actual on-ramp volume, that is,

$$l_i(k+1)=l_i(k)+T(d_i(k)-r_i(k)). \quad (11.26)$$

11.4.3 Simulations

It is worth pointing out that the traffic flow model (11.11) is merely used to generate the I/O data of the controlled freeway traffic system, but not for MFAC design.

Consider a long segment of freeway that is subdivided into 12 sections. The length of each section is 0.5 km. The desired density is set as $\rho_d = 30$ veh/km. The initial traffic volume entering section 1 is 1500 veh/h. There are two on-ramps located at sections 2 and 9 with known traffic demands, respectively, and two off-ramps located at section 1 and section 7 with unknown exiting flows, respectively. The traffic demands and the exiting flows are shown in Figure 11.17. The initial density and mean speed of each section as well as all the parameters used in this model are listed in Table 11.3.

Two cases are considered, namely, with no control (Case I) and with control (Casc II).

Case I. No control—Without any control, the traffic flow on the mainstream, coming from the inflow, on-ramps 2 and 9, is so heavy that there exists traffic congestion. The simulation results are shown in Figure 11.18. From Figure 11.18, we can observe that the density is getting higher and higher as the section number i and the time k increase and traffic congestion occurs. As a result, the traffic flow speed drops drastically and almost reaches zero.

Case II. With control—The ALINEA method and the CFDL–MFAC scheme are used for ramp metering. The feedback gain of ALINEA is set to 40 [310]. The step factors and weighting factors in MFAC scheme (11.22)–(11.24) are set to $\rho = 20$, $\eta = 0.0001$, $\mu = 0.01$, $\lambda = 0.001$, and $\varepsilon = 0.00005$. The simulation results are shown in Figure 11.19. It can be seen that, compared with the no control case, the density control performance is improved remarkably by applying ALINEA and CFDL–MFAC. In addition, the CFDL–MFAC scheme can deal with

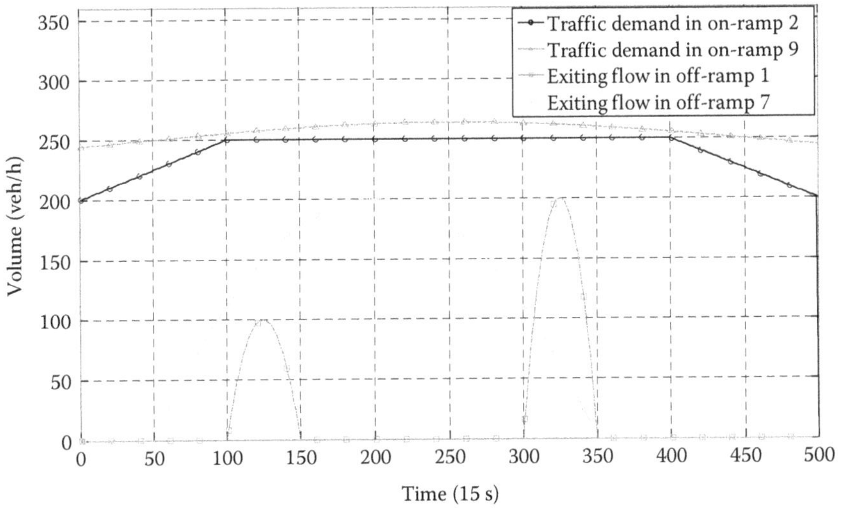

Figure 11.17 Traffic demands of on-ramps 2 and 9, and exiting flow of off-ramps 1 and 7.

Table 11.3 Initial Values and Parameters Associated with the Traffic Model

Section	*1*	*2*	*3*	*4*	*5*	*6*	*7*	*8*	*9*	*10*	*11*	*12*
$\rho_i(0)$	30	30	30	30	30	30	30	30	30	30	30	30
$v_i(0)$	50	50	50	50	50	50	50	50	50	50	50	50
Parameters	v_{free}	ρ_{jam}	l	m	κ	τ	T	γ	$q_0(k)$	$r_i(0)$	α	
	80	80	1.8	1.7	13	0.01	15	35	1500	0	0.95	

Source: Adapted from Hou ZS, Xu JX, Yan JW. *Transportation Research Part C*, 2008, 16(1): 71–97; Hou ZS, Xu JX, Zhong HW. *IEEE Transactions on Vehicular Technology*, 2007, 56(2): 466–477.

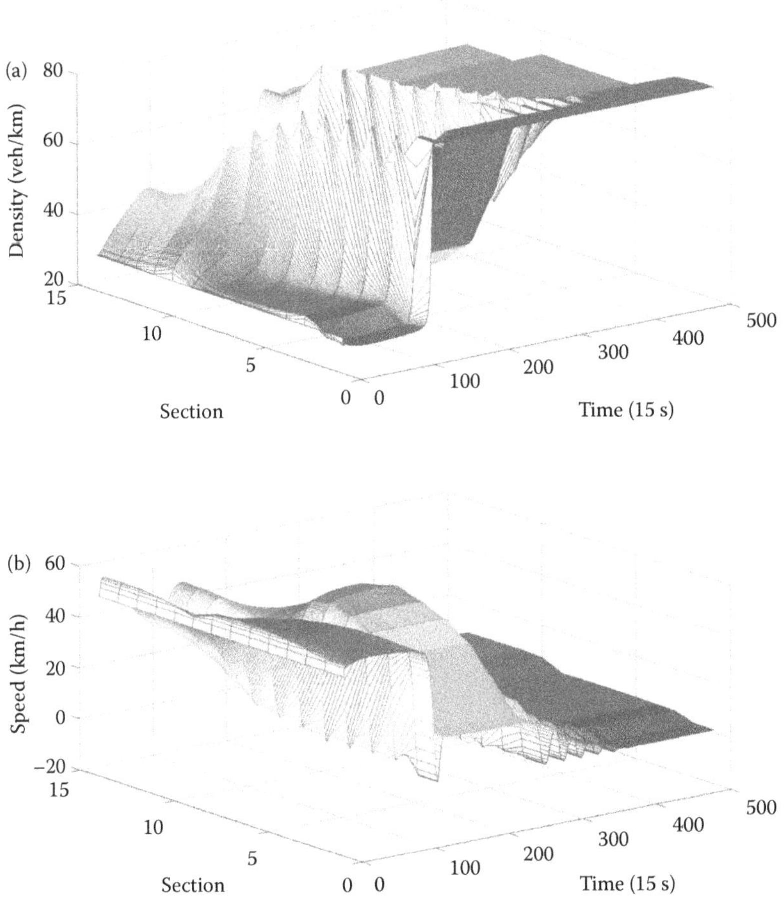

Figure 11.18 No control is applied. (a) Density. (b) Speed.

exogenous disturbances more effectively with a better control performance than ALINEA.

The simulation results show that the CFDL–MFAC scheme, which is designed for freeway traffic control systems with strong nonlinearity and uncertainty, has a better control performance.

The main advantage of MFAC based ramp metering is that the design and analysis of the control system only depend on the I/O data of the freeway traffic system without requiring any traffic model in practice. The other features are the following: First, as we all know, it is not only difficult but also time-consuming and costly to establish the macroscopic traffic model for the traffic control system and to validate the model parameters. For the MFAC approach, it does not depend on the macroscopic traffic flow model; structure and parametric uncertainties,

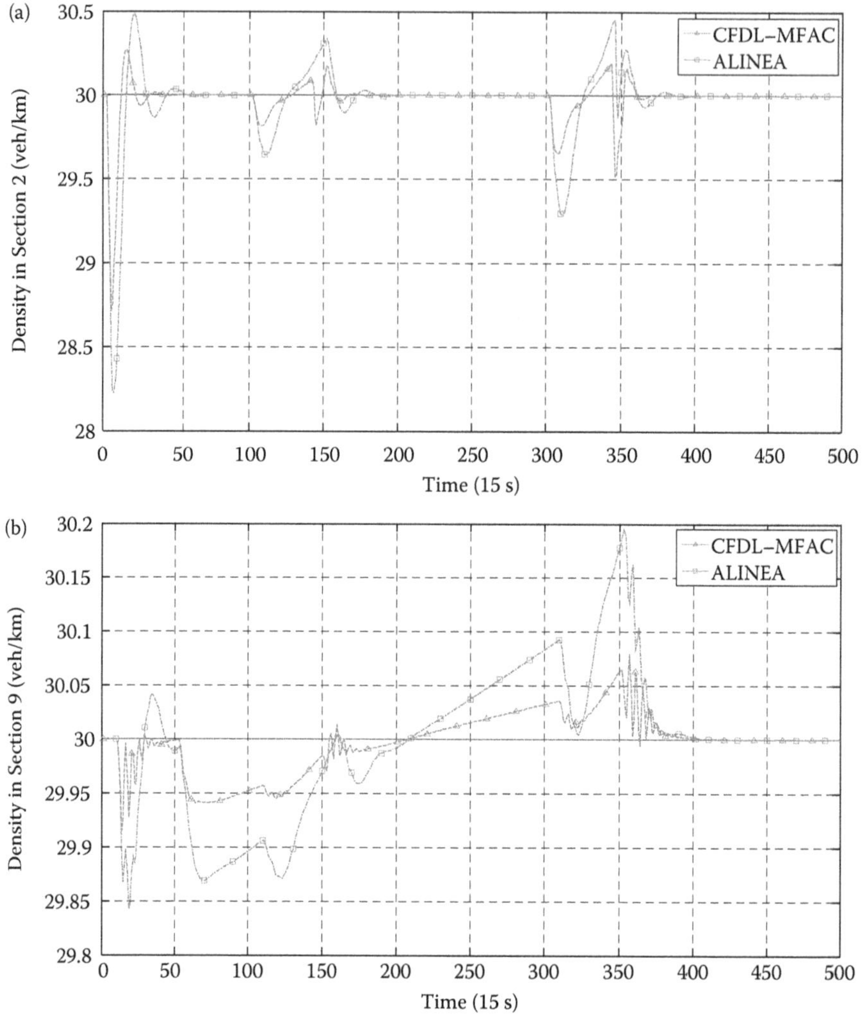

Figure 11.19 Simulation comparison between ALINEA and CFDL–MFAC. (a) Density in Section 2. (b) Density in Section 9.

whatever their variation, would not influence the performance of traffic control systems. Second, according to the theoretical proof about MFAC in Chapter 4, the CFDL–MFAC scheme can guarantee convergence of the closed-loop control system. At present, there are few control methods with stability and convergence proofs in the traffic control community. Finally, compared with the well-known ALINEA approach, CFDL–MFAC, as an adaptive control approach, can deal with time-varying parameters and disturbance.

See Refs. [209, 264, 265] for details of the applications of MFAC and its improved schemes for the freeway traffic system.

11.5 Welding Process

Arc welding is a complex process that involves interactions among materials, metallurgy, and physical chemistry. Welding quality is related to multiple technical indexes of welding technology. These indexes are coupled dynamically and overlapped statically.

There are many control methods used in the welding process, such as PID control [313], adaptive control [314], robust control [315], fuzzy control [316], neural-network-based control [317], and fuzzy neural network control [7,13,60]. Since the welding process is a multivariable, nonlinear, and time-varying process that involves many uncertain factors and constraints, an accurate mathematical model for the welding process is difficult to obtain, and the traditional model-based adaptive control and robust control approaches cannot efficiently deal with parametric uncertainties or attenuate external disturbance and parameter perturbation. Fuzzy control and neural network control are usually implemented by control experts and require a deep understanding of the controlled plant; thus the control schemes for the arc welding process are difficult to implement and maintain since their design processes are quite complex.

For engineers, a low cost controller with low online computational burden and good control performance would be the best choice to a practical welding process. In this section, the CFDL–MFAC scheme is designed for the multiple-input and single-output pulse tungsten inert gas (TIG) welding process. The simulation and applications show that the MFAC scheme has satisfactory performance and is easy to implement with low computational burden [284].

11.5.1 Experimental Setup

A diagram of the aluminum alloy pulse TIG welding system is shown in Figure 11.20. The welding system consists of a walking mechanism, a set of welding electrical units, a central controller, and a visual sensor. The system controlled by the central controller can realize functions such as arc striking, welding, arc suppression, visual sensing, and traveling.

The welding current and the welding feed speed are two very important parameters in welding beam forming. From the system control viewpoint, the backside width of the welding pool can be controlled by the welding current and the welding feed speed. Therefore, the welding speed and the welding current are considered as control input variables, and the backside width of the weld pool is the system output.

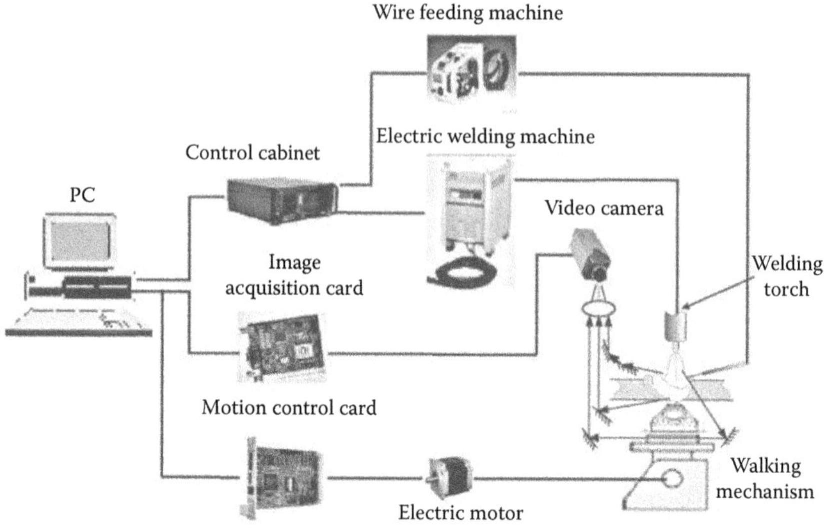

Figure 11.20 Diagram of the aluminum alloy pulse TIG welding process.

11.5.2 Control Scheme

The control objective is to regulate the welding speed and the welding current such that the backside width of the weld pool achieves the desired backside width. Since the MFAC scheme merely depends on the measured I/O data of the controlled plant without any model information about the welding process, the CFDL–MFAC scheme of this two-input and one-output arc welding process with the given desired backside width $y_d(k)$ can be designed directly as shown in Section 5.2.

$$\hat{\boldsymbol{\phi}}_c(k) = \hat{\boldsymbol{\phi}}_c(k-1) + \frac{\eta \Delta \boldsymbol{u}(k-1)(y(k) - y(k-1) - \hat{\boldsymbol{\phi}}_c^T(k-1)\Delta \boldsymbol{u}(k-1))}{\mu + \|\Delta \boldsymbol{u}(k-1)\|^2}, \tag{11.27}$$

$$\hat{\phi}_i(k) = \hat{\phi}_i(1), \quad \text{if } |\hat{\phi}_i(k)| < \varepsilon \quad \text{or} \quad \text{sign}(\hat{\phi}_i(k)) \neq \text{sign}(\hat{\phi}_i(1)), \ i = 1,2, \tag{11.28}$$

$$\boldsymbol{u}(k) = \boldsymbol{u}(k-1) + \frac{\rho \hat{\boldsymbol{\phi}}_c(k)(y_d(k+1) - y(k))}{\lambda + \|\hat{\boldsymbol{\phi}}_c(k)\|^2}, \tag{11.29}$$

where $\Delta \boldsymbol{u}(k) = \boldsymbol{u}(k) - \boldsymbol{u}(k-1)$ and $\boldsymbol{u}(k) = [I_p(k), v_f(k)]^T$. $I_p(k)$ and $v_f(k)$ denote the welding current and the wire feed speed, respectively. $y(k)$ denotes the backside width of the weld pool. $\hat{\phi}_c(1)$ is the initial value of PG estimation $\hat{\phi}_c(k)$. $\lambda > 0$, $\mu > 0$, $\eta \in (0,2]$, $\rho \in (0,1]$, and ε is a small constant.

11.5.3 Simulations

In this section, a welding process is simulated to verify the effectiveness of the CFDL–MFAC scheme for the MISO system. The dynamics of the aluminum alloy pulse TIG welding process can be described by the following ARX model [284]:

$$\begin{aligned} y(k) &= a_1 y(k-1) + a_2 y(k-2) + a_3 y(k-3) + a_4 y(k-4) + a_5 y(k-5) \\ &\quad + b_{11} I_p(k-1) + b_{12} I_p(k-2) + b_{14} I_p(k-4) \\ &\quad + b_{21} v_f(k-1) + b_{23} v_f(k-3) + b_{25} v_f(k-5), \end{aligned} \tag{11.30}$$

where $[a_1, a_2, a_3, a_4, a_5] = [1.2245, -0.7935, 0.45269, -0.23124, 0.11518]$, and $[b_{11}, b_{12}, b_{14}, b_{21}, b_{23}, b_{25}] = [0.0085696, -0.3748, 0.0039714, -0.16826, 0.0023674, -0.079501]$.

The simulation lasts for 10 s and the desired backside weld width is $y_d(k) = 5$ mm. The parameters of CFDL–MFAC scheme (11.27)–(11.29) are set to $\eta = 1$, $\mu = 1$, $\rho = 1$, $\lambda = 25$, and $\hat{\phi}_c(1) = [-0.5, 10]^T$. The simulation result is shown in Figure 11.21.

The simulation results show that the CFDL–MFAC scheme for this two-variable linear arc welding process gives good control performance. The main advantage of the CFDL–MFAC scheme is that the controller design only depends on the I/O data of the tungsten inert gas (TIG) welding process without involving its dynamic model. Refer to Refs. [284, 318] for more details.

11.5.4 Experimental Investigation

In the above section, the simulation results show that the CFDL–MFAC scheme in the MISO case can work well for the arc welding process. In this section, we will apply PID control and CFDL–MFAC to a TIG welding equipment in the laboratory to investigate the effectiveness of the CFDL–MFAC scheme and superiority to the PID controller.

Practical welding processes contain various disturbances, which can be displayed by changes at different heat dissipation conditions. Thus, a dumbbell-shaped workpiece, which dissipates heat in different styles during the welding process, is considered here in order to simulate disturbance [285]. The material and thickness of the workpiece are LD10 aluminum alloy and 4 mm, respectively. Experiments are carried out by pulsed gas tungsten arc welding on a butt joint. Owing to the fast heat dissipation of the aluminum alloy, it is difficult to penetrate at the initial welding position. Thus, to ensure penetration of the initial welding point, the welding torch should stay for a few seconds after arcing and then the welding wire begins

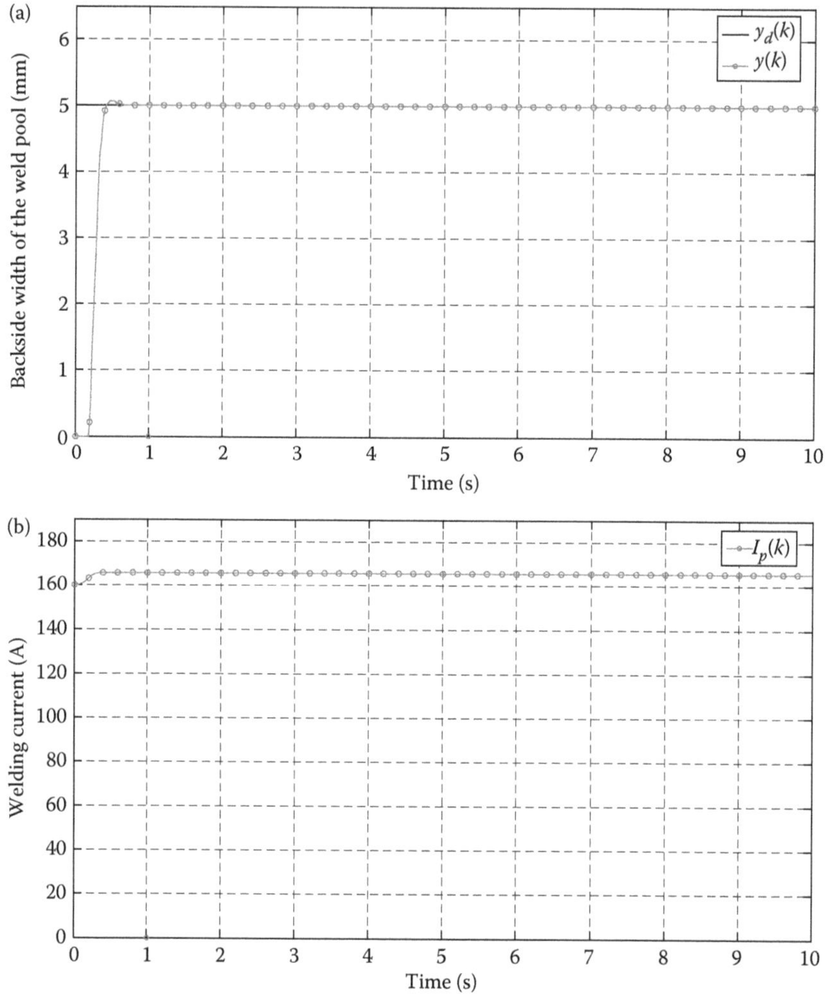

Figure 11.21 Control performance of the TIG welding process based on the CFDL–MFAC scheme. (a) Backside width of the weld pool. (b) Welding current. (c) Wire feed speed. (d) PG estimation.

to feed. The dwelling time is suggested to be 5 s by experience. The same initial values of the welding process are $I_p(1) = 220$ A and $v_f(1) = 10$ mm/s, and the desired backside width is $y_d(k) = 6$ mm. After several runs through trial and error, the PID parameters are set to $K_{p1} = 0.1$, $T_{I1} = 2.5$,$T_{D1} = 1.5$, $K_{p2} = 4$, $T_{I2} = 17$, and $T_{D1} = 0.5$. The experimental results are shown in Figures 11.22 and 11.23. It is noted that the backside width of the weld pool achieves the desired value by applying the two control methods, although the practical welding process studied here is complex and

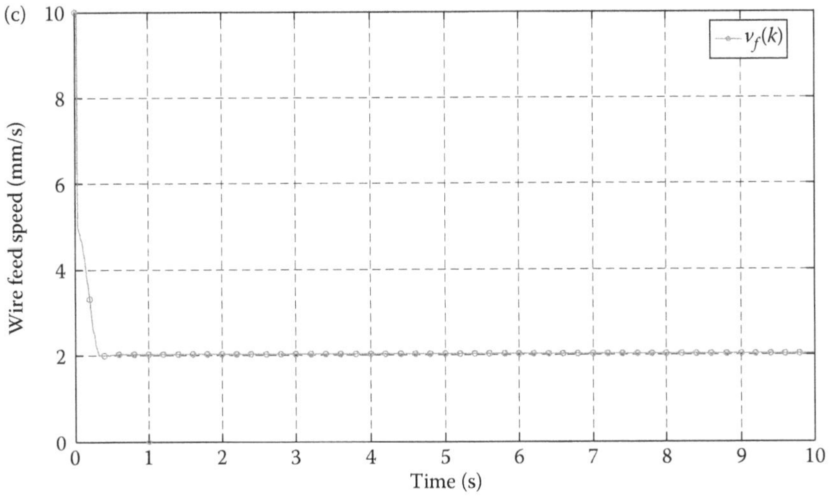

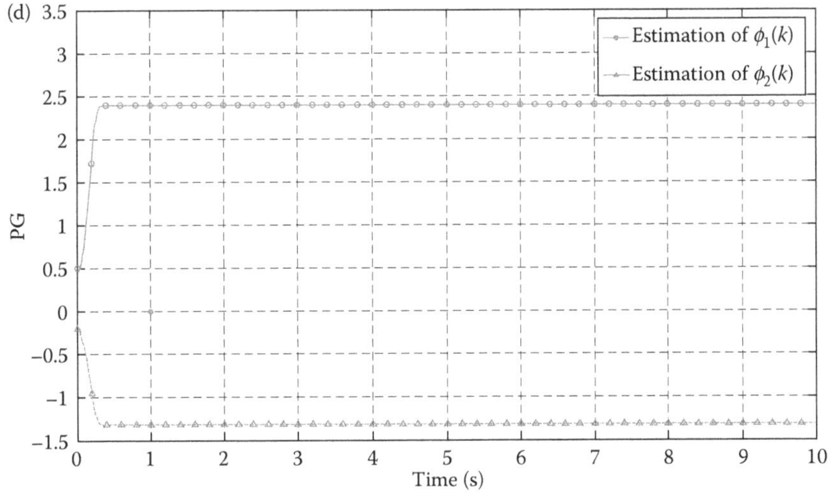

Figure 11.21 (continued) Control performance of the TIG welding process based on the CFDL–MFAC scheme. (a) Backside width of the weld pool. (b) Welding current. (c) Wire feed speed. (d) PG estimation.

with disturbance. However, it is worth pointing out that the overshoot and steady error for the MFAC scheme are less than that of PID's.

11.6 MW Grade Wind Turbine

The wind turbine blade is one of the main components of the MW grade wind turbine in a power generation system. When a major modification in technique

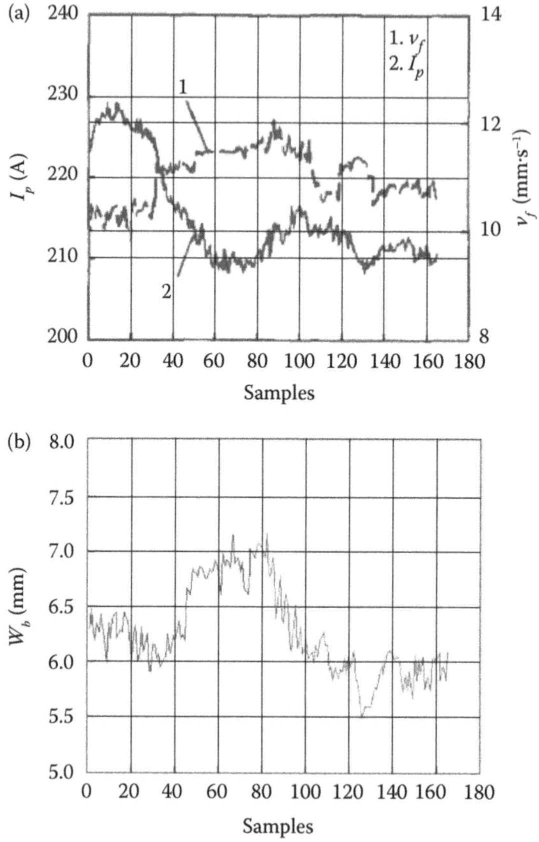

Figure 11.22 Control performance using PID. (a) Welding current and wire feed speed. (b) Backside width of the weld pool.

is made or a new wind blade is designed, a static test is needed to testify its static loading capacity, and provides necessary data for stiffness testing and structural optimization design [291].

It might be impossible to build an accurate mathematical model for the wind turbine blade since its dynamics changes during the control process in a static loading test. In addition, there exist traction force couplings among different loading nodes. For such a complex testing process with nonlinearity, time-varying parameters, and traction force decouplings, it is hard to achieve good performance by using traditional model-based control methods.

In this section, MFAC is applied to a static load testing system. The experimental results show that the traction force decoupling control problem is handled by applying the MFAC scheme effectively.

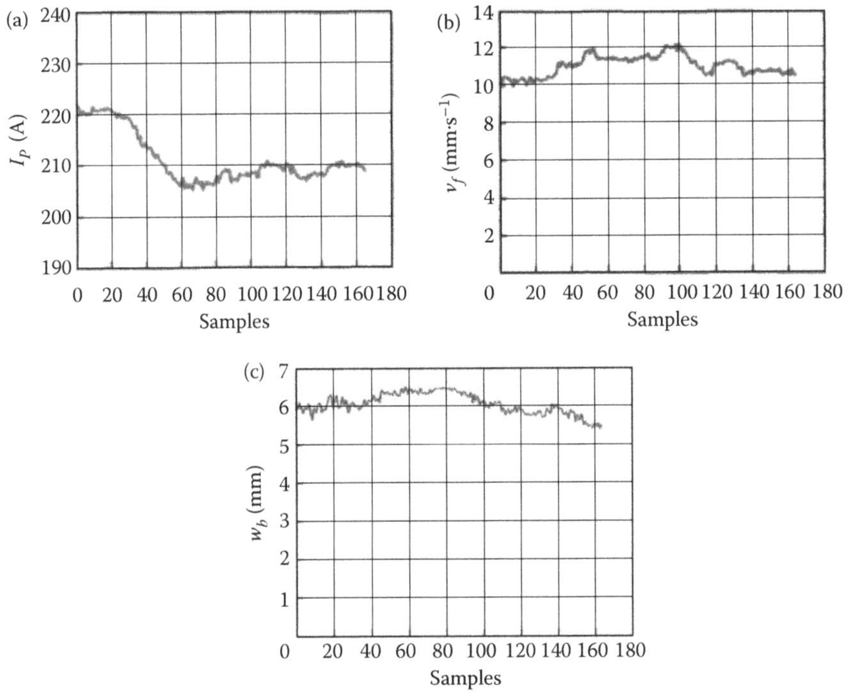

Figure 11.23 Control performance using CFDL–MFAC. (a) Welding current. (b) Wire feed speed. (c) Backside width of the weld pool.

11.6.1 Wind Turbine Blade Static Loading Control System

The static loading test bench, shown in Figure 11.24, is composed of a cylindrical fixative bracket, several loading backrests, and corresponding control systems. The cylindrical fixative bracket is fixed on the ground. A loading backrest is mainly composed of a hydraulic system and a set of pulleys. The hydraulic system is placed inside the loading backrest. The pulley position can be adjusted according to the position of the loading point node, and can be moved horizontally to keep the direction of the traction force vertical to the turbine blade. The hydraulic winch is fixed on the square steel frame. The direction of the traction force is changed via adjusting pulleys. The tested turbine blade is fixed horizontally on the cylindrical fixative bracket through a high tensile bolted flanged connection. Static loading is conducted at different loading nodes by imposing lateral traction forces by the hydraulic winches. The traction force at each loading node increases in a stepwise way after it achieves the setting value and remains steady for a while.

Three kinds of sensors, that is, the force sensor, displacement sensor, and static resistance strain gauge, are used in the static loading control system. The force

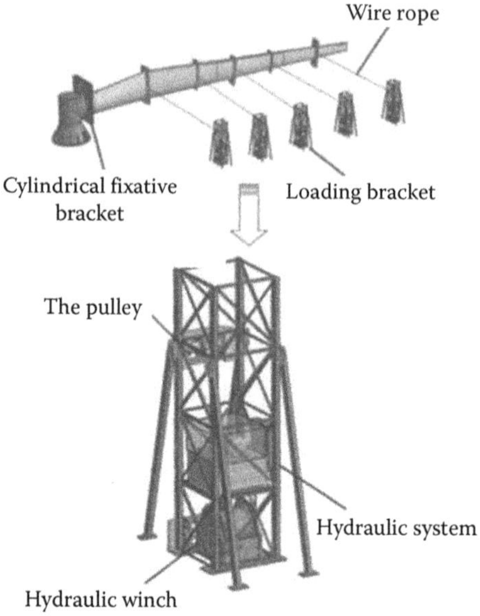

Figure 11.24 Static loading bench.

sensor is connected to the wire rope to measure real-time traction force loaded on the blade. The output of the force sensor is current and its measurement error is less than 0.3%. The displacement sensor, whose resolution is 1 mm and accuracy is ± 2 mm, is used to measure the bending deflection of the blade. The resistance strain gauge with a half-bridge circuit is stuck on the blade to realize data collection, processing, and analysis.

The MW wind turbine blade static loading system shown in Figure 11.25 consists of a computer, a master controller, and a series of local controllers. The computer is connected to the master controller through the RS485 bus, and the master controller communicates with the local controllers via the CAN bus. During the loading process, traction forces generated by hydraulic servo systems are uploaded and unloaded at different nodes on the wind turbine blade according to computer scheduling. The local controllers collect the data about the traction force and deflection via sensors, and then upload it to the computer through the RS485 bus. Meanwhile, the time–load curve and the time–deflection curve are generated automatically in the computer.

11.6.2 Control Scheme

For a given desired traction force $y_d(k)$, the CFDL–MFAC control scheme for a static loading test is designed as follows:

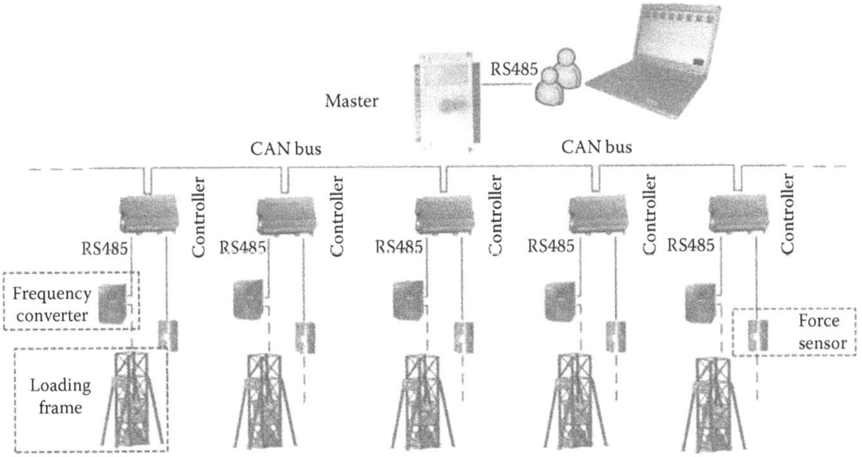

Figure 11.25 The distributed control system based on the CAN bus.

$$\hat{\phi}_c(k) = \hat{\phi}_c(k-1) + \frac{\eta\Delta u(k-1)(y(k) - y(k-1) - \hat{\phi}_c(k-1)\Delta u(k-1))}{\mu + |\Delta u(k-1)|^2}, \tag{11.31}$$

$$\hat{\phi}_c(k) = \hat{\phi}_c(1), \quad \text{if } |\hat{\phi}_c(k)| \le \varepsilon \text{ or } |\Delta u(k-1)| \le \varepsilon \text{ or } \operatorname{sign}(\hat{\phi}_c(k)) \ne \operatorname{sign}(\hat{\phi}_c(1)), \tag{11.32}$$

$$u(k) = u(k-1) + \frac{\rho\hat{\phi}_c(k)(y_d(k+1) - y(k))}{\lambda + |\hat{\phi}_c(k)|^2}, \tag{11.33}$$

where $y(k) \in R$ and $u(k) \in R$ denote the system output (traction force, kN) and the control input (static load, kN/m^2) at time instant k, respectively. $\hat{\phi}_c(k)$ is the estimation of $\phi_c(k)$; $\hat{\phi}_c(1)$ is the initial value of $\hat{\phi}_c(k)$; $\lambda > 0$, $\mu > 0$, $\eta \in (0,2]$, and $\rho \in (0,1]$ are adjustable parameters; and ε is a small positive constant.

11.6.3 Static Loading Experiment

To determine the maximum load capacity of a blade, a full-scale static loading test on the aeroblade 3.6–56.4 is carried out. The normal rated power of the wind turbine is 3.6 MW. The length of the wind blade is 56.4 m. The testing field of static

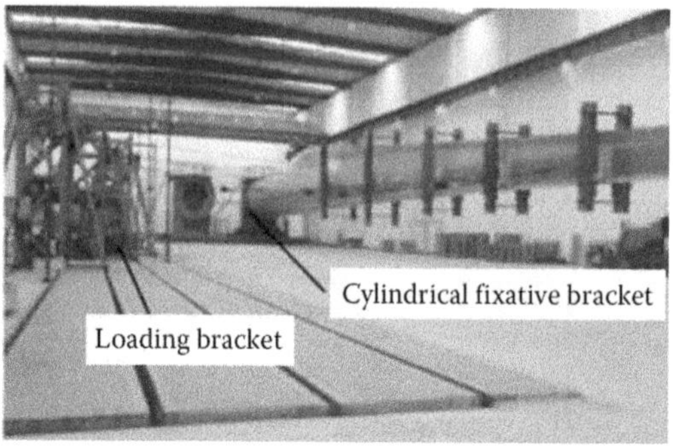

Figure 11.26 Testing field of the static loading test.

Table 11.4 Static Loading Test Data

	Node 2		*Node 3*	
Stage	*Actual Value* (kN)	*Error Rate (%)*	*Actual Value* (kN)	*Error Rate (%)*
40%	49.6	0.8	41.8	0.4
60%	74.7	0.4	62	0.6
80%	100.2	0.2	83.6	0.4
100%	125.4	0.32	103	0.32

Source: Adapted from Zhang LA et al. *China Mechanical Engineering*, 2011, 22(18): 2182–2185, 2208.

loading is shown in Figure 11.26. There are five loading nodes, seven deflection measuring points, and 40 strain measuring points in the test.

The loading process is carried out in four stages, that is, load values were set to 40%, 60%, 80%, and 100% of the maximum load force. The sampling period is $T = 0.01$s. The parameters of the MFAC scheme are set to $\rho = 1.5$, $\lambda = 0.4$, $\eta = 0.5$, $\mu = 0.1$ and the initial values of the system output and the control input are 0.

The data of the loading test at node 2# (27.9 m from the root) and node 3# (37.9 m from the root) are shown in Table 11.4 [291]. According to the results, we can observe that the traction force error is less than 1% at any stage, which satisfies the requirement of a static loading test. Although there exist strong traction couplings among different nodes, the MFAC scheme can guarantee the traction error to be within ± 4kN.

11.7 Conclusions

In this chapter, MFAC schemes are designed for five typical systems, namely, the three-tank water system, the permanent magnet linear motor, the freeway traffic system, the arc welding process, and the wind turbine machine. Extensive simulations and experiments show that MFAC is capable of controlling various unknown nonlinear plants when the plant model is unavailable. The main advantage of MFAC is that the controller design only depends on the I/O data of the controlled system, is easy to use and implement, and has very strong robustness.

Chapter 12

Conclusions and Perspectives

12.1 Conclusions

In this book, the data-driven model-free adaptive control approaches for unknown discrete-time nonlinear systems are introduced systematically. The contents of the book include the dynamic linearization approach, model-free adaptive control, model-free adaptive predictive control, and model-free adaptive iterative learning control for discrete-time nonlinear systems, with the corresponding stability analysis and typical practical applications. Moreover, some other important issues are also considered, which include model-free adaptive control for complex connected systems, modularized controller designs between model-free adaptive control and other control methods, the robustness of model-free adaptive control, and the concepts of the symmetric similarity for adaptive control system design.

MFAC is a new type of adaptive control method. The basic idea of MFAC design is implemented by building an equivalent dynamic linearization data model for the nonlinear system at each operation point first, based on some new concepts such as PPD (PG or PJM), then estimating the system's PPD (PG or PJM) online, designing the controller, and analyzing the system's performance by using only the I/O data of the controlled plant. Since the PPD (PG or PJM) in a dynamic linearization data model has no explicit relationship with the time-varying structure or time-varying parameters of the controlled plant, the MFAC method can deal with the adaptive control tasks for these systems uniformly.

The main difference between MFAC and the model-based adaptive control lie in the following three aspects: First, the plants to be controlled are different.

Traditional adaptive control is proposed for a time-invariant or slowly time-varying system with a known system structure and system orders, whereas MFAC is proposed for unknown nonlinear systems. The plant model for traditional adaptive control design is the first-principles model or the input/output identified model, whose parameters have an explicit physical meaning, whereas the plant for MFAC is the unknown discrete-time nonlinear system, which is transformed into a PPD (PG or PJM) based virtual data model without physical meaning, this data model may called the controller deigning model, which is equivalent to the original controlled plant. Second, the controller design procedures are different. The traditional adaptive control design depends on the mathematical model of the controlled plant, whereas the MFAC design depends merely on the I/O data of the controlled plant. Third, the tools for stability analysis are different. The key technical lemma or the *Lyapunov* approach is used for the stability and asymptotic convergence of the closed-loop error dynamics in traditional model-based adaptive control, whereas the generalized *Lipschitz* condition and contraction mapping approach is used to analyze the stability and monotonic convergence of the closed-loop error dynamics of the MFAC system. Finally, from the perspective of practical applications, there are a series of inevitable twinborn problems for traditional adaptive control, such as unmodeled dynamics and the robustness, accurate modeling and model reduction or controller reduction, persistent excitation condition and tracking error convergence, unknown disturbance and the upper bound of the uncertainties needed to be known for robust control design, perfect control theory in mathematics, and poor application effect in practice, whereas MFAC overcomes these difficulties of traditional adaptive control design, and all these issues in traditional adaptive control theory do not exist under the data-driven MFAC framework.

In a word, MFAC provides a novel tool to design a controller for a class of unknown discrete-time nonlinear systems, merely using the input/output data of the controlled systems, without any model dynamics involved. As a kind of data-driven control method, MFAC is an application-oriented control theory; it overcomes the dependence of modern control theory on the mathematical model of the controlled plant, and opens a new way to control theory research and practical applications.

It is worthy pointing out that, dynamic linearization method for nonlinear systems given in Chapter 3, which can transform a general discrete-time nonlinear system into a virtual equivalent time-varying data model via the concept of PPD, PG and PJM, plays a key role in controller design. Advantages of MFAC just come from this dynamic linearization method. In fact, this kind of dynamic linearization method for nonlinear systems can also be used to a general nonlinear controller which may be known or unknown. With the help of this method, the general nonlinear controller can be transformed into the controller with CFDL or PFDL or FFDL form, and then the parameter estimation algorithm for the controller parameter may be tuned on the basis of certain estimation cost function. For details, see Refs. [319, 320].

12.2 Perspectives

As discussed in the previous chapters, existing research works have laid the foundation and framework for the development of the MFAC method. However, research on model-free adaptive control is still in the initial stage from the viewpoints of both theory and applications. As one of the typical data-driven control theories and approaches, further improvement and perfection of the MFAC would be valuable for the development and application of data-driven control.

Many important issues still exist and are open for MFAC research, and the following problems may deserve further consideration and investigation:

- The issues of stability and convergence of the FFDL–MFAC schemes of the SISO, MISO, and MIMO nonlinear control plants.
- The stability and convergence of the output regulation and tracking problems of the PFDL–MFAPC and FFDL–MFAPC schemes.
- The theoretical superiority and controller evolution design problems of the adaptive control systems based on the symmetric similarity structure principle, especially the MFAC methods.
- The design and systematical analysis methods of dynamical linearization data model based ILC for nonlinear systems under nonstrictly repeatable conditions.
- The relationships among MFAC methods, model-based control methods, and other data-driven control methods, as well as how to utilize the available partial system information in the design and implementation of the data-driven control approaches.
- The robustness issues of the MFAC methods as well as the robustness enhancement approaches. Some primary works could be found in Refs. [321, 322].
- The systematic selection approach of the control input LLC and the pseudo-orders L_y and L_u in the design and implementation for the MFAC system. The systematic controller parameter tuning methods for the MFAC schemes.
- The physical meanings of pseudo-partial derivative, pseudo-gradient, pseudo-Jacobian matrix, as well as the more efficient prediction algorithms of these time-varying parameters.
- Research on the decoupling ability of MFAC for strongly coupled MIMO systems, and the corresponding results for the output regulation and tracking problems, respectively.
- The establishment of the consolidated framework for data-driven control methods. Some primary works could be found in Refs. [14, 17, 319, 320].
- The applications of MFAC and other data-driven control methods in various practical industry processes.

References

1. R. E. Kalman, A new approach to linear filtering and prediction problems, *Transactions ASME, Series D, Journal of Basic Engineering*, vol. 82, pp. 34–45, 1960.
2. R. E. Kalman, Contributions to the theory of optimal control, *Boletin de la Sociedad Matematica Mexicana*, vol. 5, pp. 102–119, 1960.
3. P. Albertos and A. Sala, *Iterative Identification and Control.* London, UK: Springer-Verlag, 2002.
4. B. D. O. Anderson, Failures of adaptive control theory and their resolution, *Communications in Information and Systems*, vol. 5, no. 1, pp. 1–20, 2005.
5. B. D. O. Anderson and A. Dehghani, Historical, generic and current challenges of adaptive control, in *Proc. Third IFAC Workshop on Periodic Control Systems*, Anichkov Palace, Russia, 2007.
6. B. D. O. Anderson and A. Dehghani, Challenges of adaptive control-past, permanent and future, *Annual Reviews in Control*, vol. 32, no. 2, pp. 123–135, 2008.
7. L. Ljung, *System Identification: Theory for the User.* Englewood Cliffs, NJ: Prentice-Hall, 1987.
8. M. Gevers, Modelling, identification and control, *Iterative Identification and Control Design*, P. Albertos and A. Sala, Eds. London: Springer Verlag, 2002, pp. 3–16.
9. R. E. Skelton, Model error concepts in control design, *International Journal of Control*, vol. 49, no. 5, pp. 1725–1753, 1989.
10. C. E. Rohrs, L. Valavani, M. Athans, and G. Stein, Robustness of continuous-time adaptive control algorithms in the presence of unmodeled dynamics, *IEEE Transactions on Automatic Control*, vol. 30, no. 9, pp. 881–889, 1985.
11. C. E. Rohrs, L. Valavani, M. Athans, and G. Stein, Robustness of adaptive control algorithms in the presence of unmodeled dynamics, in *Proc. 21st IEEE Conference on Decision and Control*, Orlando, USA, 1982, pp. 3–11.
12. T. Y. Chai and D. R. Liu (Eds.), Special Issue on Data-Based Control, Decision, Scheduling and Fault Diagnosis, *Acta Automatica Sinica*, vol. 35, no. 6, 2009.
13. T. Y. Chai, Z. S. Hou, F. L. Lewis, and A. Hussain (Eds.), Special Section on Data-Based Control, Modeling, and Optimization, *IEEE Transactions on Neural Networks*, vol. 22, no. 12, 2011.
14. Z. S. Hou and J. X. Xu, On data-driven control theory: The state of the art and perspective, *Acta Automatica Sinica*, vol. 35, no. 6, pp. 650–667, 2009.
15. J. Van Helvoort, Unfalsified Control: Data-Driven Control Design for Performance Improvement, Ph. D. thesis, Technische Universiteit Eindhoven, Eindhoven, Netherlands, 2007.

16. K. Van Heusden, Non-Iterative Data-Driven Model Reference Control, Ph. D. thesis, Ecole Polytechnique Federale de Lausanne, Lausanne, Switzerland, 2010.
17. Z. S. Hou and Z. Wang, From model-based control to data-driven control: Survey, classification and perspective, *Information Sciences*, vol. 235, pp. 3–35, 2013.
18. J. C. Spall, Multivariate stochastic approximation using a simultaneous perturbation gradient approximation, *IEEE Transactions on Automatic Control*, vol. 37, no. 3, pp. 332–341, 1992.
19. J. C. Spall, Adaptive stochastic approximation by the simultaneous perturbation method, *IEEE Transactions on Automatic Control*, vol. 45, no. 10, pp. 1839–1853, 2000.
20. J. C. Spall and J. A. Cristion, Model-free control of nonlinear stochastic systems with discrete-time measurements, *IEEE Transactions on Automatic Control*, vol. 43, no. 9, pp. 1198–1210, 1998.
21. J. C. Spall and D. C. Chin, Traffic-responsive signal timing for system-wide traffic control, *Transportation Research Part C*, vol. 5 no. 3–4, pp. 153–163, 1997.
22. F. Rezayat, On the use of an SPSA-based model-free controller in quality improvement, *Automatica*, vol. 31, no. 6, pp. 913–915, 1995.
23. Z. S. Hou, The Parameter Identification, Adaptive Control and Model Free Learning Adaptive Control for Nonlinear Systems, Ph. D. thesis, Northeastern University, Shenyang, China, 1994.
24. Z. S. Hou, *Nonparametric Models and Its Adaptive Control Theory*. Beijing: Science Press, 1999.
25. Z. S. Hou and S. T. Jin, A novel data-driven control approach for a class of discrete-time nonlinear systems, *IEEE Transactions on Control Systems Technology*, vol. 19, no. 6, pp. 1549–1558, 2011.
26. Z. S. Hou and S. T. Jin, Data driven model-free adaptive control for a class of MIMO nonlinear discrete-time systems, *IEEE Transactions on Neural Networks*, vol. 22, no. 12, pp. 2173–2188, 2011.
27. L. D. S. Coelho and A. A. R. Coelho, Model-free adaptive control optimization using a chaotic particle swarm approach, *Chaos, Solitons & Fractals*, vol. 41, no. 4, pp. 2001–2009, 2009.
28. L. D. S. Coelho, M. W. Pessôa, R. R. Sumar, and A. A. R. Coelho, Model-free adaptive control design using evolutionary-neural compensator, *Expert Systems with Applications*, vol. 37, no. 1, pp. 499–508, 2010.
29. R. M. Cao and Z. S. Hou, Nonparametric model direct adaptive predictive control for linear motor, *Control Theory & Applications*, vol. 25, no. 3, pp. 587–590, 2008.
30. K. K. Tan, T. H. Lee, S. N. Huang, and F. M. Leu, Adaptive-predictive control of a class of SISO nonlinear systems, *Dynamics and Control*, vol. 11, no. 2, pp. 151–174, 2001.
31. B. Zhang and W. D. Zhang, Adaptive predictive functional control of a class of nonlinear systems, *ISA Transactions*, vol. 45, no. 2, pp. 175–183, 2006.
32. M. G. Safonov and T. C. Tsao, The unfalsified control concept: a direct path from experiment to controller, *Feedback Control, Nonlinear Systems and Complexity*, B. A. Francis and A. R. Tannenbaum, Eds. Berlin: Springer-Verlag, 1995, pp. 196–214.
33. M. G. Safonov and T. C. Tsao, The unfalsified control concept and learning, *IEEE Transactions on Automatic Control*, vol. 42, no. 6, pp. 843–847, 1997.
34. J. Van Helvoort, B. de Jager, and M. Steinbuch, Direct data-driven recursive controller unfalsification with analytic update, *Automatica*, vol. 43, no. 12, pp. 2034–2046, 2007.

35. G. Battistelli, E. Mosca, M. G. Safonov, and P. Tesi, Stability of unfalsified adaptive switching control in noisy environments, *IEEE Transactions on Automatic Control*, vol. 55, no. 10, pp. 2424–2429, 2010.
36. S. Baldi, G. Battistelli, E. Mosca, and P. Tesi, Multi-model unfalsified adaptive switching supervisory control, *Automatica*, vol. 46, no. 2, pp. 249–259, 2010.
37. M. G. Safonov, Data-driven robust control design: unfalsified control, 2003, http://routh.usc.edu/pub/safonov/safo03i.pdf.
38. J. Sivag, A. Datta, and S. P. Bhattacharyya, New results on the synthesis of PID controllers, *IEEE Transactions on Automatic Control*, vol. 47, no. 2, pp. 241–252, 2002.
39. H. Hjalmarsson, S. Gunnarsson, and M. Gevers, A convergent iterative restricted complexity control design scheme, in *Proc. 33rd IEEE Conference on Decision and Control*, Orlando, USA, 1994, pp. 1735–1740.
40. H. Hjalmarsson, M. Gevers, and S. Gunnarson, Iterative feedback tuning-theory and applications, *IEEE Control Systems*, vol. 18, no. 4, pp. 26–41, 1998.
41. H. Hjalmarsson, Control of nonlinear systems using iterative feedback tuning, in *Proc. 1998 American Control Conference*, Philadephia, USA, 1998, pp. 2083–2087.
42. J. Sjöberg and M. Agarwal, Nonlinear controller tuning based on linearized time-variant model, in *Proc. 1997 American Control Conference*, Albuquerque, New Mexico, 1997, pp. 3336–3340.
43. J. Sjöberg, F. De Bruyne, M. Agarwal, B. D. O. Anderson, M. Gevers, F. J. Kraus, and N. Linard, Iterative controller optimization for nonlinear systems, *Control Engineering Practice*, vol. 11, no. 9, pp. 1079–1086, 2003.
44. J. Sjöberg, P. O. Gutman, M. Agarwal, and M. Bax, Nonlinear controller tuning based on a sequence of identifications of linearized time-varying models, *Control Engineering Practice*, vol. 17, no. 2, pp. 311–321, 2009.
45. H. Hjalmarsson, Iterative feedback tuning: An overview, *International Journal of Adaptive Control and Signal Processing*, vol. 16, no. 5, pp. 373–395, 2002.
46. A. Karimi, L. Miskovic, and D. Bonvin, Convergence analysis of an iterative correlation-based controller tuning method, in *Proc. 15th IFAC World Congress*, Barcelona, Spain, 2002, pp. 1546–1551.
47. L. Miskovic, A. Karimi, D. Bonvin, and M. Gevers, Correlation-based tuning of decoupling multivariable controllers, *Automatica*, vol. 43, no. 9, pp. 1482–1494, 2007.
48. A. Karimi, L. Miskovic, and D. Bonvin, Iterative correlation-based controller tuning with application to a magnetic suspension system, *Control Engineering Practice*, vol. 11, no. 6, pp. 1069–1078, 2003.
49. L. Miskovic, A. Karimi, and D. Bonvin, Correlation-based tuning of a restricted-complexity controller for an active suspension system, *European Journal of Control*, vol. 9, no. 1, pp. 77–83, 2003.
50. G. O. Guardabassi and S. M. Savaresi, Virtual reference direct design method: an off-line approach to data-based control system design, *IEEE Transactions on Automatic Control*, vol. 45, no. 5, pp. 954–959, 2000.
51. M. C. Campi, A. Lecchini, and S. M. Savaresi, Virtual reference feedback tuning: A direct method for the design of feedback controllers, *Automatica*, vol. 38, no. 8, pp. 1337–1346, 2002.
52. M. C. Campi and S. M. Savaresi, Direct nonlinear control design: The virtual reference feedback tuning (VRFT) approach, *IEEE Transactions on Automatic Control*, vol. 51, no. 1, pp. 14–27, 2006.

53. M. Nakamoto, An application of the virtual reference feedback tuning for an MIMO process, in *Proc. 2004 SICE Annual Conference*, Sapporo, Japan, 2004, pp. 2208–2213.
54. S. Yabui, K. Yubai, and J. Hirai, Direct design of switching control system by VRFT-application to vertical-type one-link arm, in *Proc. 2007 SICE Annual Conference*, Kagawa, Japan, 2007, pp. 120–123.
55. M. C. Campi, A. Lecchini, and S. M. Savaresi, An application of the virtual reference feedback tuning method to a benchmark problem, *European Journal of Control*, vol. 9, pp. 66–76, 2003.
56. F. Previdi, F. Fico, D. Belloli, S. M. Savaresi, I. Pesenti, and C. Spelta, Virtual Reference Feedback Tuning (VRFT) of velocity controller in self-balancing industrial manual manipulators, in *Proc. 2010 American Control Conference*, Baltimore, Maryland, 2010, pp. 1956–1961.
57. M. Uchiyama, Formulation of high-speed motion pattern of a mechanical arm by trial, *Control Engineering*, vol. 14, no. 6, pp. 706–712, 1978.
58. S. Arimoto, S. Kawamura, and F. Miyazaki, Bettering operation of robots by learning, *Journal of Robotic Systems*, vol. 1, pp. 123–140, 1984.
59. Y. Q. Chen and C. Y. Wen, *Iterative Learning Control–Convergence, Robustness and Applications*. Berlin: Springer Verlag, 1999.
60. K. L. Moore, *Iterative Learning Control for Deterministic Systems*. New York: Springer Verlag, 1993.
61. M. X. Sun and B. J. Huang, *Iterative Learning Control*. Beijing, China: National Defence Industry Press, 1998.
62. J. X. Xu and Y. Tan, *Linear and Nonlinear Iterative Learning Control*. Berlin Heidelberg: Springer Verlag, 2003.
63. J. X. Xu and Z. S. Hou, On learning control: The state of the art and perspective, *Acta Automatica Sinica*, vol. 31, no. 6, pp. 943–955, 2005.
64. C. J. Chen, A discrete iterative learning control for a class of nonlinear time-varying systems, *IEEE Transactions on Automatic Control*, vol. 43, no. 5, pp. 748–752, 1998.
65. T. Y. Kuc, J. S. Lee, and K. Nam, An iterative learning control theory for a class of nonlinear dynamic systems, *Automatica*, vol. 28, no. 6, pp. 1215–1221, 1992.
66. H. S. Ahn, Y. Q. Chen, and K. L. Moore, Iterative learning control: Brief survey and categorization, *IEEE Transactions on Systems, Man, and Cybernetics, Part C: Applications and Reviews*, vol. 37, no. 6, pp. 1099–1121, 2007.
67. Z. S. Hou, J. X. Xu, and J. W. Yan, An iterative learning approach for density control of freeway traffic flow via ramp metering, *Transportation Research Part C*, vol. 16, no. 1, pp. 71–97, 2008.
68. S. Schaal and C. G. Atkeson, Robot juggling: Implementation of memory-based learning, *IEEE Control Systems Magazine*, vol. 14, no. 1, pp. 57–71, 1994.
69. D. W. Aha, Editorial: Lazy learning, *Artificial Intelligence Review*, vol. 11, no. 1–5, pp. 7–10, 1997.
70. G. Bontempi and M. Birattari, From linearization to lazy learning: A survey of divide-and-conquer techniques for nonlinear control, *International Journal of Computational Cognition*, vol. 3, no. 1, pp. 56–73, 2005.
71. G. Gybenko, Just-in-time learning and estimation, *Identification, Adaptation, Learning: The Science of Learning Models from Data*, S. Bittanti and G. Picci, Eds. New York: Springer, 1996, pp. 423–434.
72. D. W. Aha, D. Kibler, and M. Albert, Instance-based learning algorithms, *Machine Learning*, vol. 6, no. 1, pp. 37–66, 1991.

73. C. G. Atkeson, A. W. Moore, and S. Schaal, Locally weighted learning for control, *Artificial Intelligence Review*, vol. 11, no. 1–5, pp. 75–113, 1997.
74. M. W. Braun, D. E. Rivera, and A. Stenman, A model-on-demand identification methodology for nonlinear process systems, *International Journal of Control*, vol. 74, no. 18, pp. 1708–1717, 2001.
75. S. Hur, M. Park and H. Rhee, Design and application of model-on-demand predictive controller to a semibatch copolymerization reactor, *Industrial & Engineering Chemistry Research*, vol. 42, no. 4, pp. 847–859, 2003.
76. G. C. Goodwin and K. S. Sin, *Adaptive Filtering Prediction and Control*. Englewood Cliffs, NJ: Prentice-Hall, 1984.
77. T. Söderström, L. Ljung, and I. Gustavsson, A theoretical analysis of recursive identification methods, *Automatica*, vol. 14, no. 3, pp. 231–244, 1978.
78. Z. G. Han, On the identification of time-varying parameters in dynamic systems, *Acta Automatica Sinica*, vol. 10, no. 4, pp. 330–337, 1984.
79. Z. G. Han, *Multi-level Recursive Method and Its Applications*. Beijing: Science Press, 1989.
80. Z. S. Hou and Z. G. Han, Weighted slow time-varying modified nonlinear system Least-Squares algorithm, in *Proc. 1993 Chinese Control and Decision Conference*, Huangshan, China 1993, pp. 268–271.
81. Z. S. Hou and Z. G. Han, Modified nonlinear system Least-Squares algorithm, *Control Theory & Applications*, vol. 11, no. 3, pp. 271–276, 1994.
82. G. H. Golub and C. F. van Loan, *Matrix Computations*. Baltimore, MD: Johns Hopkins University Press, 1983.
83. Y. M. Chen and Y. C. Wu, Modified recursive least-squares algorithm for parameter identification, *International Journal of Systems Science*, vol. 23, no. 2, pp. 187–205, 1992.
84. S. A. Billings and S. Y. Fakhour, Identification of system containing linear and static nonlinear elements, *Automatica*, vol. 18, no. 1, pp. 15–26, 1982.
85. S. A. Billings, An overview of nonlinear systems identification, in *Proc. 7th IFAC Symposium on Identification and System Parameter Estimation*, York, UK, 1985, pp. 725–729.
86. S. A. Billings and Q. M. Zhu, Rational model identification using an extended least-squares algorithm, *International Journal of Control*, vol. 52, no. 3, pp. 529–546, 1991.
87. S. Chen and S. A. Billings, Prediction error estimation algorithm for nonlinear output-affine systems, *International Journal of Control*, vol. 47, no. 1, pp. 309–332, 1988.
88. S. Chen and S. A. Billings, Recursive prediction error parameter estimator for nonlinear models, *International Journal of Control*, vol. 49, no. 2, pp. 569–594, 1989.
89. S. Chen and S. A. Billings, Representations of non-linear systems: the NARMAX model, *International Journal of Control*, vol. 49, no. 3, pp. 1013–1032, 1989.
90. F. Fraiech and L. Ljung, Recursive identification of bilinear systems, *International Journal of Control*, vol. 45, no. 2, pp. 453–470, 1987.
91. G. C. Goodwin and R. L. Payne, *Dynamic System Identification: Experiment Design and Data Analysis*. New York: Academic Press, 1977.
92. L. Ljung and T. Söderström, *Theory and Practice of Recursive Identification*. Cambridge, MA: MIT Press, 1983.
93. P. C. Young, *Recursive estimation and time-series analysis*. Berlin: Springer-Verlag, 1984.
94. Q. M. Zhu and S. A. Billings, Parameter estimation for stochastic nonlinear rational models, *International Journal of Control*, vol. 57, no. 2, pp. 309–333, 1993.

95. K. Anbunmani, L. Patnaik, and I. Serma, Self-tuning minimum variance control of nonlinear systems of the Hammerstein model, *IEEE Transactions on Automatic Control*, vol. 26, no. 4, pp. 959–961, 1981.
96. S. Svoronos, G. Stephanopoulos, and R. Aris, On bilinear estimation and control, *International Journal of Control*, vol. 34, no. 4, pp. 651–684, 1981.
97. D. J. T. Sung and T. T. Lee, Model reference adaptive control of nonlinear systems using the Wiener model, *International Journal of Systems Science*, vol. 18, no. 3, pp. 581–599, 1987.
98. T. Pröll and M. N. Karim, Real-time design of an adaptive nonlinear predictive controllers, *International Journal of Control*, vol. 59, no. 3, pp. 863–889, 1994.
99. K. R. Sales and S. A. Billings, Self-tuning control of nonlinear ARMAX models, *International Journal of Control*, vol. 51, no. 4, pp. 753–769, 1990.
100. E. Aranda-Bricaire, Ü. Kotta, and C. H. Moog, Linearization of discrete-time systems, *SIAM Journal on Control and Optimization*, vol. 34, no. 6, pp. 1999–2023, 1996.
101. J. P. Barbot, S. Monaco, and D. Normand-Cyrot, Quadratic forms and approximate feed back linearization in discrete time, *International Journal of Control*, vol. 67, no. 4, pp. 567–586, 1997.
102. J. P. Barbot, S. Monaco, and D. Normand-Cyrot, Discrete-time approximated linearization of SISO systems under output feedback, *IEEE Transactions on Automatic Control*, vol. 44, no. 9, pp. 1729–1733, 1999.
103. H. Deng, H. X. Li, and Y. H. Wu, Feedback-linearization-based neural adaptive control for unknown nonaffine nonlinear discrete-time systems, *IEEE Transactions on Neural Networks*, vol. 19, no. 9, pp. 1615–1625, 2008.
104. J. W. Grizzle, Feedback linearization of discrete-time systems, *Analysis and Optimization of Systems, Lecture Notes in Control and Information Sciences*, A. Bensoussan and J. L. Lions, Eds. Berlin: Springer, 1986, pp. 271–281.
105. J. W. Grizzle and P. V. Kokotovic, Feedback linearization of sampled-data systems, *IEEE Transactions on Automatic Control*, vol. 33, no. 9, pp. 857–859, 1988.
106. G. O. Guardabassi and S. M. Savaresi, Approximate feedback linearization of discrete-time nonlinear systems using virtual input direct design, *Systems & Control Letters*, vol. 32, no. 2, pp. 63–74, 1997.
107. G. O. Guardabassi and S. M. Savaresi, Approximate linearization via feedback—An overview, *Automatica*, vol. 37, no. 1, pp. 1–15, 2001.
108. B. Jakubczyk, Feedback linearization of discrete-time systems, *Systems & Control Letters*, vol. 9, no. 5, pp. 411–416, 1987.
109. N. Kwnaghee, Linearization of discrete-time nonlinear systems and a canonical structure, *IEEE Transactions on Automatic Control*, vol. 34, no. 1, pp. 119–122, 1989.
110. H. G. Lee, A. Arapostathis, and S. I. Marcus, Linearization of discrete-time systems, *International Journal of Control*, vol. 45, no. 5, pp. 1803–1822, 1987.
111. H. G. Lee and S. I. Marcus, Approximate and local linearizability of non-linear discrete-time systems, *International Journal of Control*, vol. 44, no. 4, pp. 1103–1124, 1986.
112. H. G. Lee and S. I. Marcus, On input-output linearization of discrete-time nonlinear systems, *Systems & Control Letters*, vol. 8, no. 3, pp. 249–259, 1987.
113. P. C. Yeh and P. V. Kokotovic, Adaptive control of a class of nonlinear discrete-time systems, *International Journal of Control*, vol. 62, no. 2, pp. 203–324, 1995.
114. L. Chen and K. S. Narendra, Identification and control of a nonlinear dynamical system based on its linearization: Part II, in *Proc. 2002 American Control Conference*, Anchorage, Alaska, 2002, pp. 382–287.

115. L. Chen and K. S. Narendra, Identification and control of a nonlinear discrete-time system based on its linearization: A unified framework, *IEEE Transactions on Neural Networks*, vol. 15, no. 3, pp. 663–673, 2004.
116. K. S. Narendra and L. Chen, Identification and control of a nonlinear dynamical system Σ based on its linearization Σ_L: Part I, in *Proc. 37th IEEE Conference on Decision and Control*, Tampa, Florida, 1998, pp. 2977–2982.
117. S. M. Savaresi and G. O. Guardabassi, Approximate I/O feedback linearization of discrete-time non-linear systems via virtual input direct design, *Automatica*, vol. 34, no. 6, pp. 715–722, 1998.
118. C. Wen and D. J. Hill, Adaptive linear control of nonlinear systems, *IEEE Transactions on Automatic Control*, vol. 35, no. 11, pp. 1253–1257, 1990.
119. Y. Xi and F. Wang, Nonlinear multi-model predictive control, in *Proc. 13th IFAC World Congress*, San Francisco, California, 1996.
120. G. A. Dumont and Y. Fu, Nonlinear adaptive control via laguerre expansion of volterra kernels, *International Journal of Adaptive Control and Signal Processing*, vol. 7, no. 5, pp. 367–382, 1993.
121. G. A. Dumont, Y. Fu, and G. Lu, Nonlinear adaptive generalized predictive control and applications, *Advances in Model-based Predictive Control*, D. W. Clarke, Ed. Oxford: Oxford University Press, 1994, pp. 498–515.
122. G. A. Dumont, C. C. Zervos, and G. L. Pageau, Laguerre-based adaptive control of PH in an industrial plant extraction stage, *Automatica*, vol. 26, no. 4, pp. 781–787, 1990.
123. J. X. Xu and Z. S. Hou, Notes on data-driven system approaches, *Acta Automatica Sinica*, vol. 35, no. 6, pp. 668–675, 2009.
124. Z. S. Hou and W. H. Huang, The model-free learning adaptive control of a class of SISO nonlinear systems, in *Proc. 1997 American Control Conference*, Albuquerque, New Mexico, 1997, pp. 343–344.
125. Z. S. Hou, On model-free adaptive control: The state of the art and perspective, *Control Theory & Applications*, vol. 23, no. 4, pp. 586–592, 2006.
126. Z. S. Hou and X. H. Bu, Model free adaptive control with data dropouts, *Expert Systems with Applications*, vol. 38, no. 8, pp. 10709–10717, 2011.
127. W. Rudin, *Principles of mathematical analysis*. New York: McGraw-Hill, 1976.
128. F. Kong and P. D. Keyser, Criteria for choosing the horizon in extended horizon predictive control, *IEEE Transactions on Automatic Control*, vol. 39, no. 7, pp. 1467–1470, 1994.
129. R. Scattolini and S. Bittanti, On the choice of the horizon in long-range predictive control: some simple criteria, *Automatica*, vol. 26, no. 5, pp. 915–917, 1990.
130. K. S. Narendra and A. M. Annaswamy, *Stable Adaptive Systems*. Englewood Cliffs, NJ: Prentice-Hall, 1989.
131. K. J. Åström and B. Wittenmark, *Adaptive Control*. Boston, MA: Addison-Wesley Longman Publishing Co., Inc, 1994.
132. L. Guo, *Time-Varying Stochastic Systems: Stability, Estimation and Control*. Changchun: Jilin Science and Technology Press, 1993.
133. K. J. Åström, Theory and applications of adaptive control-A survey, *Automatica*, vol. 19, no. 5, pp. 471–486, 1983.
134. T. Hsia, Adaptive control of robot manipulators—A review, in *Proc. 1986 IEEE International Conference on Robotics and Automation*, San Francisco, California, 1986, pp. 183–189.

135. D. E. Seborg, T. F. Edgar, and S. L. Shah, Adaptive control strategies for process control: A survey, *AIChE Journal*, vol. 32, no. 6, pp. 881–913, 1986.
136. A. Isidori, *Nonlinear Control Systems*. London, UK: Springer-Verlag, 1995.
137. G. Tao, *Adaptive Control Design and Analysis*. New York: Wiley, 2003.
138. G. Pajunen, Adaptive control of wiener type nonlinear systems, *Automatica*, vol. 28, no. 4, pp. 781–785, 1992.
139. F. C. Chen and H. K. Khalil, Adaptive control of a class of nonlinear discrete-time systems, *IEEE Transactions on Automatic Control*, vol. 40, no. 5, pp. 791–801, 1995.
140. Y. Zhang, C. Y. Wen, and Y. C. Soh, Robust adaptive control of uncertain discrete-time systems, *Automatica*, vol. 35, no. 5, pp. 321–329, 1999.
141. M. Krstic and A. Smyshlyaev, Adaptive boundary control for unstable parabolic PDEs-Part I: Lyapunov design, *IEEE Transactions on Automatic Control*, vol. 53, no. 7, pp. 1575–1591, 2008.
142. Y. Zhang, W. H. Chen, and Y. C. Soh, Improved robust backstepping adaptive control for nonlinear discrete-time systems without overparameterization, *Automatica*, vol. 44, no. 3, pp. 864–867, 2008.
143. J. Zhou and C. Y. Wen, Decentralized backstepping adaptive output tracking of interconnected nonlinear systems, *IEEE Transactions on Automatic Control*, vol. 53, no. 10, pp. 2378–2384, 2008.
144. J. Zhou, C. Y. Wen, and Y. Zhang, Adaptive backstepping control of a class of uncertain nonlinear systems with unknown backlash-like hysteresis, *IEEE Transactions on Automatic Control*, vol. 49, no. 10, pp. 1751–1759, 2004.
145. R. Findeisen, L. Imsland, F. Allgöwer, and B. A. Foss, State and output feedback nonlinear model predictive control: An overview, *European Journal of Control*, vol. 9, no. 2–3, pp. 190–206, 2003.
146. M. A. Henson, Nonlinear model predictive control: current status and future directions, *Computers & Chemical Engineering*, vol. 23, no. 2, pp. 187–202, 1998.
147. B. Kouvaritakis and M. Cannon, *Nonlinear Predictive Control: Theory and Practice*. London: The Institution of Engineering and Technology, 2001.
148. L. Magni, D. M. Raimondo, and F. Allögwer, *Nonlinear Model Predictive Control: Towards New Challenging Applications*. Berlin Heidelberg: Springer-Verlag, 2009.
149. K. S. Narendra and C. Xiang, Adaptive control of discrete-time systems using multiple models, *IEEE Transactions on Automatic Control*, vol. 45, no. 9, pp. 1669–1686, 2000.
150. K. S. Narendra, O. A. Driollet, M. Feiler, and K. George, Adaptive control using multiple models, switching and tuning, *International Journal of Adaptive Control and Signal Processing*, vol. 17, no. 2, pp. 87–102, 2003.
151. X. K. Chen, T. Fukuda, and K. D. Young, Adaptive quasi-sliding-mode tracking control for discrete uncertain input output systems, *IEEE Transactions on Industrial Electronics*, vol. 48, no. 1, pp. 216–224, 2001.
152. X. K. Chen, Adaptive sliding mode control for discrete-time multi-input multi-output systems, *Automatica*, vol. 42, no. 3, pp. 427–435, 2006.
153. B. Chen, X. P. Liu, and S. C. Tong, Adaptive fuzzy output tracking control of MIMO nonlinear uncertain systems, *IEEE Transactions on Fuzzy Systems*, vol. 15, no. 2, pp. 287–300, 2007.
154. D. V. Diaz and Y. Tang, Adaptive robust fuzzy control of nonlinear systems, *IEEE Transactions on Systems, Man, and Cybernetics, Part B: Cybernetics*, vol. 34, no. 3, pp. 1596–1601, 2004.

155. S. C. Tong, Y. M. Li, and P. Shi, Fuzzy adaptive backstepping robust control for SISO nonlinear system with dynamic uncertainties, *Information Sciences*, vol. 179, no. 9, pp. 1319–1332, 2009.
156. Y. C. Chang and H. M. Yen, Adaptive output feedback tracking control for a class of uncertain nonlinear systems using neural networks, *IEEE Transactions on Systems, Man, and Cybernetics, Part B: Cybernetics*, vol. 35, no. 6, pp. 1311–1316, 2005.
157. Y. Fu and T. Y. Chai, Nonlinear multivariable adaptive control using multiple models and neural networks, *Automatica*, vol. 43, no. 6, pp. 1101–1110, 2007.
158. S. S. Ge, G. Y. Li, J. Zhang, and T. H. Lee, Direct adaptive control for a class of MIMO nonlinear systems using neural networks, *IEEE Transactions on Automatic Control*, vol. 49, no. 11, pp. 2001–2004, 2004.
159. S. S. Ge, J. Zhang, and T. H. Lee, Adaptive MNN control for a class of non-affine NARMAX systems with disturbances, *Systems & Control Letters*, vol. 53, no. 1, pp. 1–12, 2004.
160. Q. M. Zhu and L. Z. Guo, Stable adaptive neurocontrol for nonlinear discrete-time systems, *IEEE Transactions on Neural Networks*, vol. 15, no. 3, pp. 653–662, 2004.
161. Q. Sang and G. Tao, Gain margins of adaptive control systems, *IEEE Transactions on Automatic Control*, vol. 55, no. 1, pp. 104–115, 2010.
162. K. S. Narendra and K. Parthasarathy, Identification and control for dynamic systems using neural networks, *IEEE Transactions on Neural Networks*, vol. 1, no. 1, pp. 4–27, 1990.
163. F. C. Chen, Back-propagation networks for neural nonlinear self-tuning adaptive control, *IEEE Control Systems Magazine*, vol. 10, no. 3, pp. 44–48, 1990.
164. E. I. Jury, *Theory and Application of the z-Transform Method*. New York: Wiley, 1964.
165. L. Jin, D. N. Nikiforuk, and M. M. Gupta, Adaptive control of discrete-time nonlinear systems using neural networks, *IEE Proceedings of Control Theory and Applications*, vol. 141, no. 3, pp. 169–176, 1994.
166. P. A. Cook, Application of model reference adaptive control to a benchmark problem, *Automatica*, vol. 36, no. 4, pp. 585–588, 1994.
167. M. S. Ahmed and I. A. Tasadduq, Neural-net controller for nonlinear plants: design approach through linearization, *IEE Proceedings of Control Theory and Applications*, vol. 141, no. 5, pp. 305–314, 1994.
168. V. Etxebarria, Adaptive control of discrete systems using neural networks, *IEE Proceedings of Control Theory and Applications*, vol. 141, no. 4, pp. 209–215, 1994.
169. S. Skogestad and I. Postlethwaite, *Multivariable Feedback Control: Analysis and Design*. Chichester: Wiley, 2005.
170. H. Elliot and W. A. Wolovich, A parameter adaptive control structure for linear multivariable systems, *IEEE Transactions on Automatic Control*, vol. 27, no. 2, pp. 340–352, 1982.
171. T. Asano, K. Yoshikawa, and S. Suzuki, The design of a precompensator for multivariable adaptive control-a network-theoretic approach, *IEEE Transactions on Automatic Control*, vol. 35, no. 6, pp. 706–710, 1990.
172. X. P. Liu, G. X. Gu, and K. M. Zhou, Robust stabilization of MIMO nonlinear systems by backstepping, *Automatica*, vol. 35, no. 5, pp. 987–992, 1999.
173. B. Yao and M. Tomizuka, Adaptive robust control of MIMO nonlinear systems in semi-strict feedback forms, *Automatica*, vol. 37, no. 9, pp. 1305–1321, 2001.
174. C. L. Wang and Y. Lin, Adaptive dynamic surface control for linear multivariable systems, *Automatica*, vol. 46, no. 10, pp. 1703–1711, 2010.

175. C. L. Wang and Y. Lin, Decentralized adaptive dynamic surface control for a class of interconnected nonlinear systems, *IET Control Theory & Applications*, vol. 6, no. 9, pp. 1172–1181, 2012.
176. R. Ortega, On Morse's new adaptive controller: Parameter convergence and transient performance, *IEEE Transactions on Automatic Control*, vol. 38, no. 8, pp. 1191–1202, 1993.
177. B. Chen, S. Tong, and X. Liu, Fuzzy approximate disturbance decoupling of MIMO nonlinear systems by backstepping approach, *Fuzzy Sets and Systems*, vol. 158, no. 10, pp. 1097–1125 2007.
178. W. H. Wang, Z. S. Hou, and S. T. Jin, Model-free indirect adaptive decoupling control for nonlinear discrete-time MIMO systems, in *Proc. 48th IEEE Conference on Decision and Control*, Shanghai, China, 2009, pp. 7663–7668.
179. S. T. Jin, On Model Free Learning Adaptive Control and Applications, Ph. D. thesis, Beijing Jiaotong University, Beijing, 2008.
180. W. H. Wang, Issues on Model free Adaptive Control, Ph. D. thesis, Beijing Jiaotong University, Beijing, 2008.
181. C. Y. Wen, J. Zhou, and W. Wang, Decentralized adaptive backstepping stabilization of interconnected systems with dynamic input and output interactions, *Automatica*, vol. 45, no. 1, pp. 55–67, 2009.
182. S. Gerschgorin, Uber die abgrenzung der eigenwerte einer matrix, *Izv, Akad. Nauk. USSR Otd. Fiz. -Mat. Nauk 7*, pp. 749–754, 1931.
183. L. Huang, *Linear Algebra in System and Control Theory*. Beijing: Science Press, 1984.
184. J. Zhang and S. S. Ge, Output feedback control of a class of discrete MIMO nonlinear systems with triangular form inputs, *IEEE Transactions on Neural Networks*, vol. 16, no. 1, pp. 1491–1503, 2005.
185. J. Richalet, Model predictive heuristic control: Application to industrial process, *Automatica*, vol. 14, no. 5, pp. 413–428, 1978.
186. C. R. Culter and B. L. Ramaker, Dynamic matrix control-A computer control algorithm, in *Proc. JACC*, San Francisco, USA, 1980.
187. D. W. Clarke, C. Mohtadi, and P. S. Tuffs, Generalized predictive control, *Automatica*, vol. 23, no. 2, pp. 137–160, 1987.
188. Y. Xi, *Predictive Control*. Beijing: National Defense Industry, 1993.
189. H. Chen and C. W. Scherer, Moving horizon H∞ control with performance adaptation for constrained linear systems, *Automatica*, vol. 42, no. 6, pp. 1033–1040, 2006.
190. D. Sui, L. Feng, M. Hovd, and C. Jin, Decomposition principle in model predictive control for linear systems with bounded disturbances, *Automatica*, vol. 45, no. 8, pp. 1917–1922, 2009.
191. A. L. Elshafei, G. Dumont, and A. Elaaggar, Stability and convergence analysis of an adaptive GPC based on state-space modeling, *International Journal of Control*, vol. 61, no. 1, pp. 193–210, 1995.
192. R. Rouhani and R. K. Mehra, Model algorithmic control (MAC): Basic theoretical proper, *Automatica*, vol. 18, no. 4, pp. 401–414, 1982.
193. T. W. Yoon and D. W. Clarke, Adaptive predictive control of the Benchmark plant, *Automatica*, vol. 30, no. 4, pp. 621–628, 1994.
194. P. O. M. Scokaert and D. W. Clarke, Stabilizing properties of con-strained predictive control, *IEE Proceedings of Control Theory and Applications*, vol. 141, no. 5, pp. 295–304, 1994.

195. M. Canale, L. Fagiano, and M. Milanese, Efficient model predictive control for nonlinear systems via function approximation techniques, *IEEE Transactions on Automatic Control*, vol. 55, no. 8, pp. 1911–1916, 2010.
196. D. Munoz de la Pena and P. D. Christofides, Lyapunov-based model predictive control of nonlinear systems subject to data losses, *IEEE Transactions on Automatic Control*, vol. 53, no. 9, pp. 2076–2089, 2008.
197. D. Q. Mayne, J. B. Rawlings, C. V. Rao, and P. O. M. Scokaert, Constrained model predictive control: Stability and optimality, *Automatica*, vol. 36, no. 6, pp. 789–814, 2000.
198. Z. S. Hou, The model-free direct adaptive predictive control for a class of discrete-time nonlinear system, in *Proc. 4th Asian Control Conference*, Singapore, 2002, pp. 519–524.
199. K. K. Tan, S. N. Huang, and T. H. Lee, *Applied Predictive Control*. London, UK: Springer-Verlag, 2002.
200. K. K. Tan, S. N. Huang, T. H. Lee, and F. M. Leu, Adaptive predictive PI control of a class of SISO systems, in *Proc. 1999 American Control Conference*, San Diego, California, 1999, pp. 3848–3852
201. D. W. Clarke and P. J. Gawthrop, Self-tuning control, *IEE Proceedings*, vol. 126, no. 6, pp. 633–640, 1979.
202. O. P. Palsson, H. Madsen, and H. T. Sogaard, Generalized predictive control for non-stationary systems, *Automatica*, vol. 30, no. 12, pp. 1991–1997, 1994.
203. D. A. Bristow, M. Tharayil, and A. G. Alleyne, A survey of iterative learning control: A learning-based method for high-performance tracking control, *IEEE Control Systems Magazine*, vol. 26, no. 3, no. 3, pp. 96–114, 2006.
204. Z. S. Hou, J. X. Xu, and H. W. Zhong, Freeway traffic control using iterative learning control based ramp metering and speed signaling, *IEEE Transactions on Vehicular Technology*, vol. 56, no. 2, pp. 466–477, 2007.
205. Z. S. Hou and Y. Wang, Terminal iterative learning control based station stop control of a train, *International Journal of Control*, vol. 84, no. 7, pp. 1263–1274, 2011.
206. Z. S. Hou, J. W. Yan, J. X. Xu, and Z. J. Li, Modified iterative-learning-control-based ramp metering strategies for freeway traffic control with iteration-dependent factors, *IEEE Transactions on Intelligent Transportation Systems*, vol. 13, no. 2, pp. 606–618, 2012.
207. Z. S. Hou, X. Xu, J. W. Yan, and J. X. Xu, A complementary modularized ramp metering designing approach based on iterative learning control and ALINEA, *IEEE Transactions on Intelligent Transportation Systems*, vol. 12, no. 4, pp. 1305–1318, 2011.
208. R. H. Chi and Z. S. Hou, Dual-stage optimal iterative learning control for nonlinear non-affine discrete-time systems, *Acta Automatica Sinica*, vol. 33, no. 10, pp. 1061–1065, 2007.
209. R. H. Chi and Z. S. Hou, A model-free periodic adaptive control for freeway traffic density via ramp metering, *Acta Automatica Sinica*, vol. 36, no. 7, pp. 1029–1032, 2010.
210. R. H. Chi, Z. S. Hou, and S. L. Sui, Non-parameter adaptive iterative learning control for the freeway traffic ramp metering, *Control Theory & Applications*, vol. 25, no. 6, pp. 1011–1015, 2008.
211. R. H. Chi, Z. S. Hou, L. Yu, and S. L. Sui, Higher-order model-free adaptive iterative learning control, *Control and Decision*, vol. 23, no. 7, pp. 795–798, 2008.

212. S. Huang, K. Tan, and T. Lee, Neural network learning algorithm for a class of interconnected nonlinear systems, *Neurocomputing*, vol. 72, no. 4–6, pp. 1071–1077, 2009.
213. B. Labibi, H. J. Marquez, and T. Chen, Decentralized robust output feedback control for control affine nonlinear interconnected systems, *Journal of Process Control*, vol. 19, no. 5, pp. 865–878, 2009.
214. S. S. Stankovic and D. D. Siljak, Robust stabilization of nonlinear interconnected systems by decentralized dynamic output feedback, *Systems & Control Letters*, vol. 58, no. 4, pp. 271–275, 2009.
215. R. S. Chandra, C. Langbort, and R. D. Andrea, Distributed control design with robustness to small time delays, *Systems & Control Letters*, vol. 58, no. 4, pp. 296–303, 2009.
216. C.-H. Fan, J. L. Speyer, and C. R. Jaensch, Centralized and decentralized solutions of the linear-exponential-Gaussian problem, *IEEE Transactions on Automatic Control*, vol. 39, no. 10, pp. 1986–2003, 1994.
217. D. Milutinovic and P. Lima, Modeling and optimal centralized control of a large-size robotic population, *IEEE Transactions on Robotics*, vol. 22, no. 6, pp. 1280–1285, 2006.
218. R. D'Andrea and G. E. Dullerud, Distributed control design for spatially interconnected systems, *IEEE Transactions on Automatic Control*, vol. 48, no. 9, pp. 1478–1495, 2003.
219. S. X. Ding, G. Yang, P. Zhang, E. L. Ding, T. Jeinsch, N. Weinhold, and M. Schultalbers, Feedback control structures, embedded residual signals, and feedback control schemes with an integrated residual access, *IEEE Transactions on Control Systems Technology*, vol. 18, no. 2, pp. 352–367, 2010.
220. Z. S. Hou and D. Xiong, The adaptive control system design with a model-free external loop as a compensator, in *Proc. 5th World Congress on Intelligent Control and Automation*, Hangzhou, China, 2004, pp. 444–448.
221. D. Xiong, The Modularized Control System Design with a Model-free Controller External Loop as a compensator, Master's thesis, Beijing Jiaotong University, Beijing, 2004.
222. J. W. Yan and Z. S. Hou, Convergence analysis of learning-enhanced PID control system, *Control Theory & Application*, vol. 27, no. 6, pp. 761–768, 2010.
223. Z. S. Hou and J. X. Xu, A new feedback-feedforward configuration for the iterative learning control of a class of discrete-time systems, *Acta Automatica Sinica*, vol. 33, no. 3, pp. 323–326, 2007.
224. R. Ortega and Y. Tang, Robustness of adaptive controllers—A survey, *Automatica*, vol. 25, no. 5, pp. 651–677, 1989.
225. B. B. Peterson and K. S. Narendra, Bounded error adaptive control *IEEE Transactions on Automatic Control*, vol. 27, no. 6, pp. 1161–1168, 1982.
226. K. S. Narendra and A. M. Annaswamy, A new adaptive law for robust adaptive without persistent excitation, *IEEE Transactions on Automatic Control*, vol. 32, no. 2, pp. 134–145, 1987.
227. R. Ortega, L. Praly, and I. D. Landau, Robustness of discrete time direct adaptive controllers, *IEEE Transactions on Automatic Control*, vol. 30, no. 12, pp. 1179–1187, 1985.
228. C. Y. Wen and D. J. Hill, Decentralized adaptive control of linear time-varying systems, in *Proc. IFAC 11th World Congress on Automatic Control*, Tallinn, USSR, 1990, pp. 131–136.

229. B. E. Ydstie, Transient performance and robustness of direct adaptive control, *IEEE Transactions on Automatic Control*, vol. 37, no. 8, pp. 1091–1105, 1992.
230. P. A. Ioannou and P. V. Kokotovic, Instability analysis and improvement of adaptive control, *Automatica*, vol. 20, no. 5, pp. 583–594, 1984.
231. P. A. Ioannou and K. Tsaklis, A robust direct adaptive controller, *IEEE Transactions on Automatic Control*, vol. 31, no. 11, pp. 1033–1043 1986.
232. M. Krstic, I. Kanellakopoulos, and P. V. Kokotovic, Passivity and parametric robustness of a new class adaptive systems, *Automatica*, vol. 30, no. 11, pp. 1703–1716, 1994.
233. X. H. Bu, Z. S. Hou, F. S. Yu, and F. Z. Wang, Robust model free adaptive control with measurement disturbance, *IET Control Theory & Applications*, vol. 9, no. 6, pp. 1288–1296, 2012.
234. X. H. Bu, Z. S. Hou, and S. T. Jin, The robustness of model-free adaptive control with disturbance suppression, *Control Theory & Applications*, vol. 28, no. 3, pp. 358–362, 2011.
235. J. P. Hespanha, P. Naghshtabrizi, and Y. G. Xu, A survey of recent results in networked control systems, *Proceedings of the IEEE*, vol. 95, no. 1, pp. 138–162, 2007.
236. M. Sahebsara, T. Chen, and S. L. Shah, Optimal filtering in networked control systems with multiple packet dropout, *IEEE Transactions on Automatic Control*, vol. 52, no. 8, pp. 1508–1513, 2007.
237. Y. P. Tian, C. B. Feng, and X. Xin, Robust stability of polynomials with multilinearly of dependent coefficient perturbations, *IEEE Transactions on Automatic Control*, vol. 39, no. 3, pp. 554–558, 1994.
238. Y. L. Wang and G. H. Yang, Linear estimation-based time delay and packet dropout compensation for networked control systems, in *Proc. 2008 American Control Conference*, Washington, USA, 2008, pp. 3786–3791.
239. C. C. Hsieh, P. Hsu, and B. C. Wang, The Motion Message Estimator in Networked Control Systems, in *Proc. 17th IFAC World Congress*, Seoul, Korea, 2008, pp. 11606–11611.
240. G. L. Suo and X. H. Yang, Compensation for network-induced delays and packet dropout and system stability analysis, *Control and Decision*, vol. 21, no. 2, pp. 205–209, 2006.
241. S. Y. Zhang, The symmetric and similar structure of complex control systems *Control Theory & Applications*, vol. 11, no. 2, pp. 231–236, 1994.
242. Z. S. Hou and W. H. Huang, The model-free learning adaptive control of a class of MISO nonlinear discrete-time systems, in *Proc. 5th IFAC Symposium on Low Cost Automation*, Shenyang, China, 1998, pp. 13–26.
243. Z. S. Hou and W. H. Huang, The model-free learning adaptive control of a class of nonlinear discrete-time systems, *Control Theory & Applications*, vol. 15, no. 6, pp. 893–899, 1998.
244. Z. S. Hou, Robust modeless learning adaptive control of nonlinear systems, *Control and Decision*, vol. 10, no. 2, pp. 137–142, 1995.
245. Z. S. Hou and Z. G. Han, Parameter estimation algorithm of nonlinear system and its dual adaptive control, *Acta Automatica Sinica*, vol. 21, no. 1, pp. 122–125, 1995.
246. Z. S. Hou, W. H. Huang, and Z. G. Han, On the symmetric similarity structure design of self-tuning control system: paremetric model case, *Control and Decision*, vol. 13, pp. 291–295, 1998.
247. A. Alessandri, M. Baglietto, and G. Battistelli, Receding-horizon estimation for discrete-time linear systems, *IEEE Transactions on Automatic Control*, vol. 48, no. 3, pp. 473–478, 2003.

248. G. C. Goodwin, J. A. De Doná, M. M. Seron, and X. W. Zhuo, Lagrangian duality between constrained estimation and control, *Automatica* vol. 41, no. 6, pp. 935–944, 2005.
249. R. B. Gopaluni, R. S. Patwardhan, and S. L. Shah, MPC relevant identification-tuning the noise model, *Journal of Process Control* vol. 14, no. 6, pp. 699–714, 2004.
250. I. N. M. Papadakis and S. C. A. Thomopoulos, On the dual nature of the MARC and MARI problems, in *Proc. 1991 American Control Conference*, Boston, Massachusetts, 1991, pp. 163–164.
251. D. S. Shook, C. Mohtadi, and S. L. Shah, Identification for long-range predictive control, *IEE Proceedings of Control Theory and Applications*, vol. 138, no. 1, pp. 75–84, 1991.
252. D. S. Shook, C. Mohtadi, and S. L. Shah, A control-relevant identification strategy for GPC, *IEEE Transactions on Automatic Control*, vol. 37, no. 7, pp. 975–980, 1992.
253. R. H. Chi, Adaptive iterative learning control for nonlinear discrete-time systems and its applications, Ph. D. thesis, Beijing Jiaotong University, Beijing 2006.
254. R. H. Chi and Z. S. Hou, A neural network-based adaptive ILC for a class of nonlinear discrete-time systems with dead zone scheme, *Journal of Systems Science and Complexity*, vol. 22, no. 3, pp. 435–445, 2009.
255. R. H. Chi, Z. S. Hou, and S. T. Jin, Data-weighting based discrete-time adaptive iterative learning control for nonsector nonlinear systems with iteration-varying trajectory and random initial condition, *Journal of Dynamic Systems, Measurement, and Control* vol. 134, no. 2, pp. 021016-1-10, 2012.
256. R. H. Chi, Z. S. Hou, and S. T. Jin, Discrete-time adaptive ILC for non-parametric uncertain nonlinear systems with iteration-varying trajectory and random initial condition, *Asian Journal of Control*, vol. 15, no. 2, pp. 562–570, 2013.
257. R. H. Chi, Z. S. Hou, S. L. Sui, L. Yu, and W. L. Yao, A new adaptive iterative learning control motivated by discrete-time adaptive control, *International Journal of Innovative Computing, Information and Control*, vol. 4, no. 6, pp. 1267–1274, 2008.
258. R. H. Chi, Z. S. Hou, and J. X. Xu, Adaptive ILC for a class of discrete-time systems with iteration-varying trajectory and random initial condition, *Automatica*, vol. 44, no. 8, pp. 2207–2213, 2008.
259. R. H. Chi, S. L. Sui, L. Yu, Z. S. Hou, and J. X. Xu, A discrete-time adaptive ILC for systems with random initial condition and iteration-varying trajectory, in *Proc. 17th IFAC World Congress*, Seoul, Korea, 2008, pp. 14432–14437.
260. J. X. Xu, A new periodic adaptive control approach for time-varying parameters with known periodicity, *IEEE Transactions on Automatic Control*, vol. 49, no. 4, pp. 579–583, 2004.
261. Z. S. Hou, R. H. Chi, and J. X. Xu, Reply to "Comments on 'Adaptive ILC for a class of discrete-time systems with iteration-varying trajectory and random initial condition'", *Automatica*, vol. 46, no. 3, pp. 635–636, 2010.
262. R. M. Cao and Z. S. Hou, Model-free learning adaptive control of a PH neutralisation process, *Computer Engineering and Applications*, vol. 42, no. 28, pp. 191–194, 2006.
263. J. W. Wang and Q. L. Shang, Investigation of model-free adaptive control for single-step neutralization process, *Mechanical & Electrical Engineering Magazine*, vol. 25, no. 5, pp. 96–99, 2008.
264. R. H. Chi and Z. S. Hou, A model free adaptive control approach for freeway traffic density via ramp metering, *International Journal of Innovative Computing, Information and Control*, vol. 4, no. 11, pp. 2823–2832, 2008.

265. Z. S. Hou and J. W. Yan, Model free adaptive control based freeway ramp metering with feed forward iterative learning controller, *Acta Automatica Sinica*, vol. 35, no. 6, pp. 588–595, 2009.
266. J. Ma, Z. Y. Chen, and Z. S. Hou, Model-free adaptive control of integrated roll-reducing system for large warships, *Control Theory & Applications*, vol. 26, no. 11, pp. 1289–1292, 2009.
267. J. Ma, X. H. Liu, and G. B. Li, Model-free adaptive control of test platform system of anti-rolling tank, *Ship Engineering*, vol. 28, no. 4, pp. 5–8, 2006.
268. C. Q. Li and G. S. Liu Simulation research of MFAC-PID cascade control schemes for water level of a drum boiler steam generator, *Journal of Northeast China Institute of Electric Power Engineering*, vol. 25, no. 4, pp. 11–15, 2005.
269. D. Z. Li, W. W. Ning, and W. Q. Ni, Computer simulation of a drum water level system based on double-speed sampling MFA-PID cascade control, *Boiler Technology*, vol. 38, no. 5, pp. 19–21,35, 2007.
270. J. L. Qi and G. Ma, Design of glass furnace control system based on model free adaptive control, *Journal of North China Institute of Aerospace Engineering*, vol. 20, no. 2, pp. 1–3,10, 2010.
271. W. He, An adaptive controller for magnetic suspension switched reluctance motor, *Large Electric Machine and Hydraulic Turbine*, no. 5, pp. 57–60,64, 2004.
272. Q. Zhou and W. L. Qu, Model-free learning adaptive semi-active control of structure with MR dampers, *Earthquake Engineering and Engineering Vibration*, vol. 24, no. 4, pp. 127–132, 2004.
273. P. Ma and C. H. Zhang, Model-free controller based main-steam pressure control system of power plant, *Electric Power Science and Engineering*, vol. 24, no. 10, pp. 28–30, 2008.
274. Q. M. Cheng, Y. M. Cheng, M. M. Wang, and Y. F. Wang, Simulation study on model-free adaptive control based on grey prediction in ball mill load control, *Chinese Journal of Scientific Instrument*, vol. 32, no. 1, pp. 87–92, 2011.
275. Q. M. Cheng, R. Q. Guo, X. F. Du, and Y. Zheng, A control method of multi-variable coupling control based on model-free control and its simulative application on ball mill control, *Journal of Power Engineering*, vol. 28, no. 6, pp. 891–895, 2008.
276. H. Bao, B. Y. Duan, G. D. Chen, and J. W. Mi, Control and experiment for large radio telescope, *China Mechanical Engineering*, vol. 18, no. 14, pp. 1643–1647, 2007.
277. H. Bao, B. Y. Duan, and G. Q. You, Control scheme for cable-mesh reflector adjustment on cablenet deployable antenna, *Chinese Journal of Applied Mechanics*, vol. 25, no. 1, pp. 154–157, 2008.
278. H. Bao, B. Y. Duan, and G. D. Chen, Study on control methodology of feed cabin supporting structure with flexible cable, *Machine Design and Research*, vol. 21, no. 2, pp. 64–66, 2005.
279. X. H. Shi, H. Z. Yu, and F. Qian, Slope identification based DMC adaptive prediction control method, *Chinese Journal of Scientific Instrument*, vol. 29, no. 1, pp. 152–158, 2008.
280. C. G. Liu, Z. Sui, W. Z. Fu, and M. Z. Li, Non-parametric model and adaptive control for multi-point forming process, *Control Engineering of China*, vol. 11, no. 4, pp. 306–308, 2004.
281. H. B. Wang and Q. F. Wang, Nonparametric model adaptive control for underwater towed heave compensation system, *Control Theory & Applications*, vol. 27, no. 4, pp. 513–516, 2010.

282. D. Guo, Y. L. Fu, N. Lu, and W. H. Wang, Application of model-free adaptive control in billet flash butt welding, in *Proc. 29th Chinese Control Conference*, Beijing, China, 2010, pp. 5110–5114.
283. F. L. Lv, H. B. Chen, C. J. Fan, and S. B. Chen, A novel control algorithm for weld pool control, *Industrial Robot: An International Journal*, vol. 37, no. 1, pp. 89–96, 2010.
284. F. L. Lv, J. F. Wang, C. J. Fan, and S. B. Chen, An improved model-free adaptive control with G function fuzzy reasoning regulation design and its applications, *Proceedings of the Institution of Mechanical Engineers, Part I: Journal of Systems and Control Engineering*, vol. 222, no. 8, pp. 817–828, 2008.
285. F. L. Lv, H. B. Chen, C. J. Fan, and S. B. Chen, Application of model-free adaptive control to impulse TIG welding, *Journal of Shanghai Jiaotong University*, vol. 43, no. 1, pp. 62–64,70, 2009.
286. Y. S. Li, A. G. Yao, J. B. Yang, and H. J. Wu, Experimental research on deviation control in vertical drilling based on model-free adaptive algorithm, *Exploration Engineering (Rock & Soil Drilling and Tunneling)*, vol. 36, no. S1, pp. 104–109, 2009.
287. J. F. Sun, Y. J. Feng, and X. S. Wang, A kind of macro-economic dynamic control research based on control without model, *Systems Engineering-Theory & Practice*, vol. 28, no. 6, pp. 45–52, 2008.
288. J. Jiang, Y. X. Dai, and S. Q. Peng, Development of control system for multi-wire saw, *China Mechanical Engineering*, vol. 21, no. 15, pp. 1780–1784, 2010.
289. D. J. Zhang and Y. Yao, Embedded model-free adaptive control of electrode regulator system of arc furnance, *Journal of Metallurgical Industry Automation*, vol. 33, no. S2, pp. 171–174, 2009.
290. X. P. Lu, W. Li, and Y. G. Lin, Load control of wind turbine based on model free adaptive controller, *Transactions of the Chinese Society for Agricultural Machinery*, vol. 42, no. 2, pp. 109–114,129, 2011.
291. L. A. Zhang, J. Z. Wu, Z. Q. Chen, and W. D. Wang, Design and trial of MW wind turbine blade static loading control system, *China Mechanical Engineering*, vol. 22, no. 18, pp. 2182–2185,2208, 2011.
292. Y. Chang, B. Gao, and K. Y. Gu, A model-free adaptive control to a blood pump based on heart rate, *ASAIO Journal*, vol. 57, no. 4, pp. 262–267, 2011.
293. B. Gao, K. Y. Gu, Y. Zeng, and Y. Chang, An anti-suction control for an intra-aorta pump using blood assistant index: a numerical simulation *Artificial Organs*, vol. 36, no. 3, pp. 275–282, 2012.
294. C. H. Guo and T. Q. Wang, The study of voltage source PWM rectifier based on MFAC, in *Proc. 29th Chinese Control Conference*, Beijing, China, 2010, pp. 2077–2081.
295. M. C. Campi, A. Lecchini, and S. M. Savaresi, Virtual Reference Feedback Tuning (VRFT): A new direct approach to the design of feedback controllers, in *Proc. 39th IEEE Conference on Decision and Control*, Sydney, Australia, 2000, pp. 623–629.
296. H. Hjalmarsson, From experiment design to closed-loop control, *Automatica*, vol. 41, no. 3, pp. 393–438, 2005.
297. H. Hjalmarsson, S. Gunnarsson, and M. Gevers, Model-free tuning of a robust regulator for a flexible transmission system, *European Journal of Control*, vol. 1, no. 2, pp. 148–156, 1995.
298. K. K. Tan, S. N. Huang, and T. H. Lee, Robust adaptive numerical compensation for friction and force ripple in permanent magnet linear motors, *IEEE Transactions on Magnetics*, vol. 38, no. 1, pp. 221–228, 2002.

299. Y. S. Huang and C. C. Sung, Implementation of sliding mode controller for linear synchronous motors based on direct thrust control theory, *IET Control Theory & Applications*, vol. 4, no. 3, pp. 326–338, 2010.
300. D. Naso, F. Cupertino, and B. Turchiano, Precise position control of tubular linear motors with neural networks and composite learning, *Control Engineering Practice*, vol. 18, no. 5, pp. 515–522, 2010.
301. K. K. Tan, Precision motion control with disturbance observer for pulse width-modulated-driven permanent magnet linear motors, *IEEE Transactions on Magnetics*, vol. 39, no. 3, pp. 1813–1818, 2003.
302. K. K. Tan, T. H. Lee, and H. X. Zhou, Micro-positioning of linear piezoelectric motors based on a learning nonlinear PID controller, *IEEE Transactions on Mechatronics*, vol. 6, no. 4, pp. 428–436, 2001.
303. M. Parageorgiou and A. Kotsialos, Freeway ramp metering: an overview, *IEEE Transactions on Intelligent Transportation Systems*, vol. 3, no. 4, pp. 271–278, 2002.
304. I. C. Cheng, J. B. Gruz, and J. G. Paquet, Entrance ramp control for travel rate maximization in expressways, *Transportation Research Part C*, vol. 8, no. 6, pp. 503–508, 1974.
305. L. Isaken and H. J. Payne, Suboptimal control of linear systems by augmentation with application to freeway traffic regulation, *IEEE Transactions on Automatic Control*, vol. 18, no. 3, pp. 210–219, 1973.
306. D. P. Masher, D. W. Ross, P. J. Wong, P. L. Tuan, H. M. Zeidler, and S. Petracek, Guidelines for design and operation of ramp control systems, *California: Stanford Research Institute*, 1975.
307. M. Papageorgiou, H. Hadj-Salem, and J. M. Blosseville, ALINEA: a local feedback control law for on-ramp metering, *Transportation Research Record*, vol. 1320: 58–64, 1991.
308. H. M. Zhang, S. G. Ritchie, and R. Jayakrishnan, Coordinated traffic-responsive ramp control via nonlinear state feedback, *Transportation Research Part C*, vol. 9, no. 5, pp. 337–352, 2001.
309. A. Kotsialos, Coordinated and integrated control of motor-way networks via nonlinear optimal control, *Transportation Research Part C*, vol. 10, no. 1, pp. 65–84, 2002.
310. M. Parageorgiou, H. Hadj-Salem, and F. Middleham, ALINEA local ramp metering: summary of the field results, *Transportation Research Record*, vol. 1603, pp. 90–98, 1997.
311. Y. B. Wang and M. Papageorgion, Real-time freeway traffic state estimation based on extended kalman filter: a general approach, *Transportation Research Part B*, vol. 39, no. 2, pp. 141–167, 2005.
312. M. Papageorgiou, J. M. Blosseville, and H. Hadj-Salem, Macroscopic modeling of traffic flow on the Boulevard Peripherique in Paris, *Transportation Research Part B*, vol. 23, no. 1, pp. 29–47, 1989.
313. K. A. Pietrzak and S. M. Packer, Vision-based weld pool width control, *ASME Journal of Engineering for Industry*, vol. 116, no. 1, pp. 86–92, 1994.
314. J. B. Song and D. E. Hardt, Dynamic modeling and adaptive control of the gas metal arc welding process, *ASME Journal of Dynamic Systems, Measurement, and Control*, vol. 116, no. 3, pp. 405–413, 1994.
315. Y. M. Zhang, E. Liguo, and B. L. Walcott, Robust control of pulsed gas metal arc welding, *ASME Journal of Dynamic Systems, Measurement, and Control*, vol. 124, no. 2, pp. 281–289, 2002.

316. C. H. Tsai, K. H. Hou, and H. T. Chuang, Fuzzy control of pulsed GTA welds by using real-time root bead image feedback, *Journal of Materials Processing Technology*, vol. 176, no. 1–3, pp. 158–167, 2006.
317. K. Andersen and G. E. Cook, Artificial neural networks applied to arc welding process modeling and control, *IEEE Transactions on Industry Application*, vol. 26, no. 5, pp. 824–830, 1990.
318. F. L. Lv, S. B. Chen, and S. W. Dai, A model-free adaptive control of pulsed GTAW, *Robotic Welding, Intelligence and Automation*, Lecture Notes in Control and Information Sciences, T.-J. Tarn et al., Eds. Berlin: Springer, 2007, pp. 333–339.
319. Z. S. Hou and Y. M. Zhu, Controller-dynamic-linearization based model free adaptive control for discrete-time nonlinear systems, *IEEE Transactions on Industrial Informatics*, vol. 9, no. 4, pp. 2301–2309, 2013.
320. Z. S. Hou and Y. M. Zhu, Controller compact form dynamic linearization based model free adaptive control, in *Proc. IEEE 51st Annual Conference on Decision and Control*, Maui, Hawaii, 2012, pp. 4817–4822.
321. X. H. Bu, Z. S. Hou, F. S. Yu, and Z. Y. Fu, Model free adaptive control with disturbance observer, *Control Engineering and Applied Informatics*, vol. 14, no. 4, pp. 42–49, 2012.
322. X. H. Bu, F. S. Yu, Z. S. Hou, and H. W. Zhang, Model free adaptive control algorithm with data dropout compensation, *Mathematical Problems in Engineering*, vol. 2012, pp. 1–14, 2012.

Index

I

K

L

M

N

O

P

R

S

T

U

V

W

Milton Keynes UK
Ingram Content Group UK Ltd.
UKHW031126141024
449569UK00006B/416

9 781466 594180